Mathematical Immunology of Virus Infections

Gennady Bocharov · Vitaly Volpert
Burkhard Ludewig · Andreas Meyerhans

Mathematical Immunology of Virus Infections

Springer

Gennady Bocharov
Marchuk Institute of Numerical Mathematics
Russian Academy of Sciences
Moscow
Russia

Burkhard Ludewig
Institute of Immunobiology
Kantonsspital St. Gallen
St. Gallen
Switzerland

Vitaly Volpert
Institut Camille Jordan, UMR 5208 CNRS
Centre National de la Recherche Scientifique (CNRS)
Villeurbanne
France

and

RUDN University
Moscow
Russia

Andreas Meyerhans
Parc de Recerca Biomedica Barcelona
ICREA and Universitat Pompeu Fabra
Barcelona
Spain

ISBN 978-3-030-10185-5 ISBN 978-3-319-72317-4 (eBook)
https://doi.org/10.1007/978-3-319-72317-4

Printed on acid-free paper

This Springer imprint is published by the registered company Springer International Publishing AG part of Springer Nature
The registered company address is: Gewerbestrasse 11, 6330 Cham, Switzerland

The immune system is primarily about host survival of infections and for this we also need to understand the biology of the system.

Rolf M. Zinkernagel
(from "On Immunity Against Infections and Vaccines: Credo 2004". *Scandinavian Journal of Immunology*, 2004, 60:9–13)

Preface

Understanding the cellular and molecular mechanisms that control the ability of the immune system to mount a protective response against pathogen-derived foreign antigens, but avoid a pathological response to self-antigens, is a central problem in immunology. From the new high-throughput technologies, i.e. various omics measurements, 3D visualization and immunophenotyping, we have now a static view of the numerous components of the immune system and the links between them with unprecedental resolution. However, similar to the Classical Physics of the seventeenth century before the invention of the differential calculus by Sir Isaac Newton, a dynamical systems paradigm has to be developed and enter everyday immunological research. This requires the integration of mathematical methods to complement experimentation with the aim to represent, interpret and predict the observable characteristics of infections.

Mathematics is the universal language for expressing causal and functional relationships between observations. Its mainstream developments have been inspired by the needs of Physics, Chemistry and Engineering. For the twenty-first century, it is widely expected that Biology becomes a frontier for Mathematics. The challenge is to establish an interdisciplinary dialogue between mathematicians and experimentalists so that experimentation and mathematical modelling becomes an iterative process that boosts the different disciplines. The generated models that inevitably present simplifications of the underlying biological complexity must not lose touch with reality and generate testable predictions that drive, for example, perceptions of pathogen–host interactions. The problem of how to develop, in a systematic manner, such consistent models that provide a basis for quantitative analysis and predictions raises challenges for applied mathematicians related to the formulation of genuine approaches for representing the phenotypic complexity, spatial heterogeneity, hierarchical organization and control principles inherent to the infectious disease courses and outcomes.

This book is based on several lecture courses and seminars given by us at the Lomonosov Moscow State University, University of Chester, Saarland University, University of Zurich, University Lyon 1, and University Pompeu Fabra (Barcelona). It consists of eight chapters covering basic facts on viral infections and biological systems analysis, model formulation and parameter estimation, mathematical models of experimental and human infections and multi-scale and integrative modelling approaches. We follow the route expressed in Andrew and colleagues (2007) that there is no better way to proceed with application of Mathematics to Immunology than to formulate mathematical models that correspond qualitatively to the existing theories and to form a range of models ordered according to their qualitative and quantitative consistency with the experimental and clinical observations.

Moscow, Villeurbanne, St. Gallen, Barcelona
October 2017

Gennady Bocharov
Vitaly Volpert
Burkhard Ludewig
Andreas Meyerhans

Acknowledgements

We deeply thank our colleagues and friends from various research institutes across Europe, America and Asia: Peter Aichele, Roy M. Anderson, Christopher T. H. Baker († 2017), H. T. Banks, Sergei I. Bazhan, Antonio Bertoletti, Nikolay M. Bessonov, Daniel Binder, Anass Bouchnita, Luisa Cervantes-Barragan, Valery A. Chereshnev, Jovana Cupovic, Alexander A. Danilov, John T. Edwards, Stephan Ehl, Koen Engelborghs, Jim Ferrell, Neville J. Ford, Georg A. Funk, Irina A. Gainova, Zvi Grossman, Karl-Peter Hadeler († 2017), Hans Hengartner, Willi Jäger, Tobias Junt, Arkadii V. Kim, Paul Klenerman, Philippe Krebs, Pat M. Lumb, Tatyana B. Luzyanina, Guri I. Marchuk († 2013), Martin Meier-Schellersheim, Mario Novkovic, Stewart J. Norton, Lucas Onder, Annette Oxenius, Chris A. H. Paul, William E. Paul († 2016), Fathalla A. Rihan, Dirk Roose, Tim Schenkel, Peter Seiler, Victor P. Shutyaev, Ulrich Steinhoff, Volker Thiel, Sergei Trofimchuk, Eugene E. Tyrtyshnikov, Yuri V. Vassilevski, Simon Wain-Hobson, Rolf M. Zinkernagel and Roland Züst, for productive joint research in the past which created a basis for this book.

Our special thanks goes to friends, colleagues and all the people of our laboratories from Moscow, Lyon, St. Gallen and Barcelona: Dmity Grebennikov, Rostislav Savinkov, Rufina Tretyakova, Valerya Zheltkova, Olga Orlova, Anna Savinkova, Anna Gorbatova, Daria Donecz, Jordi Argilaguet, Jordi Garcia Ojalvo, Juana Diez, Marisa and Celina Seth, Valentina Casella, Sandra Giest, Mie Kobayashi, Irene Latorre, Javier Martinez, Mireia Pedragosa, Christina Peligero, Graciela Riera, Katarina Smutna, Yasuko Yokota, Barbara Gärtner, Friedrich Grässer, Nikolaus Müller-Lantzsch († 2017), Klaus Roemer, Marlis Sauter, Martina und Urban Sester, Jean-Pierre Vartanian.

We would like to thank our editors, Jan-Philip Schmidt and Petra Jantzen, for the support and encouragement during the preparation of this book.

Finally, we thank our families for their Patience and Love.

The work on the book was supported by the Russian Science Foundation, Grants no. 15-11-00029 and 18-11-00171.

Contents

Acronyms

Ag	Antigen
Ab	Antibody
CFSE	Carboxyfluorescein succinimidyl ester
CI	Confidence interval
CTL	Cytotoxic T lymphocyte
DDE	Delay differential equation
HBV	Hepatitis B Virus
HCV	Hepatitis C Virus
HIV	Human immunodeficiency virus
hPDE	Hyperbolic partial differential equation
LCMV	Lymphocytic choriomeningitis virus
MHV	Murine hepatitis virus
MLE	Maximum likelihood estimation
ODE	Ordinary differential equation
pDC	Plasmacytoid dendritic cell
PDE	Partial differential equation
RDE	Reaction-diffusion equation

Chapter 1
Principles of Virus–Host Interaction

This chapter presents a brief overview of basic immunological concepts and ideas necessary for the development of mathematical models of immune processes during virus infections.

1.1 In Brief

Immunology as a scientific discipline studies the response of an organism to antigenic invasion, the recognition of self and non-self, and all the biological, chemical and physical aspects of immune phenomena. To protect the body against pathogens, the immune system has a repertoire of body-wide defense modalities that consist of interacting cells, humoral factors and lymphoid organs (Fig. 1.1). The latter comprise bone marrow, thymus, spleen, lymph nodes and gut that are connected by the vascular systems of blood and lymph allowing the migrating of the immune system components between the compartments and to the places of pathogenic threat. In the human immune system, there are several hundred cytokines, about 10^{20} antibody molecules and 10^{13} immuno-competent cells that migrate spatially and interact with each other either competitively or cooperatively. Furthermore, the immuno-competent cells may proliferate, differentiate, mature, age and die. They are derived from stem cells in the bone marrow and develop further in different lymphatic tissues. They function as pathogen-degrading cells (= phagocytes) (i.e. granulocytes and macrophages), antigen-presenting cells (i.e. dendritic cells (DC) and macrophages) and specific effector cells (i.e. T and B lymphocytes). The latter are responsible for cell-mediated elimination of infected cells and antibody production respectively.

Pathogens are agents that cause disease. The term is usually used in connection with an infectious microorganism. As this book focuses on the immune responses towards virus infections, only viruses will be considered. However, the principles of antiviral immune mechanisms also apply—at least partially—to other pathogens such as bacteria, helminths or fungi.

G. Bocharov et al., *Mathematical Immunology of Virus Infections*,
https://doi.org/10.1007/978-3-319-72317-4_1

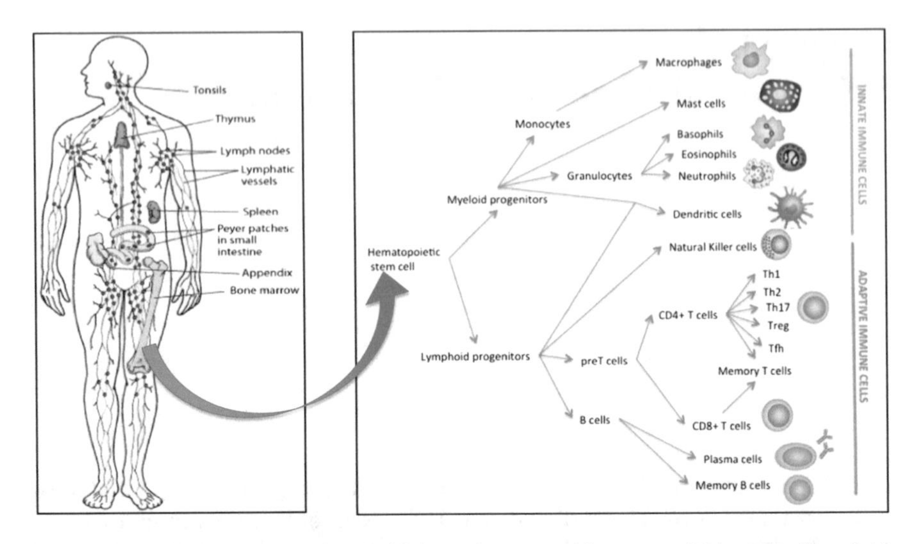

Fig. 1.1 Distribution of human lymphoid tissues/organs and immune cell hierarchy. The physiological distribution of human lymphoid tissues/organs is schematically shown in the left panel. Within primary lymphoid organs namely bone marrow and thymus, the lymphocytes are generated. Immune response initiation and lymphocyte maturation take place in the peripheral lymphoid organs comprising lymph nodes, spleen and the mucosal lymphoid tissue of the gut. Immune cells move within the human body via the lymphatic system and the blood system. The right panel shows schematically the developmental path of the major immune cell subsets. Pluripotent hematopoietic stem cells are generated in the bone marrow and give rise to myeloid and lymphoid progenitors. These then differentiate into different cell subsets with specialist effector functions. Both myeloid and lymphoid progenitors contribute to the generation of dendritic cells. Th1, T helper 1 cells; Th2, T helper 2 cells; Th17, T helper 17 cells; Treg, T regulatory cells; Tfh, T follicular helper cells; plasma cells, antibody-producing cells (antibodies Y are depicted as a brown). *Adapted from the website* http://anatomybody101.com/the-immune-responses-of-the-lymphatic-system/the-immune-responses-of-the-lymphatic-system-3d3979405d904f69a2f42156fa6a7973/. *Cells pictures were taken from King et al., 2011, Nature Reviews Immunology 11(10):685–92 and Gabrilovich et al., 2012, Nature Reviews Immunology 12(4):253–68*

Viruses are obligatory parasites that depend on host cells to multiply. They consist of a viral genome in form of ribonucleic acid (RNA) or desoxyribonucleic acid (DNA) chains that carry all necessary genetic information for virus expansion. Viral genomes are embedded within a protective protein shell named capsid that may, depending on the type of virus, be surrounded by a lipid bilayer containing the viral surface glycoproteins. These so-called envelope proteins or surface glycoproteins are the keys to enter new target cells after interacting with virus receptors on host cell surfaces. The schematic structure of a human immunodeficiency virus (HIV) particle is shown in Fig. 1.2 as an example.

To gain access to cells for multiplication, viruses need to enter a host organism. Common virus entry routes (Fig. 1.3) are the infection of cells that are exposed on skin or mucosal surfaces or the entry via injuries or intravenous inoculation as for example by insect bites. Once viruses have overcome surface barriers of an organism

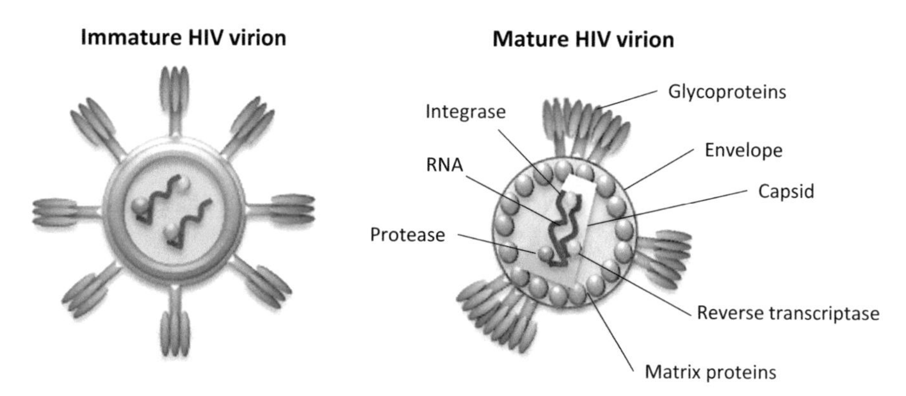

Fig. 1.2 Schematic representation of an HIV particle in the immature and mature form. A human immunodeficiency virus (HIV) particle carries two copies of single-stranded viral RNA enclosed in a conical capsid structure and is surrounded by a lipid bilayer that contains the viral glycoproteins. It has all necessary enzymes for reverse transcription of the viral RNAs into a double-stranded DNA and subsequent integration into the host cell genome. The particle is formed during budding from the cell surface of an infected cell. Particle maturation depends on the viral protease and is accompanied by glycoprotein clustering and a 1000-fold increase in viral infectivity. *The figure represents data from Chojnacki et al., Science 338, 524–528 (2012)*

and infected a host cell, they start replicating their genome and multiply. A virus life cycle from infection to progeny production may take up to 24 hours, after which hundreds to many thousand new virus particles are being released from a single infected cell. For not being overwhelmed by such explosive virus growth kinetics, organisms require multiple layers of defense mechanisms that keep viruses under control (Fig. 1.4). This significantly reduces the in vivo observable virus doubling times to about 4 hours making it comparable with the lymphocyte division time.

Protection against viruses is provided by a coordinated action of innate immunity and antigen-specific immunity. The innate immunity refers to (i) physical, chemical and microbiological barriers, (ii) responses of innate immune cells such as granulocytes, natural killer cells, macrophages and dendritic cells (DCs) that all lack the classical antigen-specific memory effect and (iii) immunologically active substances generated by innate immune cells and both stromal and parenchymal cells present in the infected organ (e.g. complement factors, cytokines, acute-phase proteins and interferons). The specific immunity is based upon the use of antigen-specific receptors on T and B lymphocytes to drive targeted effector responses against a pathogen. As this requires lymphocyte maturation and proliferation, measurable adaptive responses appear only several days post-infection. The typical dynamics of an acute infection that is resolved by an efficient immune response is shown in Fig. 1.5.

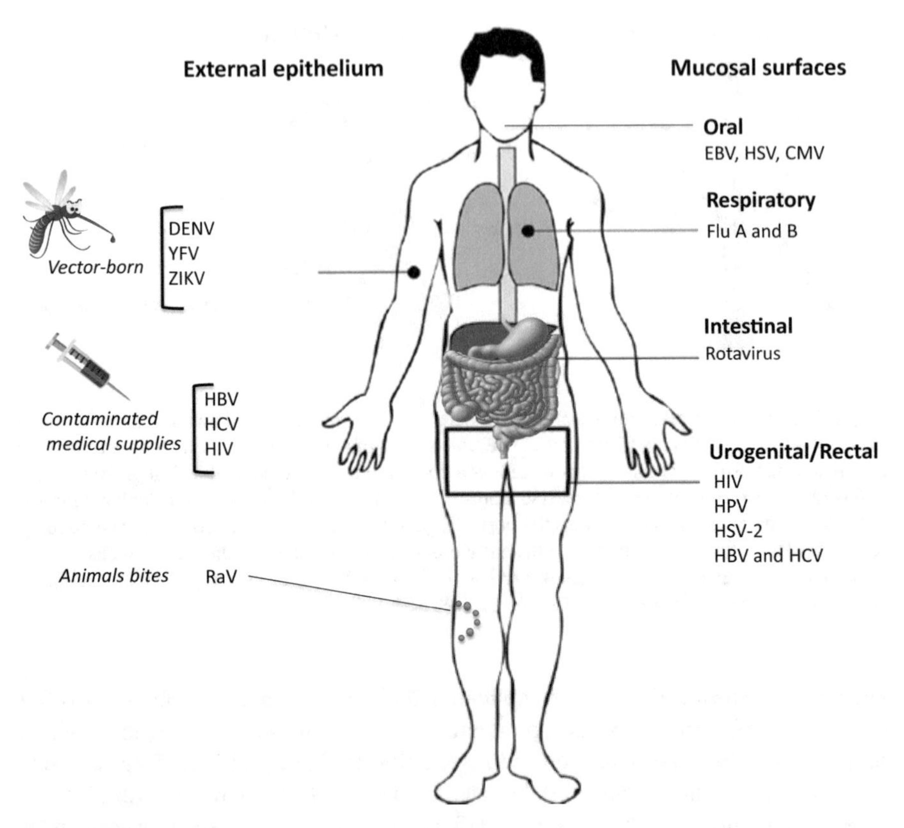

Fig. 1.3 Virus routes of infection. Viruses enter the human body by overcoming the external epithelium or mucosal surfaces. Examples are given. DENV, Dengue virus; YFV, Yellow fever virus; ZIKA, Zika virus; RaV, Rabies virus; HBV, Hepatitis B virus; HCV, Hepatitis C virus; HIV, Human immunodeficiency virus; EBV, Epstein-Barr virus; HSV, Herpes simplex virus; CMV, Cytomegalovirus; FluA, Influenza A Virus; FluB, Influenza B Virus; HPV, Human papillomavirus

1.2 Virus Recognition and Immune Responses

Viruses and other pathogens display particular molecular signatures (pathogen-associated molecular patterns) that are recognized by specific pattern recognition receptors. Signalling via these receptors leads to initiation of innate immune responses that condition the subsequent stimulation of adaptive lymphocyte responses. A general overview of these events is schematically given in Fig. 1.6. Detection of viral components and induced signalling cascades for conventional dendritic cells (cDCs) and plasmacytoid dendritic cells (pDCs), two key cell types for initiation of efficient antiviral immune responses, are shown in Fig. 1.7. The type I interferon response, a major antiviral defense component of the innate immune system, is summarized in Fig. 1.8.

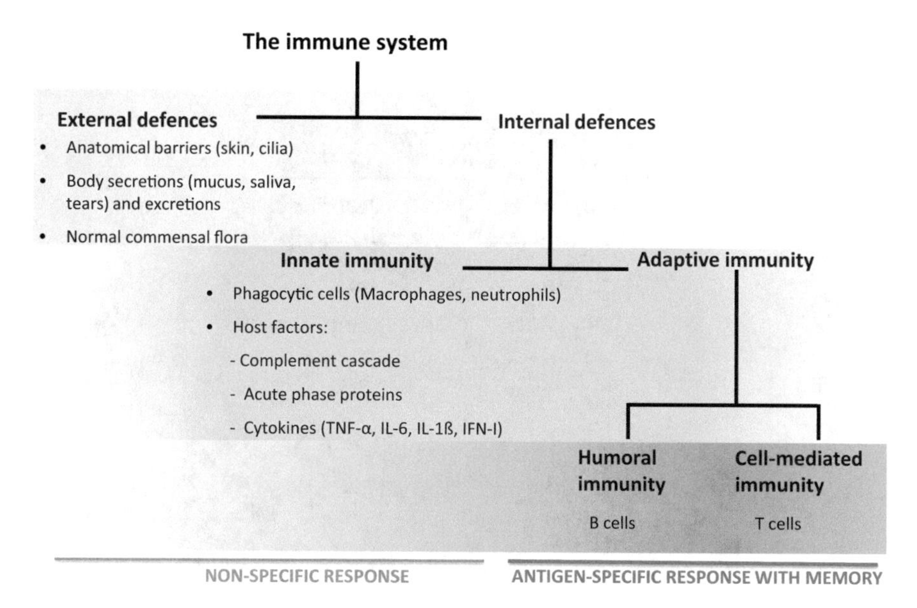

Fig. 1.4 The multiple layers of defences of the immune system. The immune system comprises external defences that prevent pathogens from entering the human body, and internal defences that provide protection against viruses that managed to overcome these barriers. External defences include host anatomic barriers like skin and cilia, body secretions like mucus, saliva and tears that contain enzymes with protective functions and the normal commensal flora that keeps pathogens at bay. Internal defences are composed of the rapid innate immune responses triggered by pattern recognition receptors (PRRs) (see Figs. 1.6 and 1.7) and the late adaptive immune responses that exhibit high pathogen specificity and memory. TNF-α, Tumour necrosis factor α; IL-6, interleukin 6; IL-1β, interleukin 1 β; IFN-I, type I interferon

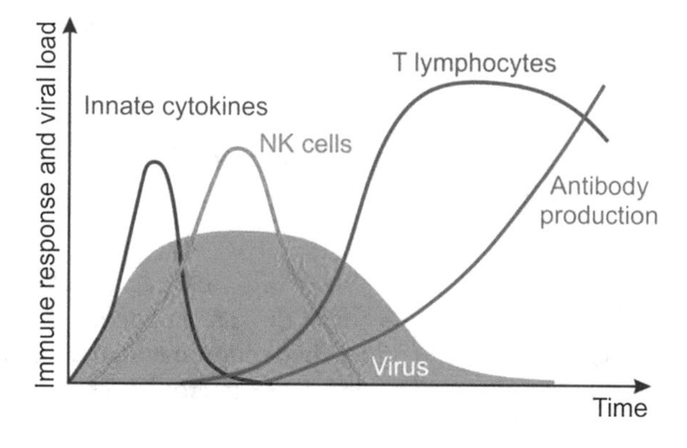

Fig. 1.5 Schematic view of the kinetics and respective immune responses of an acute virus infection. Innate cytokines like type-I interferons (IFN-α, IFN-β), TNF-α and IL-12 are produced early after a virus infection. These help activating natural killer cells (NK cells). Adaptive T- and B-cell responses (antibodies) appear subsequently and are essential to eliminate a virus infection

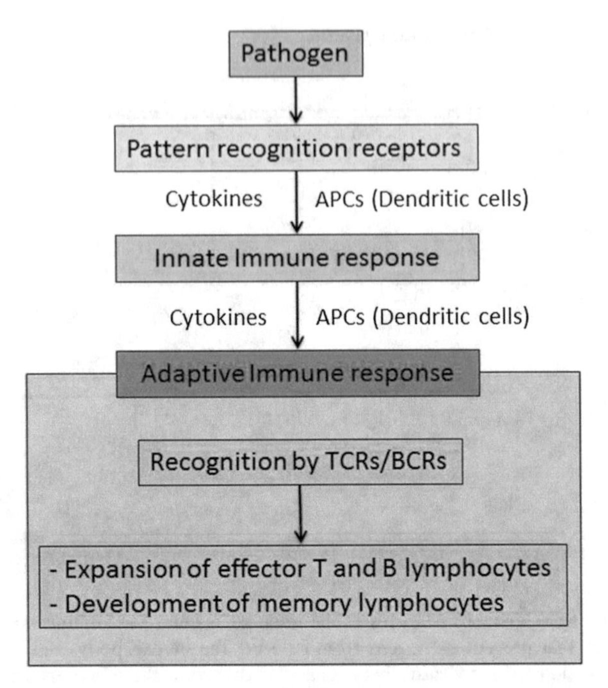

Fig. 1.6 From pattern recognition to the adaptive immune response. Viruses and other pathogens display particular molecular signatures, so-called pathogen-associated molecular patterns that are recognized by specific pattern recognition receptors (PRR) expressed by many cell types including antigen-presenting cells like dendritic cells (DCs). Pathogen recognition leads to the activation of innate immune responses as well as pathogen uptake by antigen-presenting cells (APCs), APC maturation and their migration to lymph nodes. Activated mature APCs in the lymph node can then trigger adaptive immune responses leading to pathogen-specific B- and T-cell activation, and subsequently to the development of immunological memory. TCR, T-cell receptor; BCR, B-cell receptor

1.2.1 Pattern Recognition to Initiate Innate Immune Responses

Pattern recognition receptors (PRRs) that can detect viral invaders are expressed in basically all cells in locations that are in contact with viral components like cell surfaces, vesicular structures like endosomes and the cytoplasm. They have different recognition specificities and sense microbial RNA (Toll-like receptors (=TLR) 3, 7, and 8; MDA5, RIG-I) or DNA (TLR9). Once a viral invader is detected, PRRs activate via a complex signalling cascade involving adaptor proteins, kinases and transcription factors the expression of proinflammatory cytokines including TNF-α, IL-1β and IL-6, and type I interferons (Fig. 1.7). The inflammatory response helps to combat the infection by attracting various effector molecules and cells to the invasion site. The interferons lead to activation of a myriad of interferon-induced genes (Fig. 1.8) that have diverse functions like (i) direct viral inhibition, (ii) increase

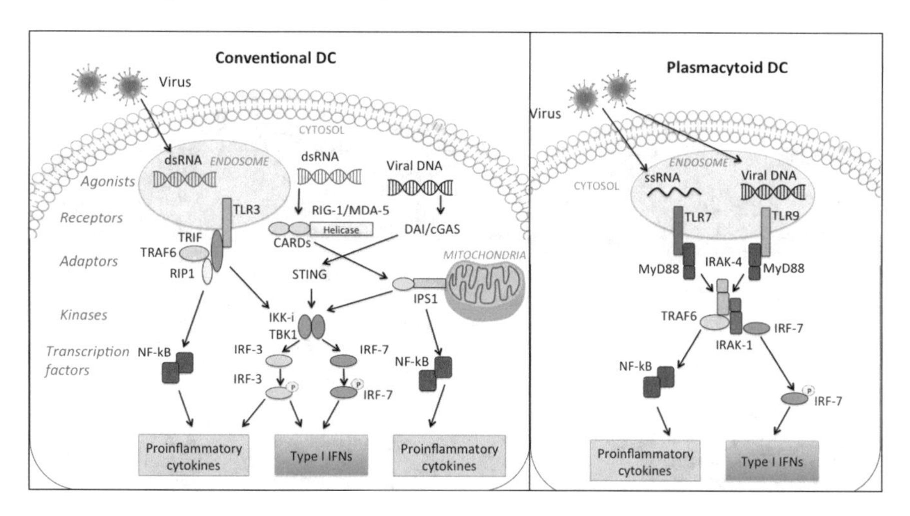

Fig. 1.7 Virus recognition by conventional and plasmacytoid dendritic cells and intracellular signal transduction. Dendritic cells are important sensors of virus infections. In conventional dendritic cells (cDCs), TLR3-dependent and RIG-I-dependent pathways operate to detect viral RNA, whereas DAI/cGAS-STING pathways detect viral DNA. Recognition of dsRNA by TLR3 in the endosomal membrane recruits TRIF to the receptor, which induces proinflammatory cytokines and type I IFNs via the RIP1/TRAF6-NF-κB pathway and the TBK1/IKK-i-IRF-3/IRF-7 pathway, respectively. Cytoplasmic dsRNA and DNA are detected via RIG-I/MDA-5 and DAI/cGAS, respectively, that signal through mitochondria-bound IPS1 or STING. In plasmacytoid dendritic cells (pDCs), TLR7 and TLR9 recognize viral ssRNA and DNA in endosomes. TLR stimulation then recruits a complex of MyD88, IRAK-4, IRAK-1, TRAF6 and IRF-7. NF-κB and phosphorylated IRF-7 translocate into the nucleus and upregulate the expression of proinflammatory cytokines and type I IFNs, respectively. *The figure is adapted and modified from Akira S et al., Cell 124, 783–801 (2006)*

of natural killer cell activity, (iii) activation of antigen-presenting cells (APCs), (iv) stimulation of chemokine expression to attract effector cells and (v) effecting cellular lifespans. Thus, PRR signalling not only initiates the immediate antiviral response but is also the basis for the subsequent adaptive immune response that is tailor-made for the infecting virus.

1.2.2 Viral Antigen Recognition by Adaptive Immune Responses

T and B lymphocytes are the major cellular immune system components that mediate specific recognition of viral invaders [12]. Each lymphocyte carries about 10^5 identical antigen receptors on its cell surface. The receptors of T lymphocytes, the T-cell receptors, can recognize virus-derived peptides in the context of cellular proteins of the major histocompatibility complex (MHC) while the B-cell receptors (= antibodies on the surface of B lymphocytes) can recognize any chemical structure

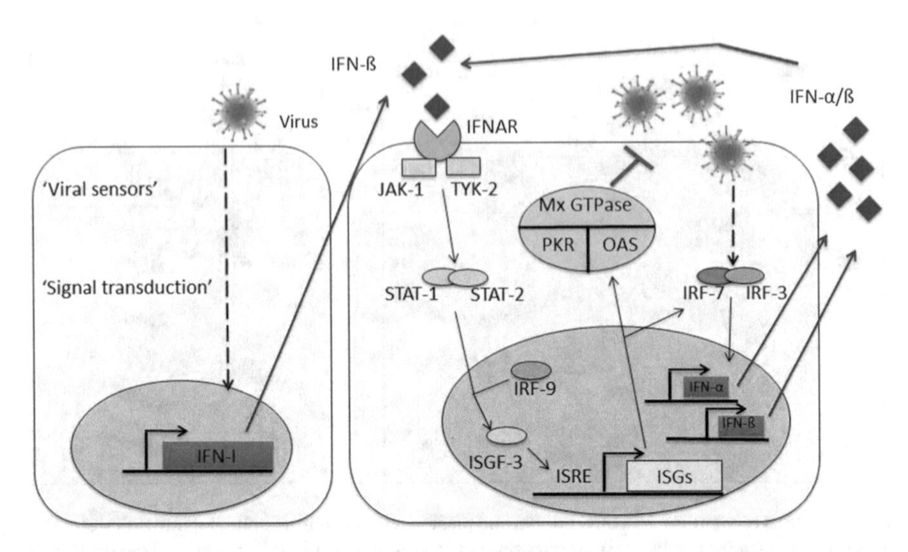

Fig. 1.8 Type I interferon production and the antiviral response. Virus infections induce IFN-I (IFN-α and IFN-β). They are secreted from the producer cell and bind to type I IFN receptor-bearing cells (IFNAR). Receptor binding activates via the JAK/STAT signalling pathway a myriad of IFN-stimulated genes (ISGs) such as MX GTPases, protein kinase R (PKR) and (2–5) oligo(A) synthetase (OAS) that exhibit antiviral activities. Some of the ISGs are interferon-regulated factors (IRFs) that can amplify the IFN response. *The figure is adapted from Haller et al., Virology 344, 119–130 (2006)*

being it protein, sugar, lipid or a stretch of nucleotides (Fig. 1.9). A population of lymphocytes with identical antigen-specific receptors is called a clone. The number of distinct lymphocyte clones represents the repertoire of specificities of the immune system. For humans, this number was estimated to range from about 10^5 to 10^7. Due to the permanent inflow of antigens and the limited lymphoid population, a selection of the lymphocyte repertoire through competition between different subsets of cells and between specific clones presumably takes place. Lymphocytes are subdivided into different cell types according to their functions in protecting a host against pathogens. $CD8^+$ cytotoxic T lymphocyte (CTL) responses represent a major mechanism in eliminating virus-infected cells. They sense virus-encoded peptides presented to their T-cell receptors in the context of MHC class I molecules and induce target cell death thus eliminating foci of virus production. The $CD4^+$ T cells have special roles as helper cells or regulatory cells; they facilitate cellular and humoral responses (the primary responses of the $CD8^+$ T cells and B cells, respectively) or inhibit responses. These functions are mainly provided by cytokines that can induce cell proliferation and differentiation, or immune system inhibition. Depending on the range of produced cytokines, $CD4^+$ T cells are classified as Th1, Th2, Th17, Tfh or Treg cells (see also Fig. 1.1). B lymphocytes are the immune cell type that produce and release antigen-specific protein molecules, the antibodies. The interaction of these humoral effector molecules with antigen can be quantitatively defined by an

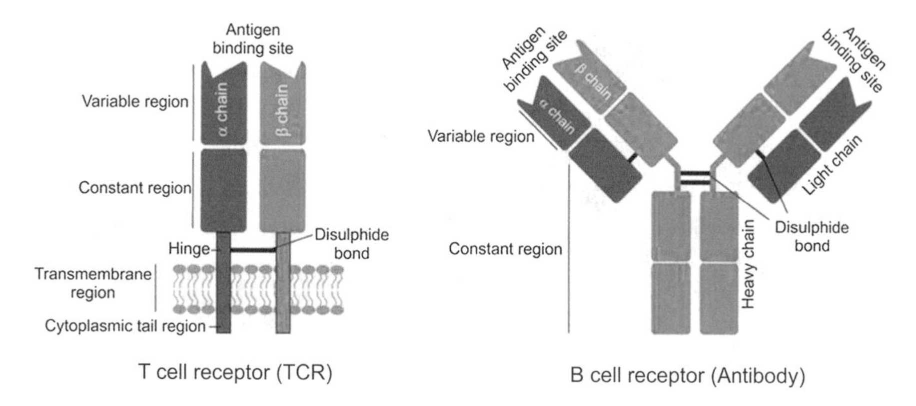

Fig. 1.9 Antigen recognition by T-cell and B-cell receptors. T cells recognize an antigen through the T-cell receptor (TCR). The TCR (left panel) is composed of two different transmembrane glycoprotein chains, α and β, connected by a disulphide bond. The extracellular portion consists of a variable and a constant domain, and the juxtaposition of the two variable domains forms the antigen-binding site that can recognize peptides presented by proteins from the major histocompatibility complex. B cells recognize an antigen through the B-cell receptor (BCR), a membrane-bound form of the immunoglobulins (right panel). The secreted form of the immunoglobulin with the same antigen specificity is the antibody produced by terminally differentiated B cells, so-called plasma cells. The antibody is composed of two hinged heavy chains (blue) and two light chains (pink) joined by disulphide bonds. The four chains are made up of a variable and a constant region. The juxtaposition of the variable region of the heavy and light chains forms two identical antigen-binding sites. The structures that may act as antigens for B cells are very flexible and can be proteins, lipids, sugars, etc.

affinity value and the chemical law of mass action. Antibodies exert their immune system function by pathogen neutralization, pathogen opsonization for subsequent phagocytosis and/or complement activation.

In order to initiate an adaptive virus-specific immune response, viral antigens have to be presented to lymphocytes in the context of antigen-presenting cells and a certain cytokine milieu that is derived from the innate response towards the invading virus. The overall outcome is then determined by a complex signal transduction and gene activation machinery at the single cell level, and various modes of communication at the cell population level. For example, a $CD4^+$ T-cell response requires activation of the cells via the TCR and a survival signal via co-stimulation which together lead to clonal T-cell expansion by proliferation (Fig. 1.10). Depending on the available cytokines, differentiation towards a T-cell subtype follows. Importantly, as signalling molecules are able to activate different genes at different concentrations, the lymphocyte response is not always proportional to the stimulus and nonlinear and bell-shaped response curves are observed under certain circumstances. The antigen-specific activation of lymphocytes leads to important changes in their protein expression profiles and their functional capabilities. As a result, three different stages of lymphocyte maturation can be identified: naive cells (cells that have not yet reacted to the antigen), antigen-activated cells and memory cells (cells that have

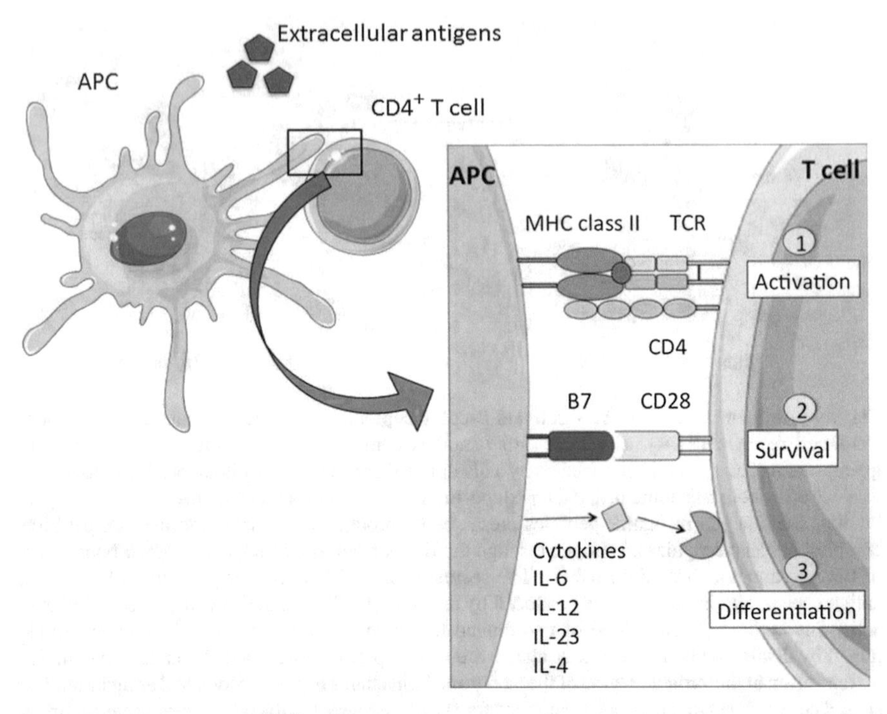

Fig. 1.10 Activation of T cells by antigen-presenting cells (APCs). After capturing and processing antigens, APCs deliver three signals that lead to the activation of naive T cells. In this example, an antigen (red circle) is presented by MHC class II proteins to the T-cell receptor (TCR) of a $CD4^+$ T cell delivering an activation signal (signal 1). To survive, the activated T cell requires further a cosignal delivered by the interaction of B7 proteins on the APC with CD28 proteins on the T-cell surface (signal 2). Depending on the nature of the antigen, APCs produce cytokines that act on T cells and induce the differentiation into specific T-cell subsets (signal 3). *Cell Pictures taken from Servier Medical Art*

encountered an antigen to which they are specific, and respond faster on re-exposure to that antigen). They are functionally different and follow distinct travelling routes within an organism. A large panel of antibodies is available today for characterizing and quantifying lymphocyte subtypes after natural or experimental virus infections.

1.3 Infection Fates with a Glimpse on the Real Complexity of Infection Immunology

The majority of viral infections can be fundamentally categorized as acute or persistent according to their temporal relationships with their hosts. Acute infections in humans are usually resolved within a few weeks by a myriad of immune system

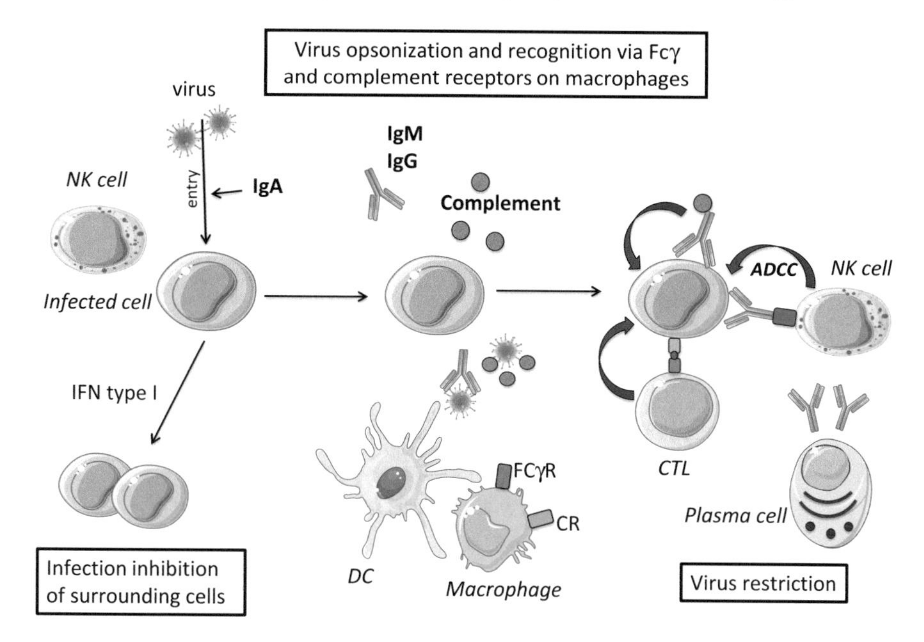

Fig. 1.11 Host defence against viruses. A host organism has multiple immune defence functions that can eliminate a virus infection. Mucosal surfaces can produce IgA antibodies that may block virus entry into a host. Interferons (IFNs) that are secreted from infected cells may inhibit further viral spread. Natural Killer (NK) cells, activated by IFNs, may recognize and kill infected cells. IgMs, IgGs and the complement system may take part in virus opsonization and elimination involving Fcγ and complement receptors (CRs) for example on macrophages. In a later stage of a virus infection, plasma cells will produce virus-specific antibodies that may neutralize a virus or initiate virus-infected cell destruction by antibody-dependent cytotoxicity (ADCC) or complement-mediated lysis. Finally, cytotoxic T cells (CTL) will efficiently kill infected cells when viral epitopes are presented on their surface by MHC class I proteins. *Cell Pictures taken from Servier Medical Art*

responses triggered upon infection or prior infection by vaccination (Figs. 1.11 and 1.12a).

In contrast, persistent infections are not resolved and, instead, develop when innate and adaptive immune responses are not sufficient to eliminate the invading virus during the primary infection phase (Fig. 1.12b). A consequence of this latter condition is the establishment of a dynamic equilibrium between virus expansion and virus-specific adaptive responses that may be maintained stably for years without major pathologic consequences or disrupted in a way that rapidly leads to overt disease. Viruses of both categories continue to threaten human health. Notable examples are the regular recurrences of influenza virus strains that cause acute infections with partly critical illness or death every year and infections with the human immunodeficiency virus (HIV) or the hepatitis B and C viruses (HBV, HCV) that can establish persistence in their hosts with different probabilities and pathogenic consequences. Whilst nearly all HIV infections lead to virus persistence, 50–80% of HCV and only about 5% of HBV infections in adults are persistent. The level of persistence of HBV-

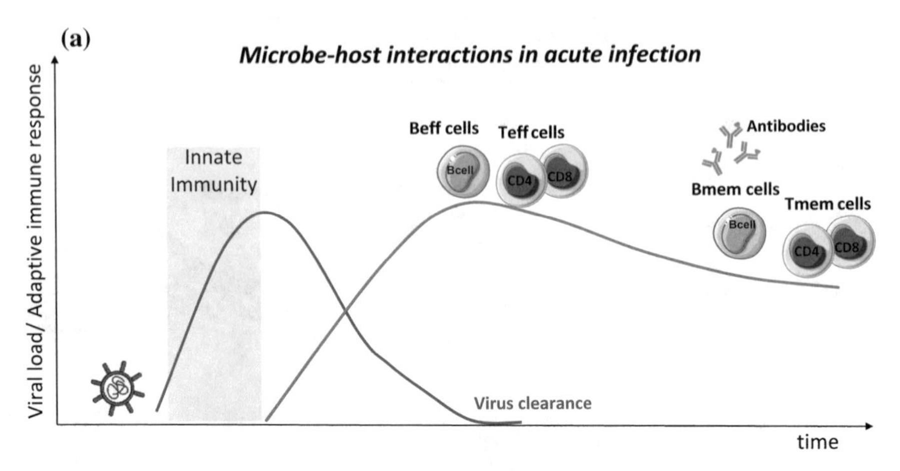

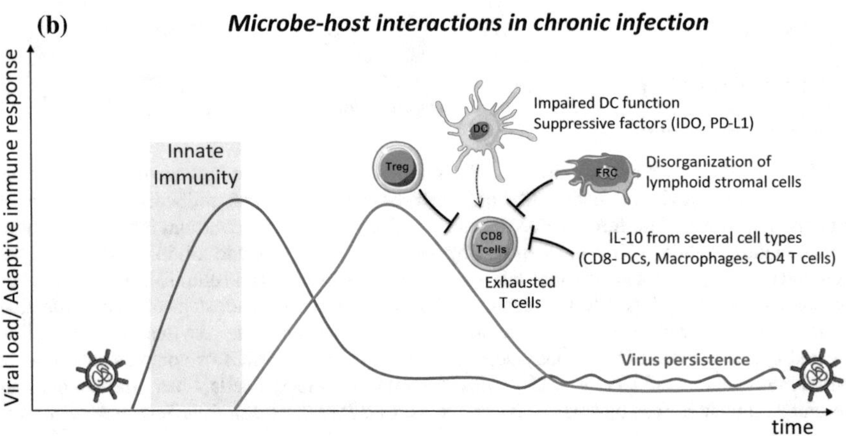

Fig. 1.12 Microbe–host interactions in acute and chronic viral infections. During the early phase of a viral infection, innate immunity slows down virus expansion through diverse mechanisms including type I interferons and NK cells. The subsequent adaptive immune response with virus-specific effector cells ($CD4^+$ helper and $CD8^+$ cytotoxic T cells, and antibody-producing B cells) may then clear an infection (1.12a; acute infection). After virus clearance, a pool of memory cells (Bmem and Tmem) remains in the host that may provide long-term protection from future infections with the same virus. However, the immune system can fail to eliminate a virus infection resulting in a long-term, persisting infection (1.12b; chronic infection). In this case, effector functions are down-regulated, i.e. in exhausted T cells and suppressive factors appear. IDO, indoleamine dioxygenase; PD-L1, programmed death ligand 1; IL-10; interleukin 10

infected newborns is massively increased to about 95% indicating that the state of the immune system is an important component in determining infection fates.

A number of viral and host factors in the early infection phase are involved in the fate decision between an acute and a persistent infection outcome [6, 8]. These include effector cell and virus expansion capacities, regulatory immune elements as

IL-10 and regulatory T cells (Tregs) and the IFN-I system. In case a host is unable to rapidly clear a viral infection, a non-functional state of pathogen-specific T cells known as exhaustion is established that helps to avoid immunopathology exerted by the simultaneous presence of virus-infected cells and virus-specific CTL [2, 5]. T-cell exhaustion is acquired via a distinct transcriptional program that leads to the permanent expression of inhibitory receptors [1, 11]. One of the main inhibitory receptors expressed on exhausted T cells is the protein programmed cell death-1 (PD1). Interactions of PD1 with its ligands PDL1 and PDL2 shut down the cells effector function and its capacity to proliferate after an antigen-specific trigger. This inhibitory system was first identified in persistent LCMV infections in mice and subsequently shown to also operate in persistent human infections like HIV and HCV, as well as in cancer [2]. Importantly, blocking the PD1–PD-ligand interaction can restore T-cell function, thus providing a promising tool for immunotherapeutic applications. Experiments in animal infection models and clinical trials with late-stage cancer patients are very encouraging [9, 10].

Besides exhaustion, T-cell functionality in persistent infections is also compromised by regulatory lymphocyte subsets, namely regulatory T cells and regulatory B cells [3, 4]. These cell subsets are a critical component in the maintenance of a balanced immune response and necessary to reduce inflammatory responses and avoid autoimmunity. They exert their function via a variety of mechanisms including the production of soluble immunosuppressive factors like IL-10 and TGF-β, and the delivery of suppressive signals by cell–cell contacts to conventional T cells and to antigen-presenting cells, thus influencing effector functions directly or indirectly. Interestingly, the PD1–PD-ligand exhaustion pathway and Tregs are linked at least in two ways. First, PD1–PDL1 interactions are involved in the generation and maintenance of induced Tregs from conventional T cells and second, PD1 can also be expressed on Tregs indicating that these suppressor cells, like their effector counterpart, are subject to downregulation. Together, this results in a regulatory circuit in which an overwhelming inflammatory response leads to effector T-cell exhaustion and generation of Tregs that are themselves blocked by exhaustion to stop excessive suppression of the immune response. While this regulatory circuit is obviously important to guarantee a balance between the defense against pathogenic intruders and the avoidance of autoimmunity, predicting the outcome of interfering with this circuit, for example, by blocking PD1–PD-ligand interactions is not straightforward. Both exhausted effector and exhausted regulatory cells may be reactivated, and the net gain of effector function will depend on the relative proliferative responsiveness and functional activities of both cell types [7]. Given the recent encouraging results from inhibiting PD1–PD-ligand interactions in late cancer patients, more anti-exhaustion immune therapies will be moving into clinical trials in the near future. Predictive mathematical models of the underlying complex immunological networks and an understanding of the respective control parameters are expected to significantly improve therapy efficacy on a rational basis.

References

1. Attanasio, J., and Wherry, E.J. (2016). Costimulatory and Coinhibitory Receptor Pathways in Infectious Disease. Immunity 44, 1052–1068.
2. Barber, D.L., Wherry, E.J., Masopust, D., Zhu, B., Allison, J.P., Sharpe, A.H., Freeman, G.J., and Ahmed, R. (2006). Restoring function in exhausted CD8 T cells during chronic viral infection. Nature 439, 682-687.
3. Li, M.O., and Rudensky, A.Y. (2016). T cell receptor signalling in the control of regulatory T cell differentiation and function. Nature reviews Immunology 16, 220-233.
4. Lund, F.E., and Randall, T.D. (2010). Effector and regulatory B cells: modulators of CD4+ T cell immunity. Nature reviews Immunology 10, 236-247.
5. Moskophidis, D., Lechner, F., Pircher, H., and Zinkernagel, R.M. (1993). Virus persistence in acutely infected immunocompetent mice by exhaustion of antiviral cytotoxic effector T cells. Nature 362, 758-761.
6. Ng, C.T., Snell, L.M., Brooks, D.G., and Oldstone, M.B. (2013). Networking at the level of host immunity: immune cell interactions during persistent viral infections. Cell host and microbe 13, 652-664.
7. Peligero, C., Argilaguet, J., Guerri-Fernandez, R., Torres, B., Ligero, C., Colomer, P., Plana, M., Knobel, H., Garcia, F., and Meyerhans, A. (2015). PD-L1 Blockade Differentially Impacts Regulatory T Cells from HIV-Infected Individuals Depending on Plasma Viremia. PLoS pathogens 11, e1005270.
8. Rouse, B.T., and Sehrawat, S. (2010). Immunity and immunopathology to viruses: what decides the outcome? Nature reviews Immunology 10, 514-526.
9. Sharma, P., and Allison, J.P. (2015). The future of immune checkpoint therapy. Science 348, 56-61.
10. Velu, V., Titanji, K., Zhu, B., Husain, S., Pladevega, A., Lai, L., Vanderford, T.H., Chennareddi, L., Silvestri, G., Freeman, G.J., et al. (2009). Enhancing SIV-specific immunity in vivo by PD-1 blockade. Nature 458, 206-210.
11. Wherry, E.J., and Kurachi, M. (2015). Molecular and cellular insights into T cell exhaustion. Nature reviews Immunology 15, 486-499.
12. Zinkernagel, R.M. (1996). Immunology taught by viruses. Science 271, 173-178.

Chapter 2
Basic Principles of Building a Mathematical Model of Immune Response

This chapter introduces a modular approach to the formulation of mathematical models of immune responses to virus infections. It presents a methodological basis for mathematical immunology of virus infections.

2.1 Systems Approach to Immunology

A 'Systems approach' in immunology [1–3] represents a framework for analysis and interpretation of complex phenomena. In biology, it is associated with the pioneer studies of von Bertalanffy [4], Mesarovich [5], Kitano [6, 7] and others. Systems biology focusses on the analyses of the structure, dynamics, design principles and control/coordination methods of biological systems (see Fig. 2.1) in order to understand how robustness, i.e. the ability to maintain a stable functioning despite various perturbations, is achieved [8, 9].

In practical terms, it is important that mechanisms that provide robustness and protect normal functions under various perturbations may also be used to maintain disease states. It is argued that a systems approach to drug and therapy design for robust diseases such as cancer, autoimmunity or diabetes should allow to identify fragilities that are hidden in the mechanisms that give rise to the robustness of the pathological states [10]. To understand the robustness/fragility, one needs to examine the stability/sensitivity of the system behaviour in light of variations in structure or parameters. Figure 2.2 summarizes the key components which have to be looked at in order to achieve a systems-level understanding of robustness. Let us start with some basic definitions from http://en.wikipedia.org.

- System, a set (complex) of interacting or interdependent components (elements) forming an integrated whole.
- Systems characteristics are structure, behaviour, interconnectivity and function(s).

G. Bocharov et al., *Mathematical Immunology of Virus Infections*,
https://doi.org/10.1007/978-3-319-72317-4_2

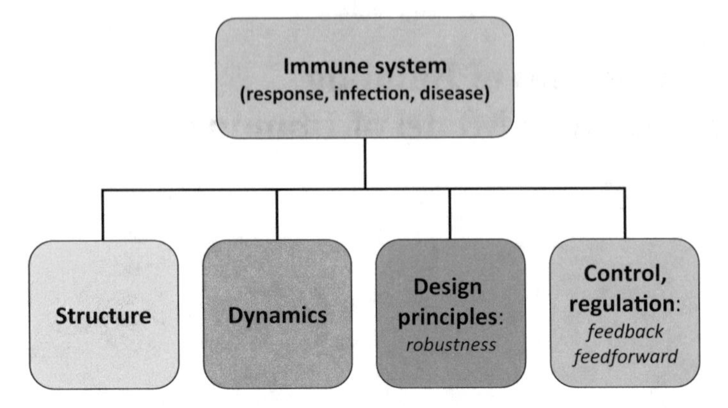

Fig. 2.1 Fundamental properties of biological systems which are the subject of systems analysis. Adapted from *Kitano, Science, 2002, 1662–1664; Csete and Doyle, Science, 2002, 1664–1669*

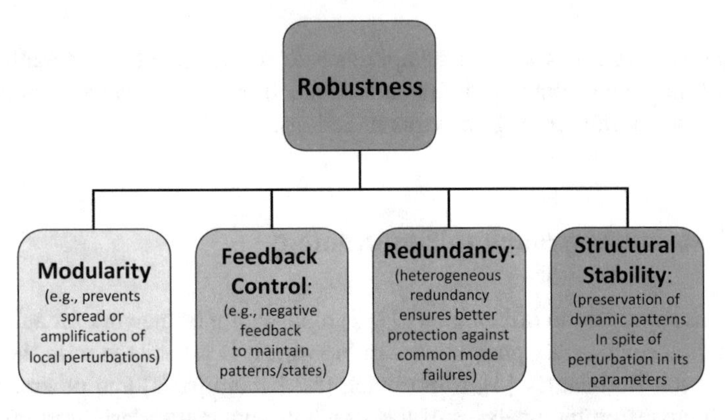

Fig. 2.2 Structural and functional features underlying robustness of biological systems. Adapted from *Kitano, Science, 2002, 1662–1664; Csete and Doyle, Science, 2002, 1664–1669*

- Systems science aims to develop interdisciplinary foundations that are applicable in a variety of areas, such as engineering, biology, medicine and social sciences.
- Systems theory is the study of the principles that can be applied to all types of systems at all levels of organization in all fields of research.
- Systems analysis is the study of sets of interacting entities.

Starting from his pioneer work in 1945 [4], Karl Ludwig von Bertalanffy stated that

> …there exist models, principles and laws that apply to generalized systems or their subclasses, irrespective of their particular kind, the nature of their component elements, and the relationships or 'forces' between them. It seems legitimate to ask for a theory, not of systems of a more or less special kind, but of universal principles applying to systems in general.

Significant development towards the concept of a system was done by Wiener [11] and Ashby [12] who pioneered the use of mathematics to study complex systems.

They invented the notion of a feedback and established a new field of mathematics—the Cybernetics. It is about the study of feedback and derived concepts such as communication and control in living organisms, machines and organizations. Its focus is on how anything (digital, mechanical or biological) processes information, reacts to information and changes or can be changed to better accomplish the first two tasks.

With the explosion of data at the subcellular level there are difficulties to interpret them in relation to the physiological behaviour of complex living organisms. A systems biology approach is a paradigm (from Noble [13]):

> ...about putting together rather than taking apart, integration rather than reduction. It requires that we develop ways of thinking about integration that are as rigorous as our reductionist programmes, but different.... It means changing our philosophy, in the full sense of the term.

Systems biology can be defined as a field of study of the interactions between the components of biological systems, and how these interactions give rise to the function and behaviour of that system. The paradigm implies the application of systems theory to the complexity of biological interactions [14] at all levels, i.e. from genes to proteins to pathways to subcellular reactions to cells to tissues to organs and finally, to the whole organisms. A systems approach is essentially based on mathematical modelling and requires the triad: Experiments + Mathematical Modelling + Theoretical concepts.

2.1.1 Theories in Immunology

Whereas the experimental immunology functions to produce observations and to understand the relationships between two or more quantities, a theoretical method ideally is needed to generate hypotheses and then to deduce the consequences to explain the cause and effect relationships between the hypotheses and the known facts [15]. Theories provide a means of making new discoveries [16]. A theory is not successful unless it is fruitful, i.e. it should enable one to deduce certain previously unknown consequences. The formation of a theory involves the construction of a model. The postulates of scientific theories must agree with the way nature is observed to behave whereas pure mathematics is not restricted by the empirical laws.

The main body of immunology is constituted by non-mathematical theories (empirically derived) such as

- Clonal selection theory (Nobel Laureate, Burnet) [17];
- Network-type (e.g. idiotypic) theory of immune regulation (Nobel Laureate, Jerne) [18];
- Spatio-temporal rules of immune response regulation (Nobel Laureate Rolf M. Zinkernagel) specifying the basic parameters in immune response phenotype regulation between activation, unresponsiveness and death [19]:

- Antigen structure,
- Mode of presentation,
- Spatial localization,
- Dose,
- Time of availability,
- Frequencies of specific T and B cells,
- Thresholds of binding avidity of T- and B-cell receptors,
- Variation of the response thresholds on cell maturation stage.

• Balance of growth and differentiation—conceptual framework of immune response regulation (Grossman and Paul) [20–23]

- Individual cells tune and update their activation thresholds,
- Immune system responds to a rapid perturbation in its homeostasis,
- Individual lymphocytes respond to a rapid change in the level of stimulation rather than a stimulation per se.

Successful theories in immunology are still in the stage of growth and revision. In particular because, data explosion in immunology due to high-throughput and advanced experimental techniques is insufficient by itself to understand the complexity and predict the immune system behaviour as a whole. Nonlinearity, threshold effects, feedback control loops, delays, compartmental organization and the redundancy inherent in the immune processes call for mathematical modelling to be a part of modern immunology and virology studies. Science and data-driven modelling analysis of the immune system is expected to allow one

• to estimate relevant parameters ('numbers game'),
• to understand the dynamic patterns of observations,
• to identify the sensitivity of dynamics to control mechanisms,
• to reduce the need for (animal) experiments and
• to optimize its performance both in vivo and in vitro.

2.2 A Mathematical Model

An insightful discussion of the role that mathematics plays in theoretical and experimental science (physics) can be found in [16]. Mathematics provides universal language for expressing causal and functional relationships between observations. The relationships can be expressed as equations. These equations can be manipulated together to derive new relations between concepts which is a subject of mathematical modelling work [24].

A mathematical model is the description of a system using mathematical concepts and language. It describes the system by a set of variables and a set of equations that establish relationships between the variables. Mathematical models can take many forms including but not limited to dynamical systems, statistical models, differential equations, game-theoretic models, etc. There exists a number of modelling issues that need to be addressed, e.g. complexity, fit to empirical data and sensitivity analysis. The model formulations should be tightly linked to a fundamental step called coordinatization (Weyl [25]), that is, the quantitative definition of observables, parameters and structures that enable the link with a biological phenotype.

2.2.1 *Basic Issues*

To formulate a mathematical model, one needs to proceed as follows:

- Generate a conceptual scheme of the system under study. One restricts the model to the most important interactions (note that, in this sense, mathematical modelling represents a reductionists' approach to studying complex systems).
- Specify modelling assumptions and select the following types of quantities
  - time and space as independent variables;
  - time- and space-dependent variables:
    population densities of cells and pathogens,
    concentrations of molecules,
  - the parameters which characterize the kinetics of the specified processes.
- Set up or derive model equations.

There is no rigorous ways to set up a mathematical model. In fact a compromise between the level of details (complexity), biological realism and tractability has to be found. The types of descriptions which can be considered for mathematical model building are shown in Fig. 2.3 [26].

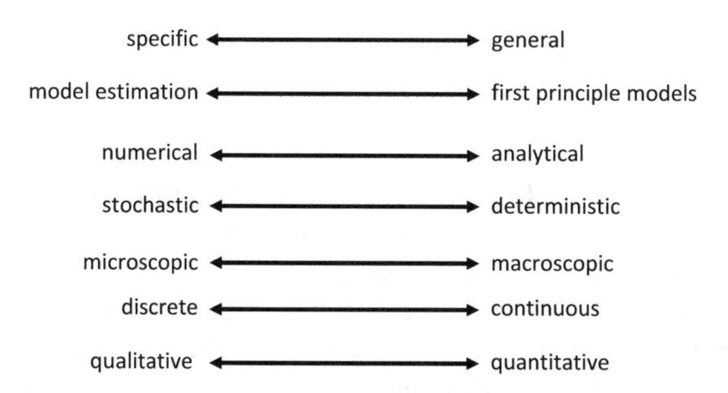

Fig. 2.3 Range of major issues to be addressed in order to formulate mathematical model of biological system. They are related to the nature of the model, level of resolution, complexity, computational implementation, etc. Reprinted from "The Nature of Mathematical Modeling" © Cambridge University Press, Neil Gershenfeld, Fig. 1.1 with permission from Cambridge University Press

2.2.2 Dynamics of Immune Responses

Lymphocytes respond to antigen stimulation through processes such as activation, division, differentiation and death. The kinetics of the cell population dynamics follows a remarkably similar pattern:

- expansion,
- contraction and
- memory persistence.

The dynamics of any population is affected by four processes. This leads to the most fundamental balance equation in population dynamics of immune responses [27]:

$$\Delta N = birth - death + immigration - emigration, \tag{2.1}$$

where ΔN is the change in the population size over some time period Δt. Even for single populations, each rate is likely to depend on the population size and the structure, and is variable in time. The critical task is to choose or specify the functional form of the processes in the model.

A mathematical description of the immune processes should address the dynamics of cellular responses to antigens (either at the population or the single cell level) occurring locally or systemically. To quantify the population dynamics of cells and molecules in one compartment[1] (defined either physically or functionally), one needs to consider the rates of (i) growth, (ii) death, (iii) immigration and (iv) emigration. The relationship can be represented by the following prototype structure of balance equation the lymphocyte population in a single compartment (e.g. spleen, lymph node):

$$\frac{d}{dt}N(t) = \begin{pmatrix} proliferation \\ or \\ \text{`}multiplication' \end{pmatrix} \pm (differentiation) - (death) \pm \begin{pmatrix} transport \\ or \\ \text{`}transfer' \end{pmatrix}. \tag{2.2}$$

In the majority of models in immunology, the building blocks (the individual terms in the differential equations) represent the following elementary processes: (i) growth of pathogens and cells; (ii) cellular and molecular interactions (e.g. antigen–antibody or receptor–ligand); (iii) activation, division and death of lymphocytes; and (iv) homeostasis in the immune system. Possible functional forms of these are given in the rest of this chapter summarizing the results from several publications [15, 28–31].

[1] A compartment is defined by a characteristic material occupying a given volume and which is kinetically homogeneous.

2.3 Elementary Building Blocks for Models

2.3.1 *Ag–Ab Interaction*

The chemical nature of interaction of antigen with antibodies was established by Heidelberger and Kendall in 1935 [32]. This implies that the chemical law of mass action can be used to model the kinetics of antigen–antibody interactions. Consider the reversible reaction of a monovalent antigen (Ag) with an antibody binding site (Ab) as depicted in Fig. 2.4:

$$Ag + Ab = \underset{k_{-1}}{\overset{k_1}{\rightleftarrows}} Ag \cdot Ab. \tag{2.3}$$

The corresponding differential equation for the rate of change of the concentration of the complex [AgAb] over time in a well-mixed compartment reads

$$\frac{d}{dt}[AgAb] = k_1[Ag][Ab] - k_{-1}[AgAb]. \tag{2.4}$$

In an equilibrium, the concentrations of the substances are related by the formula

$$\frac{[AgAb]}{[Ag][Ab]} = \frac{k_1}{k_{-1}} = K. \tag{2.5}$$

The constant K is called an intrinsic affinity of the antibody binding site for the antigen. It can be viewed as an example of the coordinatization of notion of 'specificity'.

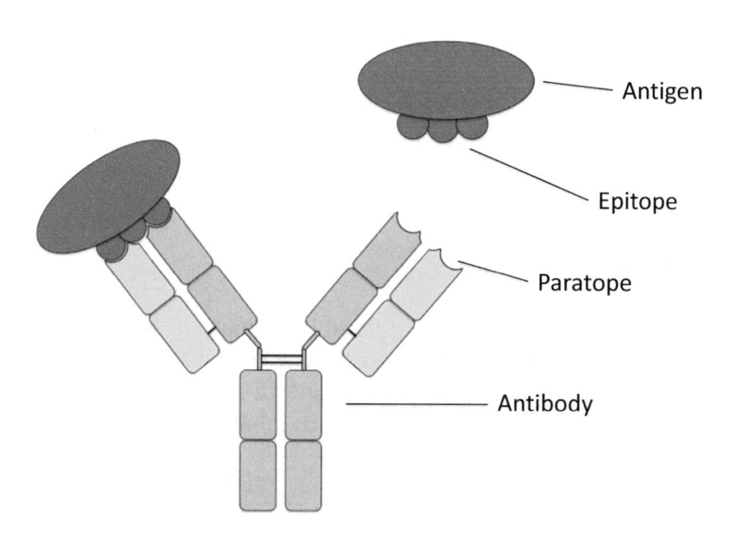

Fig. 2.4 Antigen and antibody

Note that multivalent Ab and Ag follow a more complex dynamics studied by Bell, DeLisi, Perelson, Gandolfi and others (see for References in [3, 15]).

2.3.2 *Growth Phenomena*

The process of biological species growth has common features as summarized in [33]. The material of this subsection relies on the systematic analysis of growth equations presented in the above monograph.

2.3.2.1 Exponential Growth Equation

The most basic growth equation was introduced by Malthus in his early assays on the principles of population growth already in 1798. It considers $N(t)$—the magnitude of a growing quantity as a function of time t. The rate of change of the population size

$$\frac{d}{dt}N(t) = b \cdot N(t), \tag{2.6}$$

which has the following solution:

$$N(t) = N(0)\exp^{bt}. \tag{2.7}$$

One can define a biologically relevant parameter, i.e. the 'population doubling time' in relation to the proliferation rate b

$$t_2 = \frac{\ln 2}{b}. \tag{2.8}$$

Although Eq. 2.6 is simple, it is useful in studies of short-term growth phenomena. One example of exponential cell growth is shown in Fig. 2.5 for B lymphocyte proliferation in vitro.

2.3.2.2 Logistic Growth Equation

The logistic growth equation was formulated by Verhulst in his work on limits to population growth in 1838. It can be considered as an extension of Eq. 2.6 by considering the limiting resources available for growth associated with the notion of carrying capacity C which enters the equation as follows:

$$\frac{d}{dt}N(t) = b \cdot N(t)\left(1 - \frac{N(t)}{C}\right), \tag{2.9}$$

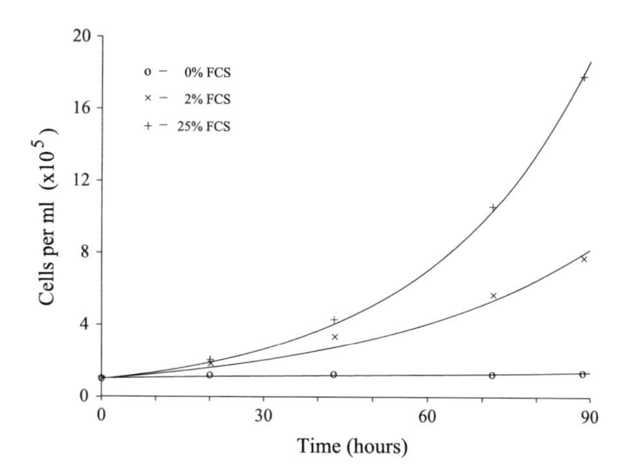

Fig. 2.5 Exponential growth of B lymphocyte in vitro and the best-fit solution of the exponential growth equation

which has the following solution:

$$N(t) = \frac{C}{1 + (C/N(0) - 1)\exp^{-bt}}. \tag{2.10}$$

Here, the parameters have the following meaning:

- b the intrinsic growth coefficient,
- C the carrying capacity,
- $d = b/C$ is called 'crowding coefficient'.

An example of virus growth in vivo, i.e. LCMV replication in spleen of C57BL/6 mice [34], consistent with the logistic model is shown in Fig. 2.6.

2.3.2.3 Confined Exponential Growth Equation

This equation describes an exponential growth confined to a limiting value bounded by C. It was first considered by Kreith 1958 and Bird et al. in 1960 to represent a source mechanism for transfer of heat or mass.

$$\frac{d}{dt}N(t) = b\,(C - N(t)), \tag{2.11}$$

which has the following solution:

$$N(t) = C - (C - N(0))\exp^{-bt}. \tag{2.12}$$

Here, the parameters have the following meaning:

- b the growth or transfer coefficient,
- C the carrying capacity.

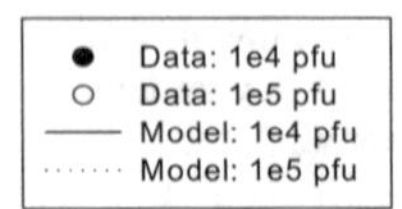

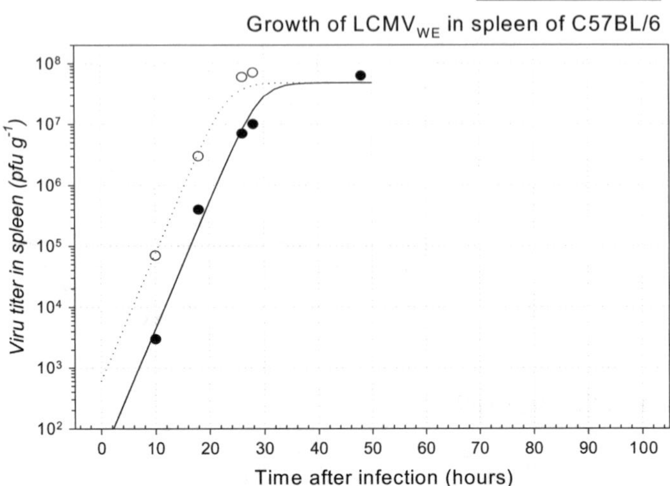

Fig. 2.6 LCMV-WE replication in spleen of C57BL/6 mice. Data (open and closed symbols) and the best-fit solution of the logistic equation. Intravenous infections with 10^4 and 10^5 pfu are shown with closed and open circles, respectively. Reprinted from Journal of Theoretical Biology, Vol. 221, Bocharov et al., Modelling the Dynamics of LCMV Infection in Mice: II. Compartmental Structure and Immunopathology, Pages 349–348, Copyright © 2003, with permission from Elsevier

The solution of the equations approaches C as $t \to \infty$. If $C = 0$, then $N(t)$ approaches 0 for increasing t. This equation is used in mathematical immunology to model the homeostasis of lymphocytes in the periphery due to the naive cell transfer from primary lymphoid organs.

2.3.2.4 Gompertz Equation

The equation formulated by Gompertz in his work on the law of human mortality in 1825 assumes that the intrinsic growth rate parameter is not constant but depends on time according to exponential growth:

$$\frac{d}{dt}N(t) = b \cdot N(t), \; N(0) = N_0, \tag{2.13}$$

$$\frac{d}{dt}b(t) = -k \cdot b(t), \; b(0) = b_0. \tag{2.14}$$

It can be formulated as a single non-autonomous equation with the explicit solution as follows:

$$\frac{d}{dt}N(t) = b_0 \exp^{-kt} \cdot N(t), \;\; N(t) = N_0 e^{\left[\frac{b_0}{k}(1-e^{-bt})\right]}. \tag{2.15}$$

Using the constant expression $C = N_0 \exp^{b_0/k}$, representing a kind of carrying capacity, the equivalent formulation of the population dynamics can be expressed in a form most close to the logistic equation

$$\frac{d}{dt}N(t) = k \cdot N(t)\,(\ln C - \ln N(t))\,. \tag{2.16}$$

The Gompertz equation provides a correct description of concave patterns of cell growth.

2.3.2.5 Time-Dependent Parameters

A generalization of the above models comes from an explicit dependence on time of the growth rate and carrying capacity parameters. These types of equations were extensively considered by Nisbet and Gurney [35] to model the dynamics of populations in fluctuating environments:

$$\frac{d}{dt}N(t) = b(t) \cdot N(t)\left(1 - \frac{N(t)}{C(t)}\right). \tag{2.17}$$

The solution can be expressed in a closed form

$$N(t) = \frac{N_0 \exp\left(\int_0^t b(\xi)d\xi\right)}{1 + N_0 \int_0^t b(\xi)/C(\xi) \exp\left(\int_0^{\xi} b(\nu d\nu)d\xi\right)}. \tag{2.18}$$

Various functional forms can be considered to account for the time dependence, e.g. of the growth rate:

- Linearly variable $b(t) = b_0(1 - at)$,
- Hyperbolically variable $b(t) = \frac{b_0}{(1+at)}$,
- Exponentially variable $b(t) = b_0 \exp^{-kt}$,
- Sinusoidally variable $b(t) = b_m + b_A \sin(2\pi/T + \phi)$.

2.3.2.6 Time-Delay Equations

In all biological systems, the present growth rate at time t depends on the size of the population at an earlier time. This assumption leads to a straightforward modification of the exponential equation [36]:

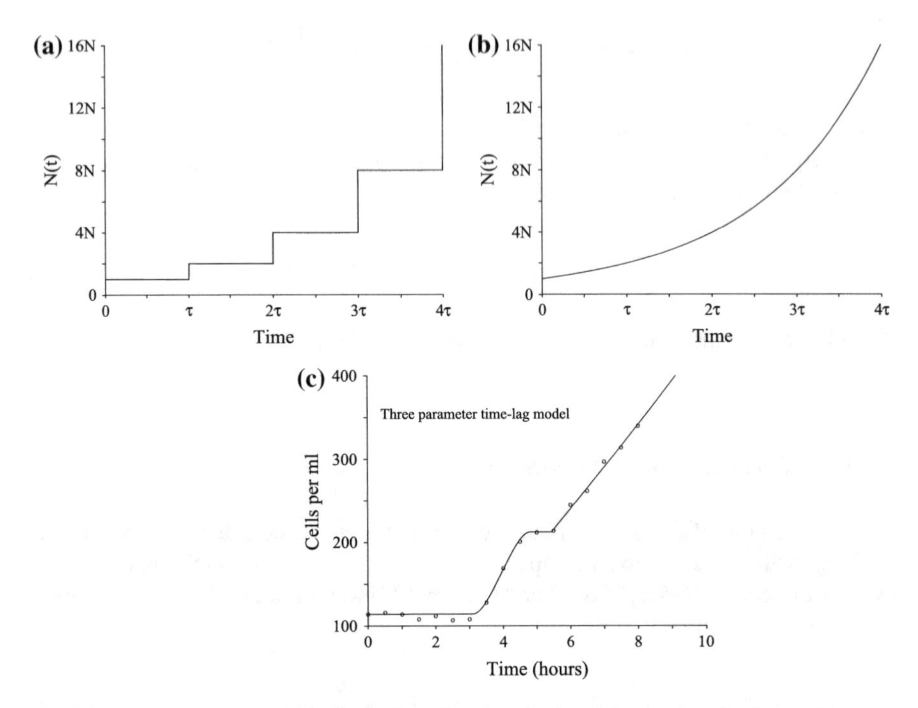

Fig. 2.7 Idealized patterns of cell growth. **a** Synchronized and **b** asynchronized growths corresponding to geometric and exponential models of proliferation, respectively. **c** Real data and the solution of the model (2.19) of cell proliferation, both characterized by the loss of synchrony

$$\frac{d}{dt}N(t) = b \cdot N(t - \tau), \ t \geq 0, \tag{2.19}$$

with τ representing reproduction/division time delay or time lag. To solve the initial value problem for the above delay differential equation (DDE), the initial data have to be specified on a time interval rather than at one time instant as in the ODE models listed above. Figure 2.7 shows the idealized patterns of cell growth and real data pattern which is consistent with a DDE-type dynamics. The outstanding feature of the cell system in vitro is that the division of cells is initially synchronized. As one can see in the figure, the synchronization is lost with time.

A broadly examined equation of population growth with a time delay is the delay logistic equation proposed by Hutchinson in 1948:

$$\frac{d}{dt}N(t) = b \cdot N(t - \tau)\left(1 - \frac{N(t)}{C}\right). \tag{2.20}$$

An outstanding feature of this scalar equation is the rich dynamics including periodic and chaotic regimes.

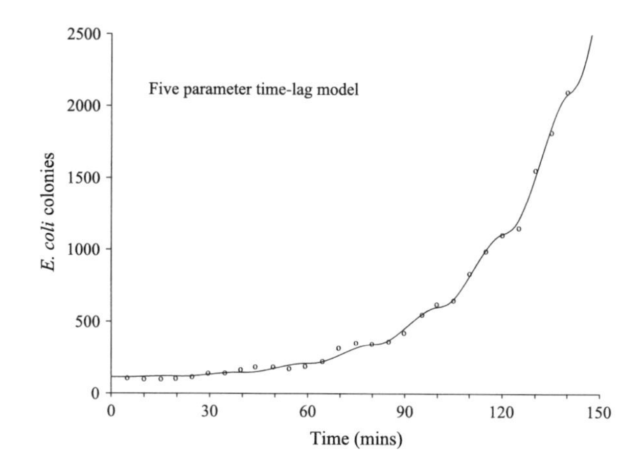

Fig. 2.8 Growth of the synchronized cell populations reproduced with a neutral DDE (2.21)

Finally, the growth of a synchronized cell populations, as shown in Fig. 2.8, can be parsimoniously represented by a neutral delay differential equation with the rate of change depending also on the rate of growth at some previous time [36]:

$$\frac{d}{dt}N(t) = a \cdot N(t) + b \cdot N(t-\tau) + c \cdot \frac{d}{dt}N(t-\tau),\ t \geq 0. \tag{2.21}$$

The numerical treatment of the neutral delay differential equations is rather demanding due to the discontinuities of solution derivatives at times defined by multiples of τ. DDEs are used quite extensively in mathematical immunology [29]. A systematic investigation of the reduction of age-structured populations dynamics models to delay differential equations is presented in [37].

2.3.3 Lymphocyte Proliferation

Induction of lymphocyte proliferation in response to antigenic stimulation is a multi-step process, which includes T-cell receptor-mediated signal transduction leading to genes up-/downregulation. The networks of protein and gene interactions in a single cell during the activation stage are in the focus of systems biology studies. At the cell population level, there exist rather few attempts to derive or infer from data-fitting constitutive relationships for the functional dependence of the antigen-induced division rate $R_{div}(Ag, N)$ of lymphocytes on the antigen concentration $Ag(t)$ and the cell density $N(t)$. Examples of possible forms can be summarized from [38, 39] as follows:

- second-order reaction kinetics, with no competition and without saturation effects:

$$R_{div} \propto b\, Ag(t)\, N(t); \tag{2.22}$$

- bounded rate growth (saturation at high antigen load):

$$R_{div} \propto b\, N(t) \frac{Ag(t)}{\theta + Ag(t)} \tag{2.23}$$

- antigen-specific resource competition:

$$R_{div} \propto b\, N(t) \frac{Ag(t)}{\theta + Ag(t) + c\, N(t)} \tag{2.24}$$

- non-specific resource competition:

$$R_{div} \propto b\, N(t) \frac{A(t)}{\theta + A(t) + c\, N(t)} \times \frac{N(t)}{1 + c\, N(t)} \tag{2.25}$$

- inhibitory lymphocyte interaction:

$$R_{div} \propto b\, N(t) \left(\frac{Ag(t)}{\theta + Ag(t)} - d\, N(t) \right). \tag{2.26}$$

The last expression includes a second term which resembles a crowding effect of cells on their division via enhancing their death.

2.3.4 Cell Death

Cell death is a key process in maintaining the homeostasis of the immune system as well as in the development of disease states (e.g. chronic infections). The lifespan of cells of the immune system is a tightly regulated process. There is a range of mechanisms responsible for cell death or elimination. Most of the mathematical models in immunology restrict the cell death description either to an exponential ('natural' death)- or a predator–prey (effector-mediated elimination)-type kinetics. The functional forms for the death rate R_{death} used in the models are

- exponential decay

$$R_{death} \propto -d\, N(t), \tag{2.27}$$

  where $d > 0$ is the intrinsic death coefficient.
- Gompertz model of cell death

$$R_{death} = -d \cdot N(t), \; N(0) = N_0, \tag{2.28}$$

$$\frac{d}{dt} d(t) = k \cdot d(t), \; d(0) = d_0, \tag{2.29}$$

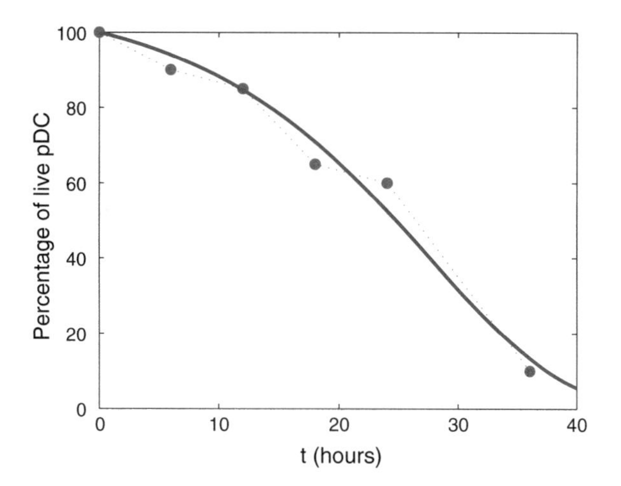

Fig. 2.9 Gompertz kinetics of cell death. Survival data of plasmacytoid dendritic cells after infection with MHV at MOI = 1 and the best-fit solution of the Gompertz model

describes the kinetics of cell survival which is characterized by a gradually increasing per capita death rate. Figure 2.9 shows an example of Gompertz-type death kinetics observed for plasmacytoid dendritic cells in vitro [40].

- the second-order kinetics due to crowding effects

$$R_{death} \propto d\, N^2(t). \tag{2.30}$$

Note that the logistic pattern of growth $N'_{death} \propto b\, N(t) - d\, N^2(t)$ may be regarded as an exponential growth mitigated by death from a second-order crowding effect.

- the effector cell-mediated elimination with or without saturation effects, respectively

$$R_{death} \propto d\ \ N(t)\, \frac{E(t)}{\theta + E(t)},\ or \propto d\, N(t)\, E(t). \tag{2.31}$$

- the fixed-time-delay description of antigen-dependent activation-induced cell death by apoptosis (see, e.g. [41])

$$R_{death} \propto d\, N(t)\, Ag(t)\, Ag(t - \tau). \tag{2.32}$$

- the division-number-dependent death rate, which assumes that there is an upper limit (the so-called 'Hayflick limit' n_{max}, that was observed for cell division *in vitro*) on the number of cell divisions n:

$$R_{death} \propto d\, \frac{(n(t)/n_{max})^m}{1 + (n(t)/n_{max})^m}, \tag{2.33}$$

where m is an integer similar to the Hill coefficient that determines the steepness of the sigmoidal dependence. It reflects the concept of replicative senescence.

2.3.5 Cell Differentiation

The activation of lymphocyte via interaction with antigen can result either in division, differentiation, anergy or death of the cell, depending on the strength of the stimulus and the availability of co-stimulatory factors. The regulation of cell proliferation versus differentiation is the least studied process in terms of a mathematical description. The commonly used approach is based on an ad hoc specification of the probability of proliferation/differentiation as a function of antigen levels: e.g.

- B lymphocytes

$$p_d = \frac{K\,Ag}{1 + K\,Ag}, \tag{2.34}$$

- T lymphocytes

$$p_p = \frac{Ag}{K + Ag}. \tag{2.35}$$

The fraction of cells following the differentiation pathway is defined as p_d or $1 - p_p$.

It is important to bear in mind that the balance of growth and differentiation is a dynamically regulated process which should be described as an emergent property of interacting heterogeneous populations of cells differing in their level of maturity [42] with a more differentiated cells exerting a negative feedback on the proliferation rate of the activated cells in a density-dependent manner.

2.3.6 Tuning of the Response

Lymphocytes are known to tune their activation thresholds. This could manifest either as a low- or high-zone tolerance phenomenon (see Fig. 2.10), or a gradual reduction of the response to the same level of stimulation (anergy). Note that an experimentally observed bell-shaped pattern of the immune response dependence on the activation level or speed of the perturbation, as shown in Fig. 2.11 (left), is consistent with the systems biology view of the need of biphasic regulatory modules Fig. 2.11 (right) underlying the functioning of the living systems [9]. The available practice with

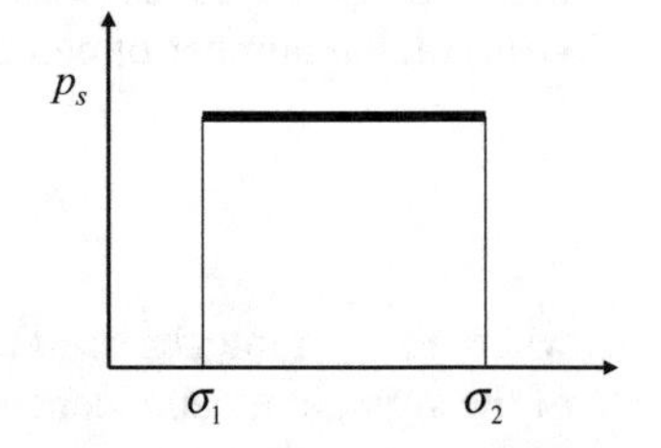

Fig. 2.10 Thresholds for low- and high-zone tolerance. Schematic view of the probability of lymphocyte activation as a function of the number of bound antigen-specific receptors

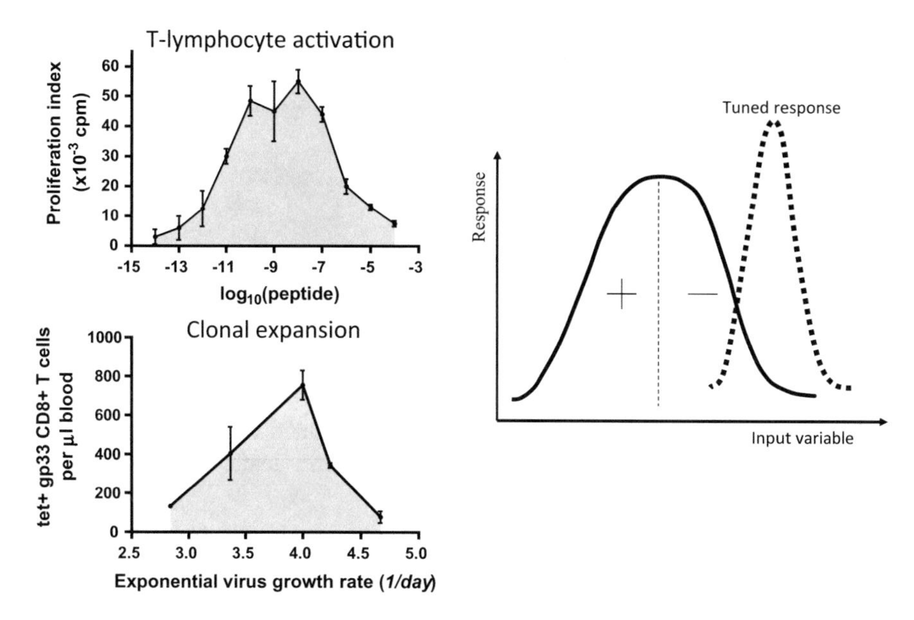

Fig. 2.11 Bell-shaped patterns in activation of immune responses for varying peptide load, and virus growth rate. Idealized form of the dose–response to be considered in equations of cell activation. Reprinted from Transfusion Medicine and Hemotherapy, Vol. 32, Bocharov, Understanding Complex Regulatory Systems: Integrating Molecular Biology and Systems Analysis, Pages 304–321, Copyright © 2005, with permission from Karger Publishers

parameterisation of these modes of response in the models can be summarized as follows:

- B lymphocytes stimulated by antigen

$$p_a = \begin{cases} 1, & \text{if } \frac{\sigma_1}{1-\sigma_1} \le K\,Ag \le \frac{\sigma_2}{1-\sigma_2} \\ 0, & \text{otherwise.} \end{cases} \tag{2.36}$$

  Here, σ_1 and σ_2 denote the threshold values for fraction of bound immunoglobulin receptors (n) needed for productive activation as shown in Fig. 2.10. The total number of receptors per B cell is about 10^5.
- T lymphocytes responding to virus infection characterized by viral load $V(t)$, see [41]

$$p_a = \frac{b \cdot V(t)}{\theta + \left(\int_0^T V(t)\,dt\right)^2}. \tag{2.37}$$

A systemic mechanism of the negative feedback of the virus infection on the immune response can be associated with the destruction of the target organ resulting in a generalized suppression of the immune processes. The mathematical models of infectious diseases proposed by G.I. Marchuk implement this paradigm via a general negative feedback function depending on the number of damaged sensitive tissue cells [28].

2.3.7 Cell Competition

The immune system needs to accommodate increasing numbers of newly generated cells within a limited space of lymphoid organ, so a selection process through competition between different subsets of cells and between specific clones takes place. The reference set of functional forms to describe the competition can be found in models of ecological systems dynamics [35, 43].

The homeostasis in the immune system is maintained by a complex regulatory network leading to cell survival, proliferation and death. This includes the following [44]:

- spatial control of cell life and death processes, which in case of two clones N_1 and N_2 could be described as follows:

$$\frac{d}{dt}N_1(t) = b_1 \cdot N_1(t)\left(1 - \frac{N_1(t) + a_{12}\,N_2(t)}{C}\right) \tag{2.38}$$

$$\frac{d}{dt}N_2(t) = b_2 \cdot N_2(t)\left(1 - \frac{N_2(t) + a_{21}\,N_1(t)}{C}\right) \tag{2.39}$$

  with a_{12}, a_{21} specifying the strength of the competition.
- competition between cells for the limited resources $R(t)$ provided by antigen-presenting cells, e.g. persisting cross-reactive antigens, self-antigens, MHC class I or II molecules and survival factors

$$\frac{d}{dt}N_1(t) = b_1 \cdot N_1(t)\frac{R(t)}{\kappa_1 + R(t)} \tag{2.40}$$

$$\frac{d}{dt}N_2(t) = b_2 \cdot N_2(t)\frac{R(t)}{\kappa_2 + R(t)} \tag{2.41}$$

$$\frac{d}{dt}R(t) = S^* - u_R \cdot R(t)\left(\frac{b_1\,N_1(t)}{\kappa_1 + R(t)} + \frac{b_2\,N_2(t)}{\kappa_2 + R(t)}\right). \tag{2.42}$$

  Here, the parameters $S^*, \kappa_1, \kappa_2, u_R$ characterize the production and consumption of the resource factors.

Other considerations could be employed to represent the inherent regulations, such as antigen/lymphocyte relationships or supportive/antagonistic relationships between components; each would bring differing details to the dynamics and the attractors, and distinguishing the optimum choice will depend on the available data.

Completing this chapter, one needs to note that modelling of the homeostatic regulation of lymphocyte clones is a practically uncovered area in mathematical immunology. A straightforward approach to model the maintenance of the i-th lymphocyte clone $N_i(t)$ within the system that controls the total number of immune cells near the carrying capacity can be described via a density-dependent term of the following type (here 10^5 is the total number of specific clones):

$$\frac{d}{dt}N_i(t) = N_i(t)\left[F\left(\sum_{i=1}^{\sim 10^5} N_i(t)\right) - d\right], \tag{2.43}$$

or as constant confined exponential growth that can be found in [29, 45].

References

1. Bocharov, G. (2005) Understanding complex regulatory systems: integrating molecular biology and systems analysis Transfusion Medicine and Hemotherapy **32**(6): 304–321.
2. Ludewig B, Stein JV, Sharpe J, Cervantes-Barragan L, Thiel V, Bocharov G. A global "imaging" view on systems approaches in immunology. *Eur. J. Immunol.* (2012); **42**(12):3116–25.
3. Mohler, R.R., Bruni, C., Gandolfi, A., A systems approach to immunology, *Proc. IEEE*, (1980) **68**: 964–990.
4. von Bertalanffy, K.L., General System Theory: Foundations, Development, Applications (1968). New York: George Braziller.
5. Mesarovic, M.D., General systems theory and biology view of a theoretician. In: Mesarovic?, D.M. (Ed.), General Systems Theory and Biology. Springer (1968).
6. Kitano, H., Systems biology: A brief overview. *Science*, **295** (2002) 1662–1664.
7. Kitano, H., Computational systems biology, *Nature*, **420** (2002) 206–210.
8. Csete, M.E., Doyle, J.C., Rerverse engeneering of biological complexity, *Science*, **295** (2002) 1664–1669.
9. Levchenko, A., Bruck, J., Sternberg, P.W., Regulatory modules that generate biphasic signal response in biological systems, *Syst. Biol.*, **1** (2004) 139–148.
10. Kitano, H., Cancer as a robust system: implications for anticancer therapy, *Nature Reviews Cancer*, **4** (2004) 227–235.
11. Wiener, N. Cybernetics: Or Control and Communication in the Animal and the Machine. (1948), Paris, (Hermann and Cie) and Camb. Mass. (MIT Press).
12. Ashby, W.R. (1956). An Introduction to Cybernetics, Chapman and Hall.
13. Noble, D. The music of life: Biology beyond the genome. (2006), Oxford: Oxford University Press.
14. Kalman, R.E., New developments in systems theory relevant to biology. In: *Systems Theory and Biology*, Springer-Verlag, Berlin (1968), Mesarović, M.D. (Editor), 222–232.
15. Bell G, Perelson AS, Pimbley G (eds): Theoretical Immunology. New York, Marcer Dekker (1978). 646 pp.
16. Constant F.W. Fundamental laws of physics (1963). Addison-Wesley, Reading, Massachusetts.
17. Burnet, F.M. The clonal selection theory of acquired immunity. *Aust. J. Sci.***20**, 6769 (1957).
18. Jerne N. The immune system. *Scientific American* (1973); 229:5260.
19. Zinkernagel, R.M., Immunology and immunity against infection: General rules, *J. Comput. Appl. Math.*, **184** (2005) 4–9.
20. Grossman, Z., What did mathematical models contribute to AIDS research? (Review of [41]) *Trends in Ecology & Evolution*, **16** (2001) 468–469.

21. Grossman, Z., Mathematical modeling of thymopoiesis in HIV infection: real data, virtual data, and data interpretation *Clin Immunol.*, **107** (2003) 137–139.
22. Grossman, Z. and Paul, W.E., Autoreactivity, dynamic tuning and selectivity, *Curr. Opin. Immunol.*, **13** (2001) 687–698.
23. Grossman, Z., Min, B., Meier-Schellersheim, M., Paul, W.E., Concomitant regulation of T-cell activation and homeostasis. *Nature Rev. Immunol.*, **4** (2004) 387–395.
24. Polderman, J.W., Willems, J.C., *Introduction to Mathematical Systems Theory. A behavioral approach*, Texts in Applied Mathematics, **26**, Springer-Verlag, New York, (1998).
25. Weyl, Hermann. The Classical Groups: Their Invariants and Representations. Princeton University Press, (2016), 336 p.
26. Gershenfeld, N.A. (2000) *The Nature of Mathematical Modelling*, Cambridge University Press, Cambridge.
27. Andrew, S.M., Baker, C.T.H., Bocharov, G.A. Rival approaches to mathematical modelling in immunology, *J. Comput. Appl. Math.*, **205** (2007) 669–686.
28. Marchuk GI. Mathematical Models in Immunology. New York, Optimization Software, Inc., (1983), 378 p.
29. Marchuk, G.I., *Mathematical models of immune response in infectious diseases*, transl. from Russian by G. Kontarev & I. Sidorov, Kluwer Press, Dordrecht (1997).
30. Merril, S.J., Computational models in immunological methods: an historical review, *Journal of Immunological Methods*, **216** (1998) 69–92.
31. Nowak, M.A., May, R.M. *Virus Dynamics. Mathematical Principles of Immunology and Virology*, Oxford University Press, Oxford (2000).
32. Heidelberger, M., Kendall, F.E., The precipitin reaction between type III pneumococcus polysaccharide and homologous antibody: III. a quantitative study and a theory of the reaction mechanism *J. Exp. Med.*, **61** (1935) 563–591.
33. Banks, R.B. (1994) *Growth and Diffusion Phenomena. Mathematical Frameworks and Applications*, Springer-Verlag, Berlin.
34. Bocharov, G., Klenerman, P., Ehl, S., Modelling the Dynamics of LCMV Infection in Mice: II. Compartmental Structure and Immunopathology. *J. Theor. Biol.*, **221** (2003) 349–378.
35. Nisbet, R.M., Gurney, W.S.C., *Modelling Fluctuating Populations*, Wiley, Chichester (1982).
36. Baker, C.T.H., Bocharov, G.A., Paul, C.A.H., Rihan, F.A., Modelling and analysis of time-lags in some basic patterns of cell proliferation, *J. Math. Biol.*, **37** (1998) 341–371.
37. Bocharov, G.A., Hadeler, K.P., Structured population models, conservation laws, and delay equations, *J. Differ. Equ.*, **168** (2000), 212–237.
38. De Boer, R.J., Perelson, A.S., Towards a general function describing T cell proliferation, *J. theor. Biol.*, **175** (1995) 567–576.
39. Borghans, J.A., Taams, L.S., Wauben, M.H.M., De Boer, R.J., Competition for antigenic sites during T cell proliferation: A mathematical interpretation of in vitro data, *Proc. Natl. Acad. Sci. USA.*, **96** (1999) 10782–10787.
40. Bocharov G, Zust R, Cervantes-Barragan L, Luzyanina T, Chiglintsev E, Chereshnev VA, Thiel V, Ludewig B. A systems immunology approach to plasmacytoid dendritic cell function in cytopathic virus infections. *PLoS Pathog.* (2010); 6(7):e1001017.
41. Bocharov, G.A., Modelling the Dynamics of LCMV Infection in Mice: Conventional and Exhaustive CTL Responses, *J. Theor. Biol.*, **192** (1998) 283–308.
42. Bocharov G, Quiel J, Luzyanina T, Alon H, Chiglintsev E, Chereshnev V, Meier-Schellersheim M, Paul W, Grossman Z. (2011) Feedback regulation of proliferation versus differentiation explains the dependence of antigen-stimulated CD4 T-cell expansion on precursor number. *Proc. Natl. Acad. Sci. USA* 108(8):3318–23.
43. Hassell M.P., *The Spatial and Temporal Dynamics of Host-Parasitoid Interactions*, Oxford University Press, Oxford (2000).
44. McLean, A.R., Rosado, M.M., Agenes, F., Vascocellos, R., Freitas, A.A., Resource competition as a mechanism for B cell homeostasis, *Proc. Natl. Acad. Sci. USA.*, **94** (1997) 5792–5797.
45. Antia, R., Pilyugin, S.S., Ahmed, R., Models of immune memory: On the role of cross-reactive stimulation, competition and homeostasis in maintaining immune memory, *Proc. Natl. Acad. Sci. USA*, **95** (1998) 14926–14931.

Chapter 3
Parameter Estimation and Model Selection

In this chapter, we illustrate a data-driven methodology to formulation and calibration of mathematical models of immune responses. The maximum likelihood approach to parameter estimation, Tikhonov regularization method and information-theoretic criteria for model ranking and selection are presented for models formulated with ODEs, DDEs and PDEs. Experimental data on CFSE-based proliferation analysis of T cells and LCMV–CTL dynamics in a low dose experimental infection of mice are used. The material of this chapter is based on our previous work published in [1–7].

3.1 General Modelling Issues

Most mathematical models of immune responses have no rigorous physical basis, i.e. the equations are not obtained from first principles (the basic laws of physics and mechanics) and therefore have no a priori claim of validity. Multiple models of differing complexity based upon different types of mathematical frameworks (and corresponding scientific hypotheses) have been proposed in order to address specific phenomena. Examples of such phenomena are the population dynamics of virus infections, the turnover of lymphocytes, the migration and homeostasis of lymphocytes, single lymphocyte regulation and the single-cell replication cycle of the virus. One can find extensive reviews covering various aspects of application of the mathematical models in immunology from the modellers' perspective [8–50].

Mathematical systems theory suggests a view of a mathematical model as an exclusion law [51]:

> ...A mathematical model expresses the opinion that some things can happen, are possible, while others cannot, are declared impossible.

G. Bocharov et al., *Mathematical Immunology of Virus Infections*,
https://doi.org/10.1007/978-3-319-72317-4_3

The above view is formalized by stating that the model selects a certain subset called *behaviour* from a *universum* of possibilities. To minimize artefacts, it is important to ensure that the repertoire of dynamics suggested by the model is not much richer than the reality.

The problem of formulating a model explaining a phenomenon specified by data sets is considered to be central to every scientific discipline. In mathematical physics, one thinks of models of physical phenomena in terms of equations, which are obtained from first principles—the basic laws of physics, mechanics, etc. The models, e.g. the diffusion equation, the transport equation can be used to describe the system behaviour. In many areas of science, modelling is, in essence, based on the restricting assumption of an isomorphism between the properties of the model and the real system [51]. This allows one to make direct inferences about the real system from the properties of the model.

In contrast to mathematical physics, mathematical modelling in immunology follows a 'systems engineering' approach. A conceptual scheme for the system is generated by a priori restricting the model to the 'most important' interactions. This defines the selection of the time- and space-dependent variables as well as the set of parameters which characterize the kinetics of the specified processes. The model equations are formulated by putting together elementary functional forms (building blocks) for the growth, death, differentiation, etc. processes rather than by deriving constitutive equations from the first principles. Various functional forms and mathematical equations can be used to build up a mathematical model in immunology. Typically, modellers borrow concepts from ecology, enzyme kinetics or epidemiology, making use of the mass action law to formulate equations and describe the dynamics.

The problem of how to develop, in a systematic manner, consistent models that provide a basis for quantitative analysis and prediction in every day immunology research raises a number of challenges. The translation process starting from observations of a particular phenomenon and scientific theories and explanations and ending with a family of mathematical models often appears to be an ill- or vaguely defined process. At the first stage, it involves the conversion of often imprecise assumptions or theories into mathematical variables and relationships between them. The next stage relies on the availability of comprehensive data measuring various aspects of the immune system. When this is available, the process involves the assessment of the accuracy, and explanatory and predictive power of a particular model and of rival models. This last stage is not routinely implemented in most modelling efforts.

The key aspects of our procedure involve the following steps:

- Assumptions about the statistical nature of the variation between one set of data and another;
- Least-squares-type fitting, using an appropriate least-squares type of objective function, related to MLE;

- The calculation of indicators, incorporating a measure of parsimony, that provide a score of the merits or demerits (reflecting the information retained or lost when the model in question is used to approximate a—in some sense—'best' model for the data) of each model.

3.2 Parameter Estimation

Those who attempt computational modelling in immunology are faced with substantial challenges due to the enormous complexity of the systemic-, cellular- and molecular-level processes underlying various immunological phenomena [52]. Model formulation is a critical element and it requires selection of essential variables and parameters and specification of causal relationships between them. The model must be related to the amount of data: small data sets support simple models with few reliably estimated parameters. However, various functional forms can be suggested for the same interaction/action, even while retaining 'simple' models.

Our objective is to present and illustrate a computational approach for (i) developing a best-fit mathematical model that provides an accurate approximation to the data, (ii) assessing the confidence in the estimates of the parameters in the model and (iii) characterizing the parsimony of the model.

The general discussion here will be conducted with reference to models based on ordinary differential equations (ODEs) and delay differential equations (DDEs) (see [1, 53], etc.). The considered ODEs and DDEs models have solutions that we denote $\mathbf{y}(t) = \mathbf{y}(t; \mathbf{p}) \in \mathbb{R}^M$, with parameter $\mathbf{p} = [p_1, p_2, \ldots, p_L]^T \in \mathbb{R}^L$; the mathematical models can be written in the following general form:

$$\begin{aligned} \mathbf{y}'(t; \mathbf{p}) &= \mathbf{f}(t, \mathbf{y}(t; \mathbf{p}), \mathbf{y}(t - \tau; \mathbf{p}); \mathbf{p}), \quad \text{for } t \in [t_0, T]; \\ \mathbf{y}(t; \mathbf{p}) &= \boldsymbol{\psi}(t; \mathbf{p}), \quad \text{for } t \in [t_0 - \tau, t_0]. \end{aligned} \tag{3.2.1}$$

We refer to the components p_ℓ of the 'parameter vector' $\mathbf{p}$ in (3.2.1) as 'parameters'. The *form* of $\mathbf{f}$ is known, so that $\mathbf{f}$ is defined precisely if $\mathbf{p}$ is specified; $\boldsymbol{\psi}$ is an initial function (possibly parameter-dependent) and for a choice of $\mathbf{p}$ the values $\mathbf{y}(\mathrm{t}_j; \mathbf{p})$ with components $y^i(\mathrm{t}_j; \mathbf{p})$ will be expected to simulate data $\{\mathbf{y}_j\}$ (observed at times[1] $\mathrm{t}_j \in [t_0, T]$) with components $\{y^i_j\}$ $(i = 1, 2, \ldots, M,\ j = 1, 2, \ldots, N)$. Specific models are given in the following sections of this chapter. In (3.2.1), $\tau \geq 0$, and if $\tau > 0$ it represents a time lag. The methodology outlined here gives, *inter alia*, an indication whether a lag parameter $\tau > 0$ can be justified. We can be asked to identify parameter values in $\boldsymbol{\psi}(t; \mathbf{p})$ as well in the equations. Throughout the chapter, L is the number of parameters, M is the dimensionality of the state vector and n denotes the sample size (the number of *scalar* observations, usually $n = NM$).

[1] With little amendment we can consider the case where different components $\{y^i_j\}_{j=1}^{N_i}$ are associated with i-dependent times $\{\mathrm{t}^i_j\}_{j=1}^{N_i}$.

3.2.1 Maximum Likelihood Approach[2]

The maximum likelihood approach allows one to find the values of model parameters that maximize the probability of obtaining exactly the observed data[3] (see [54–58]). Parameters are regarded as at our disposal; those parameters for which the likelihood is the highest are the 'maximum likelihood estimates'. We make the following assumptions.

Assumption 3.2.1 The errors in observations at successive times are independent.

Assumption 3.2.2 The errors in the observed data are assumed to have a Gaussian distribution around the vectors $\{\mathbf{y}(\mathrm{t}_j;\mathbf{p})\}_{j=1}^{N}$, that is

$$\mathbf{y}_j \sim \mathcal{N}(\mathbf{y}(\mathrm{t}_j;\mathbf{p}), \boldsymbol{\Sigma}_j),$$

where $\boldsymbol{\Sigma}_j$ is the j-th covariance matrix

Under Assumption 3.2.2, the component probability density functions are given by

$$\left\{\mathscr{H}(\mathbf{y}_j;\mathbf{p}) = \frac{1}{\sqrt{(2\pi)^M \det \boldsymbol{\Sigma}_j}} \exp\{-\frac{1}{2}[\mathbf{y}(\mathrm{t}_j;\mathbf{p}) - \mathbf{y}_j]^T \boldsymbol{\Sigma}_j^{-1}[\mathbf{y}(\mathrm{t}_j;\mathbf{p}) - \mathbf{y}_j]\}\right\}_{j=1}^{N}. \quad (3.2.2)$$

Under Assumption 3.2.1, the likelihood function is then given by

$$\mathscr{L}(\mathbf{p}) = \prod_{j=1}^{N} \mathscr{H}(\mathbf{y}_j;\mathbf{p}), \quad (3.2.3)$$

where $\mathscr{H}(\mathbf{y}_j;\mathbf{p})$ appears in (3.3.35).

Assumption 3.2.3 The errors in the components of $\mathbf{y}_j$ are assumed to be independent

Define

$$\Phi_{WLS}(\mathbf{p}) \equiv [\mathbf{y}(\mathrm{t}_j;\mathbf{p}) - \mathbf{y}_j]^T \boldsymbol{\Sigma}_j^{-1}[\mathbf{y}(\mathrm{t}_j;\mathbf{p}) - \mathbf{y}_j].$$

Under Assumption 3.2.3, we are led to define

$$\Sigma_j = \sigma^2\{\mathrm{diag}[\omega_1^{[j]}, \omega_2^{[j]}, \ldots, \omega_M^{[j]}]\}; \quad (3.2.4)$$

(where σ^2 denotes the data variance) then,

$$\Phi_{WLS}(\mathbf{p}) \equiv \sigma^{-2}\Phi_{\Omega LS}(\mathbf{p}), \quad (3.2.5a)$$

---

[2]Material of sects. 3.2.1–3.2.2 uses the results of the study Journal of Computational and Applied Mathematics, Vol. 184, C.T.H. Baker et al., Computational approaches to parameter estimation and model selection in immunology. Pages 50–76, Copyright © 2005 with permission from Elsevier.

[3]The data are regarded as fixed and assumed to have errors of a certain type.

where

$$\Phi_{\Omega LS}(\mathbf{p}) = \sum_j \|\mathrm{diag}^{-1}[\omega_1^{[j]}, \omega_2^{[j]}, \ldots, \omega_M^{[j]}][\mathbf{y}(t_j; \mathbf{p}) - \mathbf{y}_j]\|^2. \tag{3.2.5b}$$

The maximum likelihood estimate of the model parameters provides an optimal estimate of the data variance as

$$\widehat{\sigma}^2 = \frac{1}{NM}\sum_j \|\mathrm{diag}^{-1}[\omega_1^{[j]}, \omega_2^{[j]}, \ldots, \omega_M^{[j]}][\mathbf{y}(t_j, \widehat{\mathbf{p}}) - \mathbf{y}_j]\|^2 = \frac{1}{NM}\Phi_{\Omega LS}(\widehat{\mathbf{p}}), \tag{3.2.5c}$$

where $\widehat{\mathbf{p}}$ maximizes the likelihood, and then (denoting the natural logarithm by $\ell n(\cdot)$)

$$\ell n\,\mathscr{L}(\widehat{\mathbf{p}}) = -\frac{1}{2}\Big\{NM\ell n(2\pi) + NM + 2\sum_{i,j}\ell n(\omega_i^{[j]})\Big\} - \frac{1}{2}\{NM\ell n(\Phi_{\Omega LS}(\widehat{\mathbf{p}})) - NM\ell n(NM)\}. \tag{3.2.6}$$

Clearly, $\mathscr{L}(\widehat{\mathbf{p}})$ is maximized when $\Phi_{\Omega LS}(\widehat{\mathbf{p}})$ is minimized (equivalently, when $\Phi_{\mathcal{W}LS}(\widehat{\mathbf{p}})$ is minimized) and σ^2 is assigned the value $\widehat{\sigma}^2$ in (3.2.5c). We seek a best-fit parameter $\widehat{\mathbf{p}}$ for (3.2.1) for which the corresponding values $\{y^i(t_j; \widehat{\mathbf{p}})\}_{j=1:N}^{i=1:M}$, provide a 'best fit' to the given data $\{y_j^i\}_{j=1:N}^{i=1:M}$ in the sense that

$$\Phi_{\Omega LS}(\widehat{\mathbf{p}}) = \min_{\mathbf{p}} \Phi_{\Omega LS}(\mathbf{p}). \tag{3.2.7}$$

Remark 3.1 Our approach fits into a general framework as follows. A general statistical framework for parameter estimation is the Bayesian approach. Under the assumption of a uniform prior distribution of parameter values, the Bayesian approach [55, 56] reduces to a maximum likelihood estimation (MLE) [55]. The widely used least-squares technique (LSQ) is equivalent to MLE under the following set of assumptions:

(i) the observational errors are normally distributed;
(ii) equivalent positive and negative deviations from the expected values differ by equal amounts;
(iii) the errors between samples are independent.
Other powers of the deviation between model and data can be used depending on the error distribution, for example, the first power would correspond to an exponential distribution of the errors [55].

3.2.2 *Least-Squares Type Objective Functions*

As indicated above, a key element is the least-squares fitting which involves the informed selection of a least-squares objective function $\Phi_{\Omega LS}(\widehat{\mathbf{p}})$ (and this corresponds to a choice of Ω). This entails making the connection between 'the most likely' parameters and the 'best fit' parameters by ascertaining a criterion for 'best fit', that is by choosing an appropriate Ω.

In general least-squares type data fitting, one encounters in particular three types of objective functions $\Phi(\mathbf{p})$, which depend upon the given data $\{\mathrm{t}_j;\, y^i_j\}_{j=1}^N$ (for $i = 1, \cdots, M$) and the values $\{y^i(\mathrm{t}_j;\, \mathbf{p})\}_{j=1:N}^{i=1:M}$ of the solution $\mathbf{y}(t;\, \mathbf{p})$ of the parametrized model, e.g. (3.2.1). These three types (ordinary least-squares, weighted least-squares and log-least-squares objective functions) correspond to

$$\Phi_{OLS}(\mathbf{p}) = \sum_{j=1}^{N}\sum_{i=1}^{M}\left[y^i(\mathrm{t}_j;\, \mathbf{p}) - y^i_j\right]^2 = \sum_{j=1}^{N} \|\mathbf{y}(\mathrm{t}_j, \mathbf{p}) - \mathbf{y}_j\|^2; \qquad (3.2.8a)$$

$$\Phi_{WLS}(\mathbf{p}) \equiv \sigma^{-2}\Phi_{\Omega LS}(\mathbf{p}), \quad \text{where } \Phi_{\Omega LS}(\mathbf{p}) = \sum_{j=1}^{N}\sum_{i=1}^{M}\{\omega_i^{[j]}\left[y^i(\mathrm{t}_j, \mathbf{p}) - y^i_j\right]\}^2; \qquad (3.2.8b)$$

(when locating $\widehat{\mathbf{p}}$, the scaling factor σ^{-2} in the objective function is not relevant) and

$$\Phi_{LogLS}(\mathbf{p}) = \sum_{j=1}^{N}\sum_{i=1}^{M}\left[\,\ell n(y^i(\mathrm{t}_j, \mathbf{p})) - \ell n(y^i_j)\right]^2. \qquad (3.2.8c)$$

(Variants of these are possible.) To use (3.2.8c), it will be assumed that $y^i_j > 0$ and that $y^i(\mathrm{t}_j;\, \mathbf{p}) > 0$.

As observed by Gingerich [59], the objective functions (3.2.8) above correspond to maximum likelihood functions under differing assumptions. Thus, (i) (3.2.8b) (of which (3.2.8a) is a special case) follows from an assumption of arithmetic normality of observational errors, in which equivalent positive and negative deviations from expected values differ by equal amounts; (ii) (3.2.8c) is associated with an assumption of geometric normality of observational errors in which equivalent deviations differ by equal *proportions*. The use of the term 'geometric normality' refers to the errors being 'log-normal'.

If we adopt a weighted least-squares approach, as in (3.2.8b), the choice of Ω in $\Phi_{\Omega LS}(\mathbf{p})$ can be based on the natural assumption:

Assumption 3.2.4 We assume $\omega_i^{[j]} = \{y^i_j\}^{-1}$.

This implies that the variance increases with the expected value but the coefficient of variation remains constant.

3.2.3 *Uncertainty Quantification*

The general approach to characterize the reliability of parameter estimations is based upon evaluating their confidence intervals (CIs). There exist three major frameworks to evaluate CIs: a technique based on the variance–covariance matrix [56], a profile-likelihood-based method [60] and two variants of the bootstrap method—parametric and non-parametric ones [61, 62]. We review these methods and assess their relative performance by computing approximations to 95% confidence intervals for the estimated parameters of a mathematical model describing CFSE-based proliferation analysis formulated with ODEs [5].

In the following, $\Phi(\mathbf{p}^*)$ stands for the optimized least-squares function, n_d is the number of the experimental data used and n_p is the number of the estimated parameters.

3.2.4 Variance–Covariance Analysis

The variance–covariance method is based upon a parabolic approximation of the objective function around the best-fit parameter estimate $\mathbf{p}^*$. The $100 \cdot \theta\%$ confidence interval for the parameter of interest, e.g. for p_k, is approximated by the standard interval

$$CI_{p_k} = [p_k^* - \sigma_{p_k} z(\theta, n_f),\ p_k^* + \sigma_{p_k} z(\theta, n_f)], \quad k = 1, 2, ..., n_p, \tag{3.2.9}$$

where p_k^* is the best-fit parameter estimate, σ_{p_k} is the standard deviation for p_k and $z(\theta, n_f)$ is the $100 \cdot \theta$ percentage point of the Student's t-distribution with $n_f := n_d - n_p$ degrees of freedom. An estimate of the standard deviation of p_k is computed as follows. First, we construct the covariance matrix

$$\Xi(\mathbf{p}^*) = \frac{2\Phi(\mathbf{p}^*)}{n_d - n_p} H^{-1}(\mathbf{p}^*) \in \mathbf{R}^{n_p \times n_p}, \tag{3.2.10}$$

where H is the Hessian matrix,

$$H(\mathbf{p}) := \left\{\frac{\partial}{\partial \mathbf{p}}\right\}\left\{\frac{\partial}{\partial \mathbf{p}}\right\}^T \Phi(\mathbf{p}) \in \mathbf{R}^{n_p \times n_p}, \quad H_{k,m}(\mathbf{p}) = \frac{\partial^2}{\partial p_k \partial p_m}\Phi(\mathbf{p}), \tag{3.2.11}$$

with $H_{k,m}$ being the (k, m)-th element of H. The standard deviation for the k-th element of $\mathbf{p}$ is given by the corresponding diagonal element $\Xi_{k,k}$ of the covariance matrix,

$$\sigma_k = \sqrt{\Xi_{k,k}(\mathbf{p}^*)}. \tag{3.2.12}$$

3.2.5 Profile-likelihood-based Method

The profile-likelihood method provides a way for computing the confidence intervals of the maximum likelihood parameter estimates by following 'a global' behaviour of the objective function [60]. To compute approximations to the 95% CIs of the estimates, we proceed as follows. For the parameter of interest, p_k^*, we search for the interval $[p_k^{\min},\ p_k^{\max}]$ of maximal width and containing p_k^* such that

$$|\ln(\mathscr{L}(\tilde{\mathbf{p}})) - \ln(\mathscr{L}(\mathbf{p}^*))| \le \frac{1}{2}\chi^2_{1,0.95} \quad \text{whenever } p_k \in [p_k^{\min},\ p_k^{\max}]. \qquad (3.2.13)$$

In (3.2.13), $\mathscr{L}(\mathbf{p}^*)$ stands for the likelihood function,

$$\mathscr{L}(\tilde{\mathbf{p}}) := \max_{\mathbf{p}\in S(p_k)} \mathscr{L}(\mathbf{p}), \quad \text{where } S(p_k) := \Big\{[p_1, p_2, ..., p_{k-1}, p, p_{k+1}, ..., p_{n_p}]|_{p \text{ fixed}}\Big\},$$

and $\chi^2_{1,0.95} = 3.841$ is the 0.95th quantile of the χ^2-distribution for one degree of freedom.

3.2.6 Bootstrap Method

The bootstrap technique is a computationally intensive method for estimating the mean and standard error on the basis of samples generated from small data sets (say of size n) [62]. The new data sets, being subsets of available data, are randomly drawn from original observations set of data points and fit as if they were independent observations. Therefore, the method works by resampling randomly, for example, with a uniform probability equal to $1/n$, the observed sample values to model the unknown population of observations. The whole process of random sampling with replacement is repeated M times to generate M data sets. For these data sets, one computes the best-fit parameter estimates $\mathbf{p}_m^*$, $m = 1, 2, ..., M$, which provide an estimate of the 'true' standard error by taking the standard deviation of the M values of $\mathbf{p}_m^*$. The standard interval

$$CI_{p_k} = [\mathscr{E}(p_k^*) - \sigma_{p_k} z(\theta),\ \mathscr{E}(p_k^*) + \sigma_{p_k} z(\theta)], \quad k = 1, 2, ..., n_p, \qquad (3.2.14)$$

approximates the $100 \cdot \theta$ bootstrap confidence interval. Here $\mathscr{E}(p_k^*)$ and σ_{p_k} are the mean and standard deviation, respectively, of the estimates of the parameter p_k^* that are found by fitting the bootstrap resamples of the original data and $z(\theta)$ is the $100 \cdot \theta$-th percentile of a normal deviate. The value $z(0.95) \approx 1.96$ is used for approximating the 95% CI in conjunction with a large number of resamplings M.

3.2.7 Example of Computational Analysis of CFSE Proliferation Assay

CFSE-based tracking of the lymphocyte proliferation using flow cytometry is a powerful experimental technique in immunology allowing for the tracing of labelled cell populations over time in terms of the number of divisions cells have undergone (see

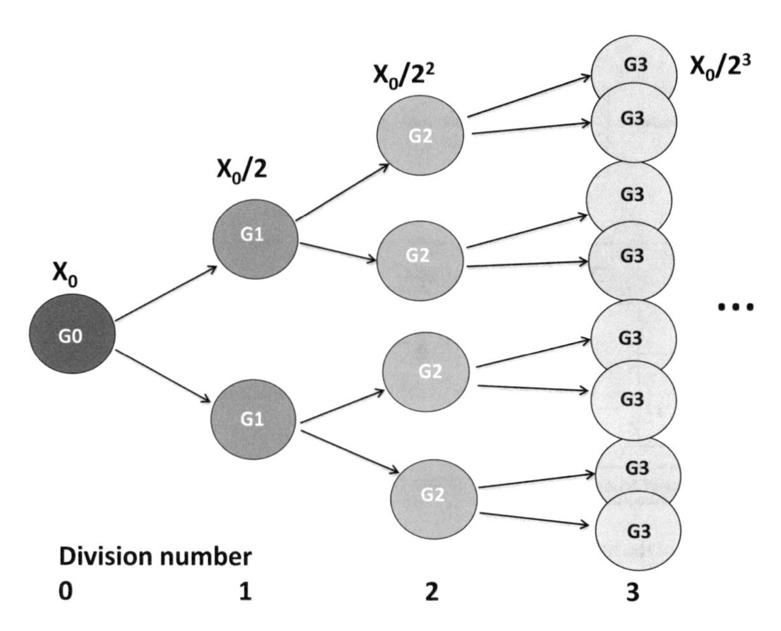

Fig. 3.1 Schematic view of the CFSE-based tracking of the lymphocyte proliferation. The division of cells is assumed to reduce the amount of CFSE label in daughter cells as compared to the mother cell by half. Cells with the same number of completed divisions j contain the same amount of the label equal to $\frac{X_0}{(2^j)}$

Fig. 3.1). Interpretation and understanding of such population data can be greatly improved through the use of mathematical modelling.

3.2.7.1 ODE Model of CFSE-labelled Cell Growth

We apply a heterogenous linear compartmental model, formulated by a system of ordinary differential equations [5]. This model allows the division number-dependent rates of cell proliferation and death, and describes the rate of changes in the numbers of cells having undergone j divisions as shown in Fig. 3.2.

In our setting, the compartments represent the cell populations, which have made a specified number of divisions. In the equations, $N_j(t)$ and $D(t)$ denote the population sizes at time t of live lymphocytes having undergone j divisions and dead but not disintegrated lymphocytes, respectively. The heterogenous compartmental model assumes that the per capita proliferation and death rates of T lymphocytes, α_j, respectively, β_j, depend on the number of divisions lymphocytes have undergone. The rates of change of $N_j(t)$ and $D(t)$ with time are represented by the following set of equations:

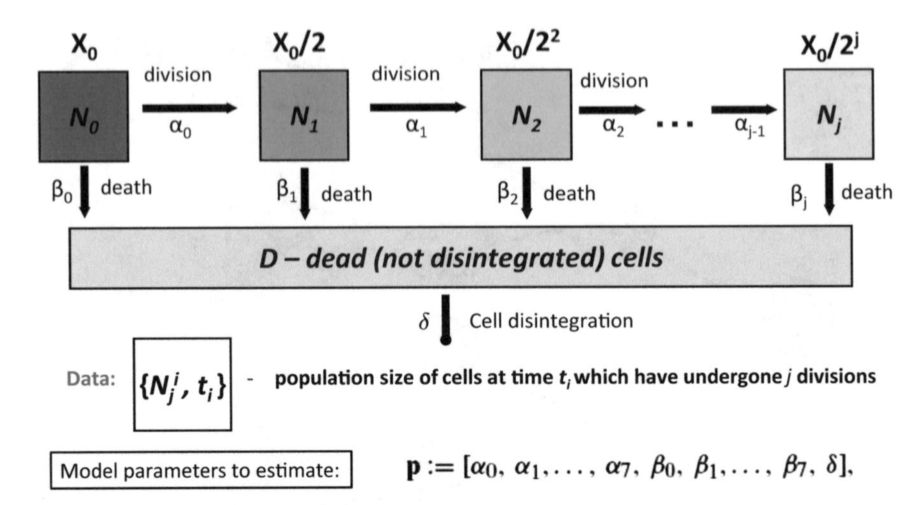

Fig. 3.2 Scheme of the heterogenous linear compartmental model of CFSE-labelled cell proliferation. Basic assumption: the level of CFSE expression (x) on an individual cell takes on a discrete set of values, related to the number of divisions the cell has undergone => each cell generation displays some mean CFSE intensity

$$
\begin{aligned}
\tfrac{dN_0}{dt}(t) &= -(\alpha_0 + \beta_0)N_0(t), \\
\tfrac{dN_j}{dt}(t) &= 2\alpha_{j-1}N_{j-1}(t) - (\alpha_j + \beta_j)N_j(t), \quad j = 1, \ldots, J, \\
\tfrac{dD}{dt}(t) &= \textstyle\sum_{j=0}^{J} \beta_j N_j(t) - \delta D(t).
\end{aligned}
\tag{3.2.15}
$$

The first term on the right of equations for $N_j(t)$ represents the cell birth (influx from previous compartment because of division), while the last term on the right represents cell loss (outflux from the compartment) due to division and death. In the equation for dead cells, δ denotes the specific (fractional) decay rate of dead lymphocytes due to disintegration and phagocytosis.

Assuming that the population sizes at time t_0 are specified by initial data $N_j(t_0)$ and $D(t_0)$, and the condition $\alpha_j + \beta_j \neq \alpha_i + \beta_i$ is fulfilled for $i \neq j$, the solution of the model is expressed in the form

$$
\begin{aligned}
N_j(t) &= \sum_{s=1}^{j} \left\{ 2^s N_{j-s}(t_0) \prod_{m=j-s}^{j-1} \alpha_m \sum_{i=j-s}^{j} \exp^{-c_i(t-t_0)} \prod_{k=j-s, k\neq i}^{j} (c_k - c_i)^{-1} \right\} + N_j(t_0)\exp^{-c_j(t-t_0)}, \\
&\qquad j = 0, 1, \ldots J, \quad t \geq t_0, \\
D(t) &= \sum_{j=0}^{J} \beta_j \left\{ \sum_{s=1}^{j} 2^s N_{j-s}(t_0) \prod_{m=j-s}^{j-1} \alpha_m \sum_{i=j-s}^{j} \frac{\exp^{-c_i(t-t_0)} - \exp^{-\delta(t-t_0)}}{\delta - c_i} \prod_{k=j-s, k\neq i}^{j} (c_k - c_i)^{-1} \right\} \\
&\quad + \sum_{j=0}^{J} \beta_j N_j(t_0) \frac{\exp^{-c_j(t-t_0)} - \exp^{-\delta(t-t_0)}}{\delta - c_j} + D(t_0)\exp^{-\delta(t-t_0)}, \quad t \geq t_0,
\end{aligned}
\tag{3.2.16}
$$

where $c_j := \alpha_j + \beta_j$. The availability of the closed-form solution to the model reduces the computational treatment of the model-driven data analysis.

A simplified version of the heterogenous model can be obtained if we assume that the proliferation and death rates of cells, α and β, do not depend on the number of divisions cells have undergone. For practical examples, we refer to cell kinetics studies [63–66]. The corresponding 'homogenous compartmental model' is defined by system (3.2.15) with $\alpha_i = \alpha,\ \beta_i = \beta,\ i = 0, 1, \ldots, J$, and the same initial data as for the heterogenous model. The solution of this model for $t \geq t_0$ is given by

$$N_j(t) = \exp^{-c(t-t_0)} \sum_{i=0}^{j} (2\alpha)^i \frac{(t-t_0)^i}{i} N_{j-i}(t_0), \quad j = 0, 1, \ldots, J,$$

$$D(t) = \beta \sum_{j=0}^{J} \sum_{i=0}^{j} (2\alpha)^i N_{j-i}(t_0) \left\{ \frac{(-1)^i}{(\delta - c)^{i+1}} \left(\exp^{(\delta-c)(t-t_0)} \sum_{k=i}^{0} \frac{((t-t_0)(c-\delta))^k}{k!} - 1 \right) \right\} \exp^{-\delta(t-t_0)} + D(t_0) \exp^{-\delta(t-t_0)}, \tag{3.2.17}$$

where $c := \alpha + \beta$.

3.2.7.2 Data of CFSE-labelled Cell Proliferation

The experimental data set that we analyse is derived from measuring the kinetics of PHA-induced human T lymphocyte proliferation in vitro from day 3 to day 7. Specified characteristics are as follows: (i) the total number of live cells, $N(t_i)$, (ii) the total number of dead but not disintegrated cells, $D(t_i)$, and (iii) the number of cells divided j times, $N_j(t_i), i = 0, 1, \ldots, 4,\ j = 0, 1, \ldots, 7$. Table 3.1 presents the set of CFSE data analysed.

Table 3.1 Quantitative dynamics of human peripheral blood T lymphocytes following stimulation with PHA in vitro. At various times, CFSE profiles were obtained by flow cytometry. The total numbers of live, $N(t_i)$, and dead, $D(t_i)$, lymphocytes and the distribution of lymphocytes with respect to the number of divisions they have undergone, $N_j(t_i),\ j = 0, 1, \ldots, 7$, were followed from day 3 to day 7 at the indicated times $t_i,\ i = 0, 1, \ldots, 4$

Time days t_i	Total number of live cells $N(t_i)$	Total number of dead cells $D(t_i)$	Numbers of cells w.r.t. the number of divisions (j) they undergone $N_j(t_i)$							
			0	1	2	3	4	5	6	7
3	1.4×10^5	1.6×10^4	29358	22876	43372	39970	5208	98	14	0
4	2.5×10^5	2.4×10^4	16050	12600	22650	57025	96350	46950	2500	25
5	4.4×10^5	6.0×10^4	14476	14784	25344	58652	141460	156290	32076	440
6	5.0×10^5	1.2×10^5	13500	12150	24150	55000	137850	188950	69450	2150
7	5.7×10^5	1.3×10^5	13509	12198	21603	51927	140560	232160	96102	3420

3.2.7.3 Confidence in Best-Fit Estimates

We search for a vector of best-fit parameters, $\mathbf{p}^*$, for which the model solution $N_j(t;\mathbf{p}^*)$, $D(t;\mathbf{p}^*)$ is closest according to the least-squares criterion to the given experimental data at the time points of the measurements, i.e. the solution fits the data in an optimal way. The vector $\mathbf{p}$, $\mathbf{p} \in \mathbf{R}^L$ has the components

$$\mathbf{p} := [\alpha_0,\ \alpha_1, \ldots,\ \alpha_7,\ \beta_0,\ \beta_1, \ldots,\ \beta_7,\ \delta],$$

and

$$\mathbf{p} := [\alpha,\ \beta,\ \delta]$$

for the heterogenous and the homogenous model, respectively. Our data set consists of cell numbers $N_j^i := N_j(t_i)$ and $D^i := D(t_i)$ measured at times t_i, cf. Table 3.1, where $t_0 = 72,\ t_1 = 96,\ t_2 = 120,\ t_3 = 144,\ t_4 = 168$ (h). The values N_j^0 and D^0 are used as the initial data $N_j(t_0)$ and $D(t_0)$ for the models.

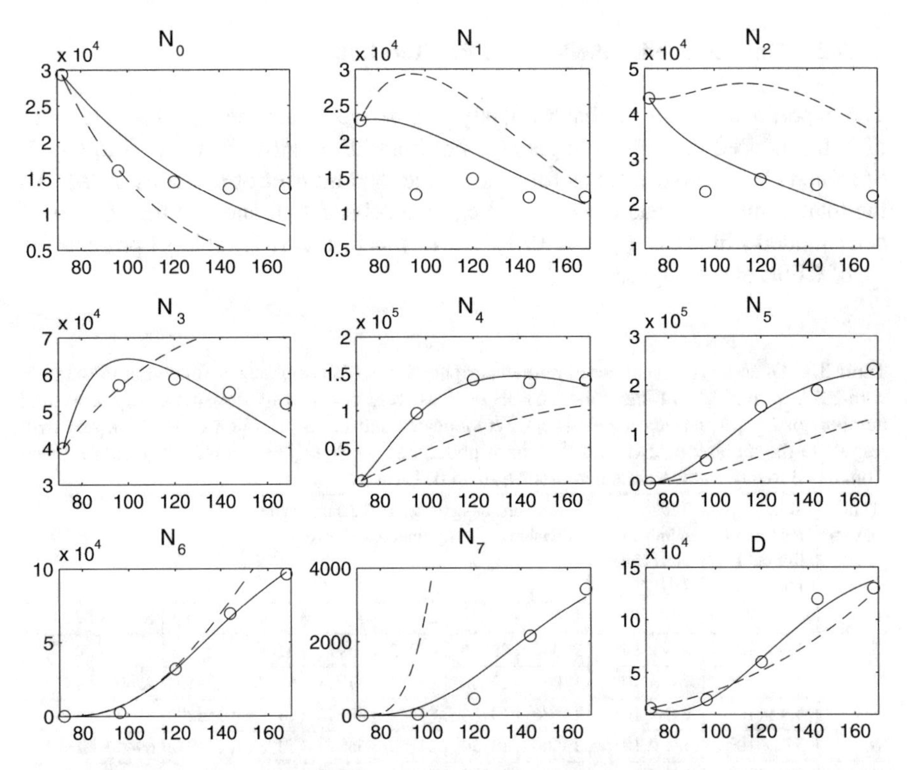

Fig. 3.3 Experimental data (circle) and the best-fit solutions of the heterogenous (solid lines) model (3.1) and the homogenous (dashed lines) model

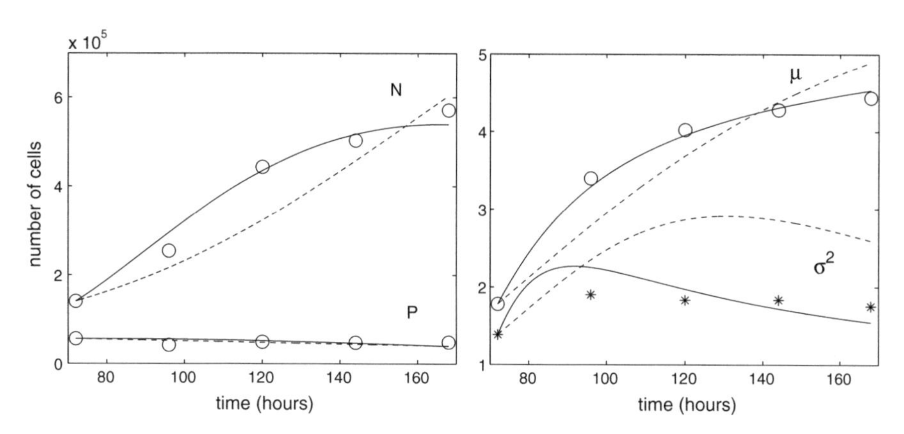

Fig. 3.4 The kinetics of the human peripheral blood T cell population growth. Left: The total number of live lymphocytes N over all divisions and the number of precursors P that would have generated the current lymphocyte population in the absence of death, estimated from the data in Table 3.1 (circle) and predicted by the heterogenous (solid line) model (3.2.15) and the homogenous (dashed line) model for the best-fit parameter values. Right: The behaviour of the mean division number $\mu(\circ)$ and its variance σ^2 (asterisk) suggested by the data. The solid and dashed curves correspond to $\mu(t)$ and $\sigma^2(t)$ computed using the best-fit solutions of the heterogenous and homogenous models, respectively

Figure 3.3 shows the experimental data and the solutions of the two models corresponding to the best-fit parameter estimates. The kinetics of cells which have undergone more than two divisions is consistently reproduced by the heterogenous model. In contrast, the data characterizing the kinetics of the first two divisions appear to be a problem for both models. This discrepancy might be related to a large observation error for the number of T cells that have undergone one or two divisions. Indeed, the decline in the number of undivided cells is not accompanied by an increase in the number of cells that have divided once or twice, which seems to be counter-intuitive. As shown in Fig. 3.4 (left), the growth of the total cell population N slows down after day 5. This concave pattern is consistently captured by the heterogenous model, whereas the homogenous model predicts a biased dynamics. The numbers of precursors P estimated from the data and predicted by the models are close to each other. Figure 3.4 (right) shows that the mean number of divisions cell populations have undergone (μ) is predicted reliably by the heterogenous model. However, the evolution of the variance (σ^2) in the mean division number over time is not precisely reproduced. The homogenous model gives a poor fit of μ and σ^2.

Parameter estimation results obtained using the ordinary least-squares approach for both models are summarized in Table 3.2. The best-fit estimates of the heterogenous model parameters suggest that the lymphocyte proliferation and death rates are not constant but vary essentially with the division number in a non-monotone way, Fig. 3.5.

The best-fit estimates of parameters β_j, $j = 0, 1, 2, 3, 6, 7$, are close to zero, taking numerical values ranging between 10^{-15} and 10^{-11} (hours^{-1}). Biologically,

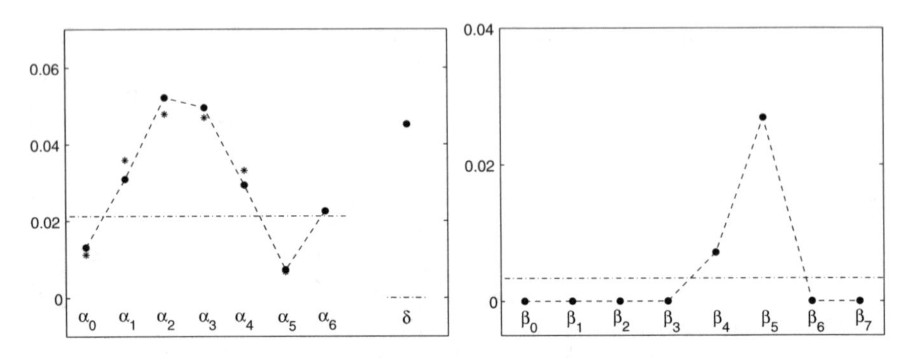

Fig. 3.5 Dependence of the cell turnover parameters on the division number. The best-fit estimates of the division rate (left) and the death rate (right) parameters of the heterogenous (circle) and homogenous (dot-dashed curves)models. The best-fit value of α_7, $\alpha_7^* \approx 1.37$ is not shown. The best-fit estimates of α_j, $j = 0, ..., 5$, for the parameterized version of the model with the proliferation rate described by the function $\alpha_j = z_1 - z_2(z_3 + j)^2$ are indicated by asterisks

Table 3.2 Best-fit parameter estimates of the heterogenous and homogenous models obtained using the ordinary least-squares approach and the data set in Table 3.1

Heterogenous model	Division rate (1/hour)		Death rate (1/hour)		Disintegration rate (1/hour)	Doubling time (hours)
α_0	1.31×10^{-2}	β_0	2.33×10^{-15}	δ	4.52×10^{-2}	52.8
α_1	3.10×10^{-2}	β_1	7.98×10^{-13}			22.4
α_2	5.21×10^{-2}	β_2	5.56×10^{-13}			13.3
α_3	4.95×10^{-2}	β_3	1.54×10^{-14}			14.0
α_4	2.94×10^{-2}	β_4	7.12×10^{-3}			23.5
α_5	7.28×10^{-3}	β_5	2.69×10^{-2}			95.3
α_6	2.26×10^{-2}	β_6	7.07×10^{-15}			30.6
α_7	1.37	β_7	6.249×10^{-11}			0.5
Homogenous model						
α	2.13×10^{-2}	β	3.35×10^{-3}	δ	5×10^{-18}	32.5

these small values would imply zero death rate of the proliferating cells with division number age from zero to three, six and seven. We have performed computational analysis of the confidence intervals for the vector of 11 parameters

$$\mathbf{p} = [\alpha_0, \alpha_1, \alpha_2, \ldots, \alpha_7, \beta_4, \beta_5, \delta], \tag{3.2.18}$$

of the heterogenous model (3.2.15) with $\beta_j = 0$, $j = 0, 1, 2, 3, 6, 7$, as fixed ad hoc.

The computed estimates of 95% CIs ($z \approx 2.06$) for the best-fit parameters of model (3.2.15) are presented in Table 3.3 and shown in Fig. 3.6. The intervals appear to be quite narrow for all parameters except α_6 and α_7. The estimated CIs indicate that data sets covering seven divisions (J), such as presented in Table 3.1, are informative

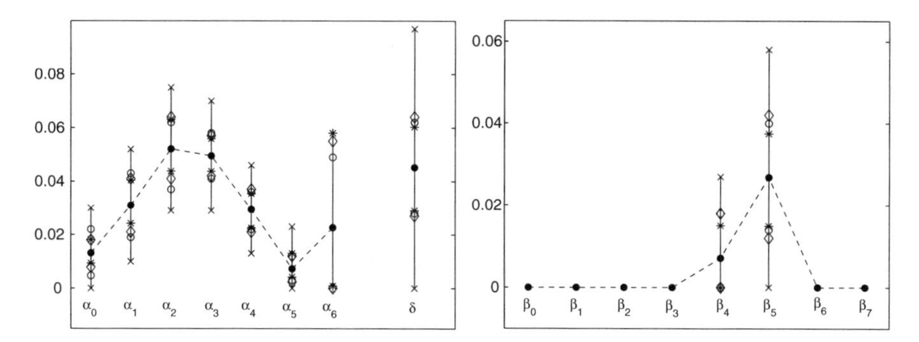

Fig. 3.6 Comparison of the approximations to 95% confidence intervals for the best-fit parameter values of the heterogenous model computed by three methods: the variance–covariance method (diamond), the profile-likelihood-based method (asterisk) and two versions of the bootstrap method non-parametric (times) and parametric (open circle) approaches. The best-fit parameter values are marked by filled circle

enough to estimate reliably the proliferation and death rates of the first six $(J-1)$ successive divisions.

The estimated 95% confidence intervals by a profile-likelihood-based method, cf. Table 3.3 and Fig. 3.5, were computed using a numerical algorithm presented in [60]. The profile-based CIs, except CI_{α_7}, are rather close to the variance–covariance-based ones. The fact that they turned out to be somewhat narrower for most of the parameters indicates that the objective function grows a bit faster than the parabolic one. The profile-likelihood-based method does not provide an estimate of the upper limit of the CI_{α_7}. Its value tends to infinity as the iterations of the computational algorithm continue. The reason is as follows. Using the relationship between the maximum likelihood and least-squares objective function, the expression (3.2.13) is equivalent to

$$|\ln(\Phi(\tilde{\mathbf{p}})) - \ln(\Phi(\mathbf{p}^*))| \leq \frac{1}{n_d}\chi^2_{1,0.95} \quad \text{whenever } p_k \in [p_k^{\min},\ p_k^{\max}], \qquad (3.2.19)$$

where

$$\Phi(\tilde{\mathbf{p}}) := \min_{\mathbf{p}\in S(p_k)} \Phi(\mathbf{p}).$$

Then, using $\ln(\Phi(\mathbf{p}^*)) \approx 20.96$, the final expression for computing the 95% CI_{p_k} is equivalent to

$$\Phi(\tilde{\mathbf{p}}) \in [\Phi(\mathbf{p}^*),\ 1.41 \times 10^9] \quad \text{whenever } p_k \in [p_k^{\min},\ p_k^{\max}]. \qquad (3.2.20)$$

The value of $\Phi(\tilde{\mathbf{p}})$ stays below the threshold 1.41×10^9 as $\alpha_7 \to +\infty$.

The bootstrap approach can be used in two ways, i.e. following either a non-parametric or parametric bootstrap. We used a non-parametric bootstrap approach [61] as follows. The original observations are available for days 3 to 7 and characterize

Table 3.3 Computational estimates of 95% confidence intervals for the best-fit parameter values of the heterogenous model (3.2.15) approximated by the variance–covariance, profile-likelihood and bootstrap methods

p	Best-fit values	Estimates of 95% confidence intervals			
		Variance–covariance method	Profile-likelihood method	Bootstrap method	
				Non-parametric	Parametric
α_0	1.31×10^{-2}	$[0.79,\ 1.8] \times 10^{-2}$	$[0.94,\ 1.8] \times 10^{-2}$	$[0,\ 2.9] \times 10^{-2}$	$[0.51,\ 2.1] \times 10^{-2}$
α_1	3.10×10^{-2}	$[2.1,\ 4.1] \times 10^{-2}$	$[2.4,\ 4.0] \times 10^{-2}$	$[1.1,\ 5.1] \times 10^{-2}$	$[2.0,\ 4.2] \times 10^{-2}$
α_2	5.21×10^{-2}	$[4.1,\ 6.4] \times 10^{-2}$	$[4.4,\ 6.3] \times 10^{-2}$	$[3.1,\ 7.4] \times 10^{-2}$	$[3.8,\ 6.7] \times 10^{-2}$
α_3	4.95×10^{-2}	$[4.2,\ 5.7] \times 10^{-2}$	$[4.4,\ 5.6] \times 10^{-2}$	$[3.0,\ 6.9] \times 10^{-2}$	$[4.1,\ 5.8] \times 10^{-2}$
α_4	2.94×10^{-2}	$[2.1,\ 3.7] \times 10^{-2}$	$[2.4,\ 3.5] \times 10^{-2}$	$[1.4,\ 4.5] \times 10^{-2}$	$[2.3,\ 3.6] \times 10^{-2}$
α_5	7.28×10^{-3}	$[0.24,\ 1.2] \times 10^{-2}$	$[0.42,\ 1.3] \times 10^{-2}$	$[0,\ 2.2] \times 10^{-2}$	$[0.32,\ 1.1] \times 10^{-2}$
α_6	2.26×10^{-2}	$[0,\ 5.5] \times 10^{-2}$	$[0.09,\ 5.8] \times 10^{-2}$		$[0,\ 4.8] \times 10^{-2}$
α_7	1.37	$[0,\ 6.8]$	$[0.016,\ \infty)$		$[0,\ 2.7]$
β_4	7.12×10^{-3}	$[0,\ 1.8] \times 10^{-2}$	$[0,\ 1.51] \times 10^{-2}$	$[0,\ 2.6] \times 10^{-2}$	$[0,\ 1.7] \times 10^{-2}$
β_5	2.69×10^{-2}	$[1.2,\ 4.2] \times 10^{-2}$	$[1.0,\ 3.8] \times 10^{-2}$	$[0,\ 5.6] \times 10^{-2}$	$[1.4,\ 3.9] \times 10^{-2}$
δ	4.52×10^{-2}	$[2.7,\ 6.4] \times 10^{-2}$	$[2.9,\ 6.0] \times 10^{-2}$	$[0,\ 9.4] \times 10^{-2}$	$[2.9,\ 6.1] \times 10^{-2}$

the numbers of live and dead cells. One can treat the measurements of live and dead lymphocytes at a given time as independent, whereas the distribution of cells over the division number is a single entity which cannot be split. Therefore, we considered five measurement times (days 3 to 7) for the live and dead cells as separate data points, which represent altogether the original sample set of size $2n_t$, $n_t = 5$. The resampled data sets were generated by choosing the measurement times randomly from the original set using a uniform probability distribution.

Let the m-th resampling procedure select the following set of measurement days $\{t_0^N, t_1^N, t_2^N, t_3^N, t_4^N\}$ for N and $\{t_0^D, t_1^D, t_2^D, t_3^D, t_4^D\}$ for D. The day set is further ordered to ensure that $t_0^N \le t_1^N \le t_2^N \le t_3^N \le t_4^N$ and $t_0^D \le t_1^D \le t_2^D \le t_3^D \le t_4^D$. The resamples consisting of only one day of measurement, i.e. the measurement day represented n_t times, are discarded from analysis as being not informative. The bootstrapping analysis involves the following heuristics:

- construct samples of measurements by selecting the measurement times and corresponding cell data at the sampled days;

- take the best-fit parameter vector as an initial guess for fitting the resampled data;
- compute the best-fit estimate $\tilde{\mathbf{p}}^*$ using the objective function

$$\tilde{\Phi}(\mathbf{p}) = \sum_{i=1}^{4}\Big(\sum_{j=0}^{7}(N_j^{t_i^N} - N_j(t_i^N;\mathbf{p}))^2 + (D^{t_i^D} - D(t_i^D;\mathbf{p}))^2\Big), \qquad (3.2.21)$$

where $N_j^{t_i^N}$ and $D^{t_i^D}$ are the original data at the resampled times t_i^N and t_i^D, respectively. If it happens that $t_0^D < t_0^N$, we compute the model predictions for N_j for all times t_i^D for which $t_i^D < t_0^N$, $i = 0, \ldots$.

Repeating the above bootstrapping procedure M times, we compute a set of best-fit vectors $\tilde{\mathbf{p}}_m^*$, $m = 1, \ldots, M$, for M bootstrap samples. The standard deviation of the element p_k of the vector $\tilde{\mathbf{p}}_m^*$ can be approximated from the above set in a usual way,

$$\sigma_{p_{k,M}} = \Big(\frac{\sum_{m=1}^{M}|\tilde{p}_{k,m}^* - \hat{p}_k^*|^2}{M-1}\Big)^{1/2}, \quad \text{where} \quad \hat{p}_k^* := \frac{\sum_{m=1}^{M}\tilde{p}_{k,m}^*}{M}, \qquad (3.2.22)$$

cf. [61], Eq. (2.4). If

$$\lim_{M\to\infty}\sigma_{p_{k,M}} =: \sigma_{p_k}$$

exists, then σ_{p_k} is the bootstrap estimate of the standard deviation for p_k. In this case, a standard bootstrap 95% CI for p_k is given by

$$CI_{p_{k,M}} = [\hat{p}_k^* - 1.96\sigma_{p_{k,M}},\ \hat{p}_k^* + 1.96\sigma_{p_{k,M}}]. \qquad (3.2.23)$$

It is important that M is large enough to achieve convergence of $\sigma_{p_{k,m}}$ as $m \to M$. However, as it was noticed in [61], the bootstrap is not generally reliable for small sample sizes regardless of how many resamples M are used.

The computed 95% CIs for the parameters of the heterogenous model are presented in Table 3.3 and Fig. 3.6. These results are based on $M = 4000$ resamplings. The bootstrap CIs for all parameters appear to be much broader than the ones predicted with variance–covariance and profile-likelihood-based methods.

Applying a parametric variant of the bootstrap method to estimate the 95% CIs for the best-fit values of the heterogenous model parameters, new data samples were generated by perturbing the original data as follows:

$$\begin{aligned} \tilde{N}_j^i &= N_j^i + \sigma^*\mathcal{N}(0,1), \\ \tilde{D}^i &= D^i + \sigma^*\mathcal{N}(0,1), \quad i = 0, 1, \ldots, 4, \quad j = 0, 1, \ldots, 7, \end{aligned} \qquad (3.2.24)$$

where N_j^i and D^i are the data from Table 3.1 at time t_i, $\mathcal{N}(0, 1)$ is a normally distributed random variable with zero mean and variance equal to one, and $\sigma^* \approx 5.9 \times 10^3$, where

$$\sigma^{*^2} = \frac{1}{n_d}\Phi(\mathbf{p}^*), \tag{3.2.25}$$

which follows from the optimality condition $\partial(\ln(\mathcal{L}(\mathbf{p}^*; \sigma)))/\partial\sigma^2 = 0$. The maximum likelihood estimate of the variance σ^{*^2} is rather large compared to the values of N_7^i. Hence, the parametric resampling (3.2.24) can generate biologically improper values of $\tilde{N}_7^i$, e.g. negative, sharply decreasing values. Such inconsistent data for $\tilde{N}_7^i$ were filtered out by ignoring the sequences that do not increase monotonically. The above difficulty indicates that the variance in the components of the observed state space vector might not be equal. Rather, σ^{*^2} should be dependent on the number of divisions cells have undergone and therefore, estimated separately.

For each perturbed data sample, we computed the bootstrap estimates of the best-fit parameters by minimizing the following objective function:

$$\tilde{\Phi}(\mathbf{p}) = \sum_{i=1}^{4}\Big(\sum_{j=0}^{7}(\tilde{N}_j^i - N_j(t_i;\mathbf{p}))^2 + (\tilde{D}^i - D(t_i;\mathbf{p}))^2\Big). \tag{3.2.26}$$

Similar to the non-parametric bootstrap, we generated 4000 data samples. The bootstrap set of the best-fit parameter values was used to calculate, by (3.2.22) and (3.2.23), the standard deviation of the estimators to approximate the 95% CIs, cf. Table 3.3 and Fig. 3.6. The estimated CIs are consistent with those given by the variance–covariance and profile-likelihood methods.

The presented results show that the three techniques give rather close estimates of the 95% confidence intervals of the identified parameters of the considered heterogenous model.

3.3 Regularization of Parameter Estimation

The parameter estimation task discussed above represents an example of an inverse problem. A fundamental question governing the solution of an inverse problem is whether it is well posed in the sense of Hadamard [67, 68]: an inverse problem, defined on certain metric spaces, is well posed if its solution exists in the given space, the solution is unique and the problem is stable in these spaces. An ill-posed function coefficient inverse problem is typically characterized by the lack of uniqueness of its solution (a few parameter sets yield a solution of the model that is close to the data within the noise level in the data) and by a high sensitivity to noise in the measured data (instabilities), cf. [68–77]. Ill-posedness of a function coefficient inverse problem might be caused by (i) modelling errors, (ii) errors in the experimental data and (iii) a high dimension (L) of the finite-dimensional space used to approximate the unknown function (which is not justified by the information content of the data). This important issue will be addressed for the CFSE proliferation assay when a

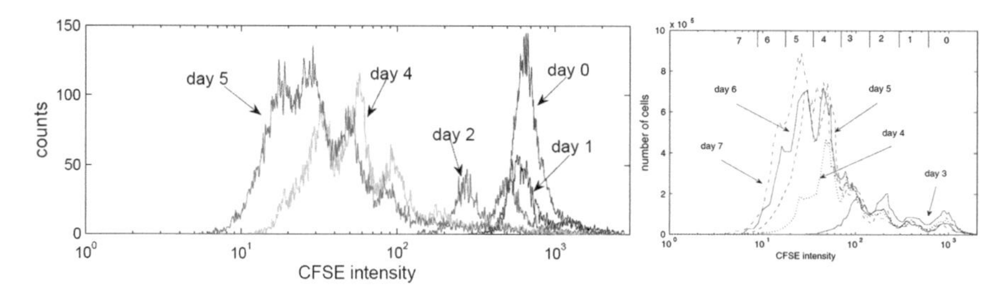

Fig. 3.7 Examples of the original CFSE histograms of PHA-stimulated T cells at days 0,1,2,4,5. Left: cell counts distributions. Right: cell number rescaled distributions suitable for parameter estimation

more complex description of the cell growth is used, i.e. the model formulated as a distributed parameter system.

In a standard experimental approach, the CFSE fluorescence histograms (a representative example is shown in Fig. 3.7) are used to evaluate the fractions of T cells that have completed certain number of divisions. This type of 'mean fluorescence intensity' data can be obtained either manually or by using various deconvolution techniques implemented in programmes, such as ModFit (Verity Software), CellQuest (Becton Dickinson) and CFSE Modeler (ScienceSpeak). The corresponding computer-based procedures require setting of the spacing between generations, i.e. marking the CFSE fluorescence intensities that separate consecutive generations of dividing cells. Note that when the starting population of cells exhibits a broad range of CFSE fluorescence, the division peaks can be not easily identifiable, making conventional division tracking analysis problematic. Therefore, this mapping procedure can often be vaguely defined.

Mathematical modelling of CFSE-labelled cell proliferation is an area of active research (see recent reviews [78–81]). From the viewpoint of the cell turnover parameters estimation, the models can be subdivided into two categories: the 'division-structured models' and the 'label-structured models'. The first one requires preprocessing of the histogram data to quantify the cell populations differing in the number of completed cell cycles. Note that decomposition of the histograms into cell generations is subject to bias due to the lack of clear peak separation, the CFSE label degradation during the experiment, the presence of noise, etc. The second group of the models deals directly with the CFSE histograms, and, therefore, eliminates the need for histogram analysis.

3.3.1 Distributed Parameter Model

We present the label-structured cell population model proposed in [6] to quantitatively describe the dynamics of CFSE-labelled cell populations and to estimate the rates of cell division, death, label decay and the label dilution factor from CFSE proliferation assays. Consider a population of cells that are distinguishable from each other in terms of their fluorescence intensity (UI) with the distribution function $n(t,x)$ ($cell/UI$), characterizing the number of cells at time t which have a label intensity between x and $x+dx$. The total number of cells in a system is then given by $N(t) = \int_{x_{\min}}^{x_{\max}} n(t,x)\mathrm{d}x$. We assume that (i) at the beginning of the experiment the cells are labelled with CFSE giving rise to an initial distribution function; (ii) some proliferation inducing stimuli are provided and (iii) the level of stimulation remains uniform during the experiment. We assume constant (i.e. independent of the CFSE intensity level) rates of cell death and label decay, consistent with the CFSE data set used, cf. [6]. Under these assumptions, the cell population dynamics is described by the following population balance hyperbolic PDE model:

$$\frac{\partial n}{\partial t}(t,x) - v\frac{\partial n}{\partial x}(t,x) = -(\alpha(x)+\beta)n(t,x) + 2\gamma a(\gamma x)n(t,\gamma x), \quad t > t_0, \tag{3.3.27}$$

where

$$a(\gamma x) = \begin{cases} \alpha(\gamma x), & x_{\min} \le x \le x_{\max}/\gamma \\ 0, & x_{\max}/\gamma < x \le x_{\max}. \end{cases}$$

The model consists of the following terms:

- The advection term $v\partial n(t,x)/\partial x$ describes the natural decay of the CFSE fluorescence intensity of the labelled cells with the rate v ($UI/hour$).
- The term $-(\alpha(x)+\beta)n(t,x)$ describes the local disappearance of cells with the CFSE intensity x due to the division associated label dilution and cell death, with the coefficients $\alpha(x) \ge 0$ and $\beta \ge 0$ characterizing the proliferation and death rates, respectively. The unit of $\alpha(x)$ and β is $1/hour$. The precise dependence of $\alpha(x)$ on x is not known a priori, and it is estimated from the data.
- $2\gamma\alpha(\gamma x)n(t,\gamma x)$ represents the birth of two cells due to division of the mother cell with the label intensity γx. The factor 2 accounts for the doubling of numbers and the coefficient γ ($1 < \gamma \le 2$) accounts for the difference in the size of the CFSE intervals to which daughter and mother cells belong.

The initial data for model (3.3.27) specify the distribution of cells at time t_0,

$$n(t_0,x) =: n_0(x), \quad x \in [x_{\min}, x_{\max}]. \tag{3.3.28}$$

For the data set considered in this paper, the observations start at $t_0 = 72$ h (day 3) after the onset of stimulation. The initial function $n_0(x)$ is determined by an interpolation of the vector $\mathbf{n}_0 := \{n_{0,j}\}_{j=1}^{M_0}$ obtained as described below. To translate the flow cytometry counts data to cell numbers considered in the mathematical model, the following transformation is used:

$$n_{i,j} = \frac{c_{i,j}N_i}{F_i}, \quad F_i = \int_{s_{\min}}^{s_{\max}} \tilde{c}_i(s)\mathrm{d}s, \quad i = 0,1,\ldots,M, \quad j = 1,\ldots,M_i, \tag{3.3.29}$$

where N_i and F_i denote the total number of cells and cell counts, respectively, at time t_i, and $\tilde{c}_i$ is a continuous approximation of the vector $\mathbf{c}_i$ defined on the mesh $\mathbf{s}_i$. Figure 3.7 (right) shows the resulting set of five ($M = 4$) histograms of CFSE-labelled cell distributions that we used to identify the model parameters.

The boundary condition

$$n(t, x_{\max}) = 0, \quad t > t_0, \tag{3.3.30}$$

reflects the absence of cells with CFSE intensity above the given maximal value $x_{\max}$ for all $t > t_0$.

The CFSE histograms obtained by flow cytometry use the base 10 logarithm of the marker expression level. Therefore, for numerical analysis, we reformulate model (3.3.27) using the $\log_{10}$-transformed coordinate $z = \lg x$:

$$\frac{\partial n}{\partial t}(t, z) - \nu(z)\frac{\partial n}{\partial z}(t, z) = -(\alpha(z) + \beta)n(t, z) + 2\gamma a(z + \lg \gamma)n(t, z + \lg \gamma), \quad t > t_0, \tag{3.3.31}$$

where $\nu(z) = v/10^z \ln 10$. The initial and boundary conditions now read as

$$n(t_0, z) =: n_0(z), \quad z \in [z_{\min}, z_{\max}]; \qquad n(t, z_{\max}) = 0, \quad t > t_0. \tag{3.3.32}$$

3.3.2 *Distributed Parameter Estimation*

The population balance model (3.3.31) depends on the unknown function $\alpha(z)$ and the parameters $v,\ \beta,\ \gamma$. The identification of these unknowns from the observed CFSE histograms, using some measure of closeness of the model solution to the observations, represents an inverse problem. This problem is characterized by a finite set of observations and an infinite-dimensional space of the unknowns to be estimated. To transform an optimization in a function space into a finite-dimensional parametric optimization problem, we parameterize the unknown function. In this section, we present the parametrization of $\alpha(z)$ used in this study and outline our choice of the cost function and the numerical methods used to solve the direct and inverse problems.

3.3.2.1 Parameterization of Unknown Rate Functions

To obtain a biologically consistent approximation of the birth rate $\alpha(z)$, we use the fact, following from CFSE histograms, that at any time t there are no divisions of cells with the marker intensity below some level $z \leq z^*(t)$ and hence $\alpha(z) = 0$ for $z \in [z_{\min}, z^*(t)]$. The CFSE histogram data at days 4 to 7 are such that $z^*(t)$ is (nearly) independent of time, and hence, we can set $z^*(t) = z^*$. To fulfil this aspect in computations, we set $\alpha(z) = 0$ for $z \in [z_{\min}, z^*]$ with $z^* = 1.15$ for the considered data set.

Our previous studies [5, 6] showed that the birth rate of cells depends on the number of divisions that cells have undergone and that the rate is a bell-shaped function. We approximate the function $\alpha(z)$ using piecewise cubic Hermite splines,

$$\alpha_L(z) = \sum_{j=0}^{L} a_j \phi_j(z), \quad z \in [z^*, z_{\max}], \tag{3.3.33}$$

where $\phi_j(z)$ is a piecewise cubic polynomial defined on the mesh $\mathscr{Z} := [z^* = \tilde{z}_0, \tilde{z}_1, \ldots, \tilde{z}_L = z_{\max}]$ with the property $\phi_j(\tilde{z}_j) = 1,\ \phi_j(\tilde{z}_i) = 0,\ i \neq j$, and hence $\alpha_L(\tilde{z}_j) = a_j,\ j = 1, \ldots, L,\ \alpha_L(\tilde{z}_0) = a_0 = 0$. Note that the approximation of a function by Hermite splines is a general representation of a smooth function which ensures continuity of its first derivative.

One can consider two types of the parametrization (3.3.33):

1. The nodes of the mesh $\mathscr{Z}$ are equally spaced and a priori fixed so that only the elements of the vector $\mathbf{a} := [a_1, \ldots, a_L]$ are estimated.
2. The internal nodes of the mesh $\mathscr{Z}$, i.e. the elements of the vector $\tilde{\mathbf{z}} := [\tilde{z}_1, \ldots, \tilde{z}_{L-1}]$, are estimated together with the elements of the vector $\mathbf{a}$.

Note that while fixed uniform meshes for the parametrization of an unknown function are often used in practice, e.g. [69, 82], the authors are not aware of the use of estimated mesh points in similar problems. The above parametrization of the function $\alpha(z)$ reduces the original infinite-dimensional function identification problem to a finite-dimensional parameter estimation problem for the following vector of parameters:

$$\mathbf{p} = [\mathbf{a},\ \upsilon,\ \beta,\ \gamma] \in \mathbf{R}^{L+3} \quad \text{or} \quad \mathbf{p} = [\mathbf{a},\ \tilde{\mathbf{z}},\ \upsilon,\ \beta,\ \gamma] \in \mathbf{R}^{2L+2}. \tag{3.3.34}$$

We refer to them as minimization problem 1 and minimization problem 2, respectively.

We present a regularization procedure for the second more complex parameter estimation problem.

3.3.2.2 Regularized Maximum Likelihood Approach

The objective is to find the elements of the vector $\mathbf{p}$ such that the corresponding model solution computed at time t_i at the points of the data mesh $\mathbf{s}_i$, i.e.

$$\mathbf{n}(t_i, \mathbf{s}_i; \mathbf{p}) := [n(t_i, s_{i,1}; \mathbf{p}), \ldots, n(t_i, s_{i,M_i}; \mathbf{p})],$$

is quantitatively consistent with the data on the CFSE-labelled cell distribution available at time t_i,

$$\mathbf{n}_i := [n_{i,1}, \ldots, n_{i,M_i}], \quad i = 0, 1, \ldots, M.$$

For an optimal estimation of the parameter values, we seek to maximize the likelihood that the data agree with the model. To apply the maximum likelihood approach, we assume that (i) the observational errors, i.e. the residuals defined as a difference between observed and model-predicted values, are normally distributed; (ii) the errors in observations at successive times are independent; (iii) the errors in cell counts for consecutive label bins are independent; (iv) the variance σ^2 of observation errors is the same for all observation times and independent of the label expression level. Then, the corresponding likelihood function is

$$\mathscr{L}(\mathbf{p};\sigma)=\prod_{i=0}^{M}(2\pi\sigma^2)^{-M_i/2}\exp\Big\{-\frac{1}{2}\Big(\mathbf{n}(t_i,\mathbf{s}_i;\mathbf{p})-\mathbf{n}_i\Big)\sigma^{-2}\Big(\mathbf{n}(t_i,\mathbf{s}_i;\mathbf{p})-\mathbf{n}_i\Big)^T\Big\}. \tag{3.3.35}$$

The maximization of the log-likelihood function

$$\ln(\mathscr{L}(\mathbf{p};\sigma))=-0.5\Big(n_d\ln(2\pi)+n_d\ln(\sigma^2)+\sigma^{-2}\Phi(\mathbf{p})\Big) \tag{3.3.36}$$

is equivalent to the minimization of the ordinary least-squares function $\Phi(\mathbf{p})$,

$$\Phi(\mathbf{p})=\sum_{i=0}^{M}\|\mathbf{n}(t_i,\mathbf{s}_i;\mathbf{p})-\mathbf{n}_i\|_2^2=\sum_{i=0}^{M}\sum_{j=1}^{M_i}(n(t_i,s_{i,j};\mathbf{p})-n_{i,j})^2, \tag{3.3.37}$$

provided that σ^2 is assigned the maximum likelihood value ${\sigma^*}^2=\Phi(\mathbf{p}^*)/n_d$, where $\mathbf{p}^*$ is the vector which gives a minimum to $\Phi(\mathbf{p})$ and $n_d:=\sum_{i=1}^{M}M_i$ is the total number of scalar measurements fitted by the model solution ($M=4$ for the data used).

Note that the following issue, related to the presence of measurement errors, has to be taken into account in any data-fitting procedure, cf., e.g. [68]. Any model solution, consistent with the data set within the noise level, can be considered as an acceptable one. In other words, the best-fit vector $\mathbf{p}^*$ (i.e. the minimizer of $\Phi(\mathbf{p})$) does not necessarily provide a biologically most consistent parameter estimate since the best data fitting may imply a good fitting of the error in the data (the overfitting problem). Hence, the use of a priori information about experimental data (e.g. noise level) and an information of the qualitative nature concerning the physical meaning of the solution (e.g. boundness, smoothness) is important in a data-fitting procedure. For $\Phi(\mathbf{p})$ defined by (3.3.37), we can use the l_2-norm to measure the cumulative contribution of the noise to the data,

$$\delta_\varepsilon=\Big(\sum_{i=1}^{4}\|\varepsilon\mathbf{n}_i\|_2^2\Big)^{1/2}=\varepsilon\Big(\sum_{i=1}^{4}\sum_{j=1}^{M_i}n_{i,j}^2\Big)^{1/2}, \tag{3.3.38}$$

where ε is the relative level of the measurement errors, e.g. $\varepsilon=0.2$ in case of 20% noise in the data. The errors in the data measurements that we consider were

Table 3.4 Results on the solution of the minimization problem 2 with $\alpha(z)$ approximated by $\alpha_L(z)$, $L = 3, 4, 6$. For all computed minima $M_{L,i}$: the total number of the estimated model parameters n_p, the estimated model parameters υ, β, γ, the corresponding value of $\Phi(\mathbf{p})$, the condition number τ and the Akaike index μ are presented

L	$M_{L,i}$	n_p	υ	β	γ	$\Phi(\mathbf{p})$	τ	μ
3	$M_{3,1}$	8	0.050	0.0212	1.92	7.89×10^{11}	374	8675
	$M_{3,2}$		0.118	0.0218	1.99	7.94×10^{11}	2080	8677
4	$M_{4,1}$	10	0.088	0.0207	1.90	6.16×10^{11}	2395	8601
	$M_{4,2}$		0.118	0.0174	1.94	7.28×10^{11}	8605	8654
6	$M_{6,1}$	14	0.057	0.0211	1.92	5.78×10^{11}	1291	8573
	$M_{6,2}$		0.129	0.0176	1.94	6.88×10^{11}	3783	8645

within 15–20% (S. Ehl, personal communication), and hence δ_ε ranges in the interval $[\delta_{0.15}, \delta_{0.2}]$. Since $\Phi(\mathbf{p})$ is the squared norm of the sum of the residuals, the value of interest is δ_ε^2, which can be confronted with the value of $\Phi(\mathbf{p})$ at a computed minimum. For the data used, $\delta_\varepsilon^2 \in [6.3, 11.3] \times 10^{11}$.

We use a non-uniform mesh $\mathscr{Z} = [1.15 = \tilde{z}_0, \tilde{z}_1, \ldots, \tilde{z}_L = z_{\max}]$ with its internal nodes, i.e. the elements of the vector $\tilde{\mathbf{z}} = [\tilde{z}_1, \ldots, \tilde{z}_{L-1}]$, included in the list of parameters to be estimated. We keep the ordering of the mesh points $\tilde{z}_k$ not allowing them to jump over each other during the minimization procedure. In fact, we solve a minimization problem with inequality constraints of a special form (Fig. 3.8).

The results on the estimation of the parameters of model (3.3.31) using the parametrization (3.3.33) of $\alpha(z)$ with $L = 3, 4, 6$ and estimated nodes of the mesh $\tilde{\mathbf{z}}$ are presented in Table 3.4 and Figs. 3.8 and 3.9. For $L = 3$, the larger number of the estimated parameters (8 against 6 in case of the uniform mesh $\mathscr{Z}$) leads to multiple minima of $\Phi(\mathbf{p})$, cf. Table 3.4 and Fig. 3.8 (left). Hence, the minimization problem 2 is ill-posed for all values of L. A large uncertainty in the right part of the estimated $\alpha_L(z)$, observed for the minimization problem 1, is also present as confirmed by the results presented in Fig. 3.8 (right) for $L = 4$. Note that small values of the computed 95% CIs for the estimated parameters which determine $\alpha_3(z)$ corresponding to the minimum $M_{3,2}$, cf. Figure 3.8 (left), are due to a specific location of the estimated nodes $\tilde{z}_1$ and $\tilde{z}_2$ and their very small CIs.

The minimization approach 2 provides a better fit to the data. Indeed, the value of $\Phi(\mathbf{p})$ at the computed minima is reduced significantly, by about 50%, for $L = 3$ and by $23 - 30\%$ and $17 - 25\%$ for $L = 4$ and $L = 6$, respectively. Obviously, this is due to a larger number of the estimated parameters, as compared to the corresponding minimization problem 1. Regularization of the ill-posed minimization problem 2 is the subject of the following section.

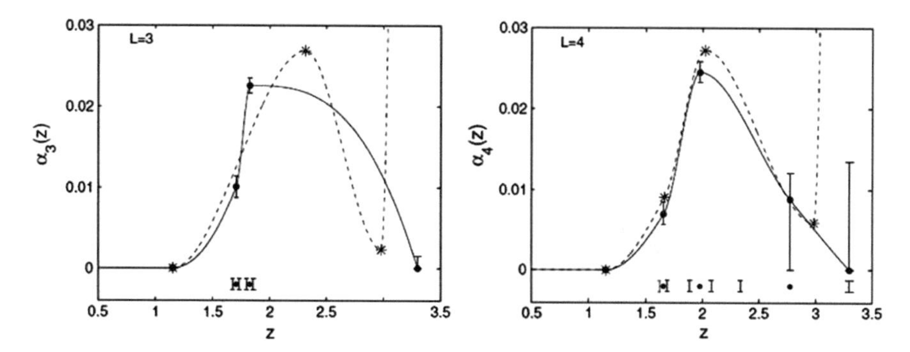

Fig. 3.8 The estimated functions $\alpha_3(z)$ and $\alpha_4(z)$. Bullets and stars indicate the pairs $\{\tilde{z}_j, a_j\}_{j=0}^{L}$ with estimated $\{\tilde{z}_j\}_{j=1}^{L-1}$ and $\{a_j\}_{j=1}^{L}$. The dashed and solid curves indicate $\alpha_L(z)$ corresponding to the minima $M_{L,1}$ and $M_{L,2}$, respectively. The estimated values $a_3 \approx 0.58$ and $a_4 \approx 0.61$ in case of the minima $M_{3,1}$ and $M_{4,1}$ are not shown. Approximations to 95% CIs for the estimated parameters $\{\tilde{z}_j\}_{j=1}^{L-1}$ and $\{a_j\}_{j=1}^{L}$ are computed for the minima $M_{3,2}$ and $M_{4,2}$

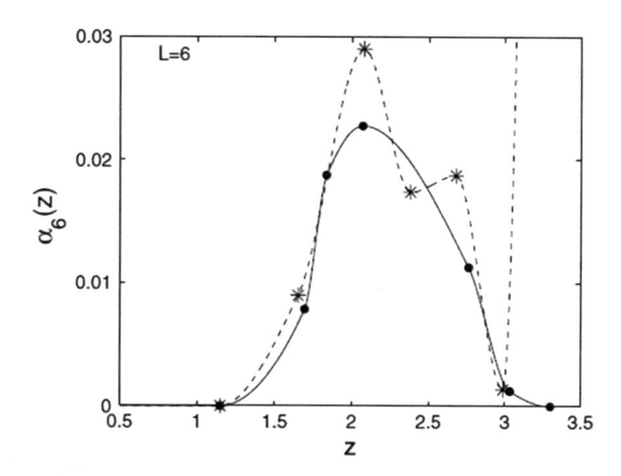

Fig. 3.9 The estimated functions $\alpha_6(z)$. Bullets and stars indicate the pairs $\{\tilde{z}_j, a_j\}_{j=0}^{6}$ with estimated $\{\tilde{z}_j\}_{j=1}^{5}$ and $\{a_j\}_{j=1}^{6}$. The dashed and solid curves indicate $\alpha_6(z)$ corresponding to the minima $M_{6,1}$ and $M_{6,2}$, respectively. The estimated value $a_6 \approx 0.32$ in case of the minimum $M_{6,1}$ is not shown

3.3.3 Regularization of the Parameter Estimation

A common approach to deal with ill-posed problems is to transform the original problem to a well-posed problem which is close, in a certain sense, to the original one. This approach leads to some forms of regularization of the original ill-posed problem. We use the well-known Tikhonov regularization framework [68, 71]. One of the main ideas of Tikhonov regularization is to use, as much as possible, supplementary information concerning the solution. In the context of ill-posed inverse problems, this implies the use of a priori information about the experimental data (e.g. noise level) and the qualitative nature of the solution (e.g. boundness, smoothness) consistent with its physical meaning. This allows one to restrict the space of computed solutions and to obtain a meaningful solution of the well-posed problem. According to the Tikhonov

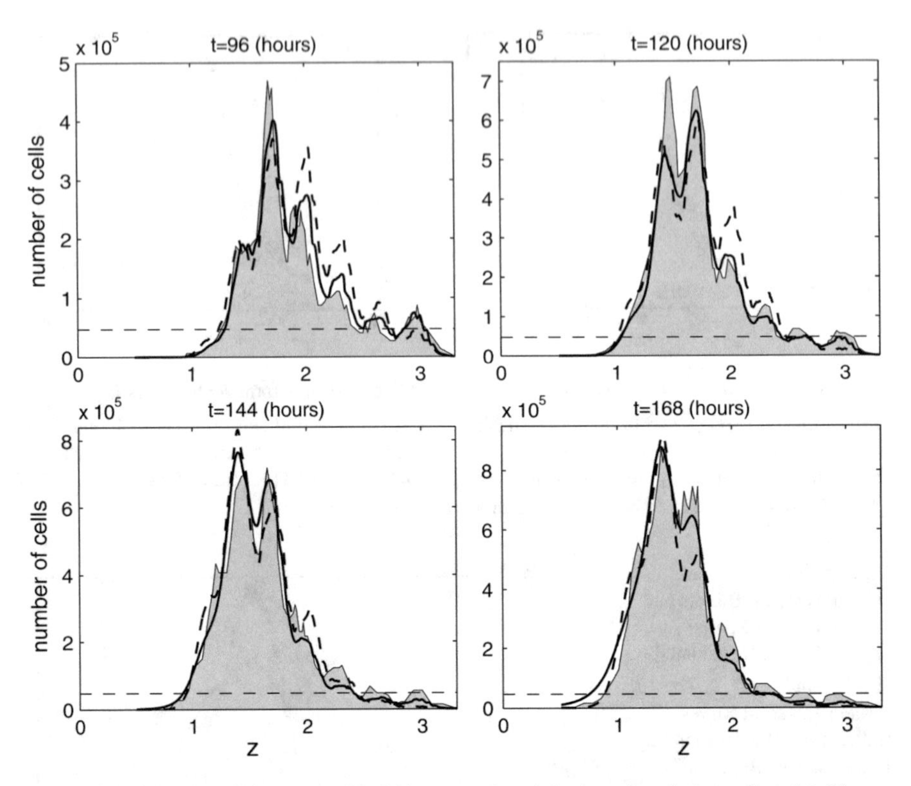

Fig. 3.10 Experimental data (grey filled histograms) and the best-fit solutions of model (5) corresponding to (i) the minimization problem 1 with $\alpha(z)$ approximated by $\alpha_3(z)$ (dashed curves) and (ii) the regularized minimization problem 2 (solid curves). The dashed horizontal lines indicate the maximum likelihood estimate of the data standard deviation $\sigma^* = 4.7 \times 10^4$ corresponding to $\Phi(p) = 7 \times 10^{11}$

regularization procedure, the regularized solution $\mathbf{p}_\lambda$ is obtained as the minimizer of the following combination of the objective function $\Phi(\mathbf{p})$ and a weighted penalty functional $\Omega(\mathbf{p})$:

$$\mathbf{p}_\lambda = \arg\min_{\mathbf{p}}\{J(\mathbf{p};\lambda) := \Phi(\mathbf{p}) + \lambda\Omega(\mathbf{p})\}, \tag{3.3.39}$$

where the scalar $\lambda > 0$ is the regularization parameter that controls the trade-off between the two objectives. The non-negatively valued penalty functional $\Omega(\mathbf{p})$ reflects desirable constraints on the solution $\mathbf{p}$, i.e. its choice is prompted by the nature of the problem. We seek regularized solutions $\mathbf{p}_\lambda$ such that $\alpha_L(z)$ is a non-oscillatory function. Therefore, we consider

$$\Omega(\mathbf{p}) = \int_{z^*}^{z_{\max}} \left(\frac{\mathrm{d}^2\alpha_L(z)}{\mathrm{d}z^2}\right)^2 \mathrm{d}z. \tag{3.3.40}$$

This type of $\Omega(\mathbf{p})$ implies that the function $\alpha_L(z)$ is approximated by functions having the smallest integrated squared curvature.

The regularization parameter λ should be chosen with care. According to Tikhonov regularization [68], λ must be adjusted to the noise level δ in the data, $\lambda = \lambda(\delta)$. One of the methods to choose $\lambda(\delta)$ is the discrepancy principle [68, 83, 84]: λ^* is determined as the minimizer of the functional $Q(\mathbf{p}_\lambda)$,

$$\lambda^* = \arg\min_\lambda \{Q(\mathbf{p}_\lambda) := |\Phi(\mathbf{p}_\lambda) - \delta^2|\}, \tag{3.3.41}$$

with δ being the corresponding norm of the error in the data. Note that although the rigorous justification of the Tikhonov regularization with the regularization parameter chosen by the discrepancy principle (e.g. convergence of the regularized solution to the exact one as the noise level in the data vanishes) is established for linear ill-posed problems, this method is widely used for nonlinear problems and, in particular, for nonlinear least-squares problems.

The discrepancy method, based on an embedded minimization problem, is computationally rather expensive. In [68], the following algorithm for computing $\lambda(\delta)$ is proposed: using

$$\lambda_k = \lambda_0 q^k, \quad q > 0, \tag{3.3.42}$$

we compute $\mathbf{p}_{\lambda_k}$ for $k = 0, 1, \ldots, K$, where K is such that λ_K gives a minimum to $Q(\mathbf{p}_\lambda)$ with a reasonable accuracy. Using this algorithm, one needs to solve $K + 1$ minimization problems of the form (3.3.39).

We apply the Tikhonov regularization procedure to the minimization problem 2 with $\alpha(z)$ approximated by $\alpha_4(z)$. We consider $L = 4$ for the following reason. For any value of λ, the value of $\Phi(\mathbf{p}_\lambda)$ at the computed minimum is larger than the one without regularization. Moreover, the larger L is (i.e. if more freedom for the shape of $\alpha_L(z)$ is given), the stronger regularization is necessary. However, this leads to a significant increase in the value of $\Phi(\mathbf{p}_\lambda)$ at the computed minimum and hence to a decrease of the quality of the data fitting. As one can see in Table 3.4, the increase of L from 3 to 4 reduces $\Phi(\mathbf{p})$ by about 20%, while $\Phi(\mathbf{p})$ changes negligibly when increasing L from 4 to 6. A similar behaviour is observed for the value of the Akaike index. Hence, the use of $L = 4$ seems to be the most suitable for the parametrization of $\alpha(z)$ with the ensuing regularization.

Using the discrepancy method, we found that the minimization problem (3.3.39) with $\alpha(z)$ approximated by $\alpha_4(z)$ has a unique solution $\mathbf{p}_\lambda$ when λ is about 5×10^{12} and larger. For $\lambda = 5 \times 10^{12}$, the corresponding model solution fits the data with an accuracy within the noise level in the data: $\Phi(\mathbf{p}_\lambda) \approx \delta^2_{0.166}$. The regularized function $\alpha_4(z)$ is shown in Fig. 3.11 (left).

For comparison, we computed the regularized solution $\mathbf{p}_\lambda$ for larger values of λ: $\lambda = 10^{13}$ and $\lambda = 4 \times 10^{13}$, cf. Table 3.5 and Fig. 3.11 (left). The corresponding regularized functions $\alpha_4(z)$ do not differ much. However, the contribution of the term $\lambda\Omega(\mathbf{p}_\lambda)$ to the value of the minimized functional $J(\mathbf{p}_\lambda; \lambda)$, cf. (3.3.39), grows with growing λ, cf. Table 3.5, and reaches half of the value of $J(\mathbf{p}_\lambda, \lambda)$ at the computed

Table 3.5 Results on the solution of the regularized minimization problem 2 with $\alpha(z)$ approximated by $\alpha_4(z)$. For the given values of the regularization parameter λ: the estimated model parameters υ, β, γ, the corresponding values of $\Phi(\mathbf{p}_\lambda)$, δ^2 and $\lambda\Omega(\mathbf{p}_\lambda)/\Phi(\mathbf{p}_\lambda)$ are presented

λ	υ	β	γ	$\Phi(\mathbf{p}_\lambda)$	δ^2	$\lambda\Omega(\mathbf{p}_\lambda)/\Phi(\mathbf{p}_\lambda)$
5×10^{12}	0.132	0.0167	1.93	7.79×10^{11}	$\delta^2_{0.166}$	0.15
10^{13}	0.127	0.0170	1.94	8.08×10^{11}	$\delta^2_{0.169}$	0.22
4×10^{13}	0.113	0.0176	1.95	9.49×10^{11}	$\delta^2_{0.184}$	0.45

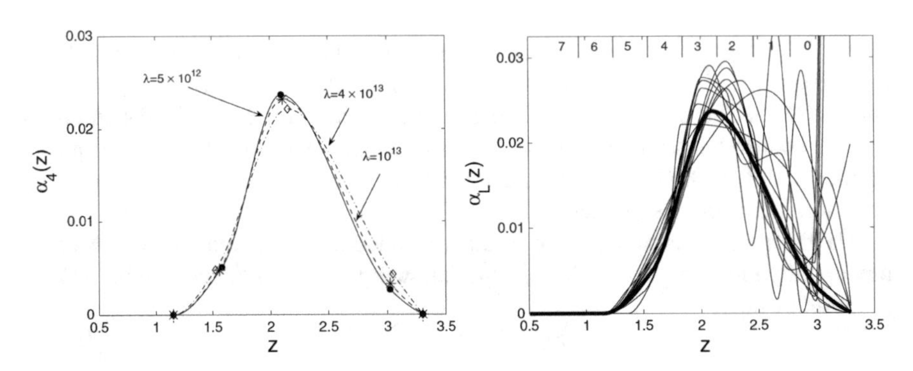

Fig. 3.11 Left: The regularized functions $\alpha_4(z)$ estimated using different values of the regularization parameter λ. Bullets, stars and diamonds indicate the pairs $\{\tilde{z}_j, a_j\}_{j=0}^4$ with estimated $\{\tilde{z}_j\}_{j=1}^3$ and $\{a_j\}_{j=1}^4$. Right: The estimated regularized function $\alpha_4(z)$ (thick curve) and all $\alpha_L(z)$ corresponding to the minima estimated via minimization 1 and 2 approaches (thin curves). The vertical lines at the top of the figure indicate intervals of the CFSE intensity z corresponding to the generations of cells which have undergone $0, 1, \ldots, 7$ divisions

minimum for $\lambda = 4 \times 10^{13}$. The latter implies too strong regularization, while the value $\lambda\Omega(\mathbf{p}_\lambda)/\Phi(\mathbf{p}_\lambda) = 0.16$ for $\lambda = 5 \times 10^{12}$ is acceptable.

Since the impact of the term $\lambda\Omega(\mathbf{p}_\lambda)$ in the minimized functional $J(\mathbf{p}_\lambda; \lambda)$ is minimal for $\lambda = 5 \times 10^{12}$, this value of λ can be accepted as the optimal one. The regularized function $\alpha_4(z)$ for $\lambda = 5 \times 10^{12}$ and all estimated $\alpha_L(z)$ corresponding to the minima listed in Tables 3.4 and 3.5 are shown in Fig. 3.11 (right). We observe that the right part of the estimated $\alpha(z)$ is much more affected by the regularization procedure than the left part. Figure 3.10 shows the best-fit of the experimental data by the model solution corresponding to the regularized solution $\mathbf{p}_\lambda$ of the minimization problem (3.3.39) with $\lambda = 5 \times 10^{12}$. This figure clearly indicates that the model solution is consistent with the CFSE histogram data.

3.3.4 *Cell Growth Model with Asymmetry and Time Lags*

The interpretation of CFSE proliferation assays relies on the assumption that the label is divided equally between the daughter cells upon cell division. However, recent experimental studies indicate that division of cells is not perfectly symmetric, and

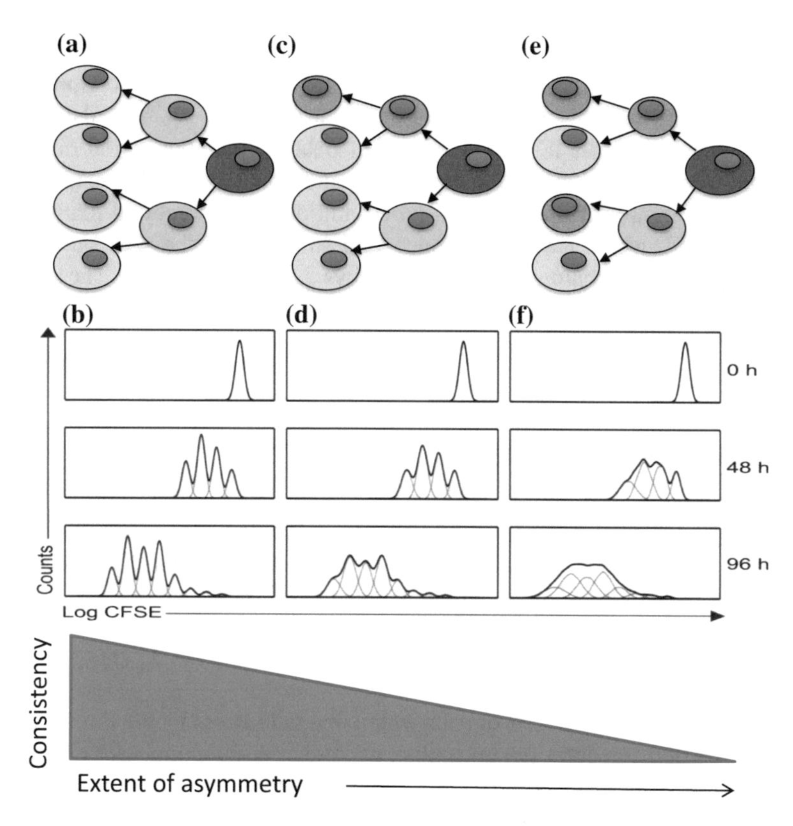

Fig. 3.12 Impact of asymmetry in T cell division impinges on fluorescent protein partition between daughter cells. **a**, **b** Symmetric cell division with equal distribution of the fluorescent dye between daughter cells **a** and modelled time course analysis of T cell proliferation as determined by flow cytometry (**b**, solid black lines). Dashed red lines in **b** indicate the evolution of CFSE intensity of the cohorts (generations) of cell which differ in the number of completed divisions with the assumption of symmetric division. **c**, **d** Asymmetric cell division with low asymmetry (**c**) and modelled flow cytometric time course analysis of CFSE dilution (**d**, solid black lines) that corresponds to an asymmetry 46 and 54% (**d**, dashed red lines describe the CFSE distributions for cell cohorts differing in terms of the completed divisions). **e**, **f** T cells dividing with high asymmetry (**e**) and corresponding model-generated flow cytometric CFSE dilution patterns (**f**, solid black lines) with asymmetry values of 42% and 58% describing the behaviour of the T cells in this setting (**f**, dashed red lines describe the cell cohorts corresponding to different generations)

there is unequal distribution of protein between sister cell pairs. The uneven partition of protein or mass to daughter cells can lead to an overlap in the generations of CFSE-labelled cells with straightforward consequences for the resolution of individual generations as shown in Fig. 3.12.

Numerous mathematical models developed for the analysis of CFSE-based proliferation incorporate the premise that the CFSE fluorescence intensity is halved in the two daughter cells. For quantitative analysis of CFSE-labelled cell proliferation which is characterized by poorly resolved peaks of cell generations in flow cytometric histograms, an extension of the label-structured analysed in the previous section is needed to take into account the cell division asymmetry. We formulate the corresponding model in the form of a system of delay hyperbolic partial differential equations. This model considering asymmetry and time lag in cell division belongs to a powerful family of the so-called label- and division-structured mathematical models [85–89], which have some conceptual similarity to the age- and division-structured model (Bernard et al. 2003, Biophys J. 84(5):3414–24). These models allow for a direct reference to the histograms of cell distribution and are more robust to a poor peak resolution of cell generations.

3.3.5 Division-Structured DDE Model

Let $N_i(t)$ be the total population size of live T lymphocytes (cells) which have undergone i divisions at time t as shown in Table 3.6. We split the population of cells in each generation i in two subpopulations as follows:

$$N_i(t) = N_i^r(t) + N_i^c(t), \quad i = 0, 1, \ldots, i_r - 1, \quad N_{i_r}(t) = N_{i_r}^r(t). \tag{3.3.43}$$

The N_i^r-subpopulation consists of cells which divided i times by the time t and they are not in the division cycle (resting cells), while the N_i^c-subpopulation contains the cells which are in the process of the $i+1$ division (cycling cells). These dividing cells will complete the $(i+1)$-th division at time $t + \tau_{i+1}$ if their i-th division happened at time t, where τ_{i+1} is the duration of the $i+1$ round of division. The last expression for $N_{i_r}(t)$ represents the simplifying assumption that the last (i_r-th) generation of cells which can be experimentally traced with CFSE consists of T cells which do not divide. The schematic illustration of the mathematical model is given in Fig. 3.13. The rates of change of $N_i^r(t)$ and $N_i^c(t)$ with time are represented by the following set of delay differential equations:

$$\begin{aligned}
\frac{dN_0^r(t)}{dt} &= -(\alpha_0 + \beta_0)N_0^r(t), \\
\frac{dN_i^r(t)}{dt} &= -(\alpha_i + \beta_i)N_i^r(t) + 2\alpha_{i-1}N_{i-1}^r(t - \tau_{i-1}), \quad i = 1, 2, \ldots, i_r, \\
\frac{dN_i^c(t)}{dt} &= \alpha_i(N_i^r(t) - N_i^r(t - \tau_i)), \quad i = 0, 1, \ldots, i_r - 1,
\end{aligned} \tag{3.3.44}$$

with initial conditions: $N_0^r(\xi) = 0,\ \xi \in [-\tau_0, 0)$, $N_0^r(0) = N^0$ is the given number of cells at the start of the experiment (cf. Table 3.6), $N_i^r(\xi) = 0,\ \xi \in [-\tau_i, 0],\ i = 1, 2, \ldots, i_r$, $N_i^c(\xi) = 0,\ \xi \in [-\tau_i, 0],\ i = 0, 1, \ldots, i_r - 1$.

The first term on the right of equations for $N_i^r(t)$ represents the cell loss (outflux from the compartment) due to division and death, while the last term on the right represents the cell birth (influx from the previous compartment due to cell division).

Table 3.6 In vitro proliferation data of MHV s598-605-specific CD8$^+$ T lymphocytes following co-culture with dendritic cells pulsed with the s598 peptide at a concentration of 10^{-5} M. At the indicated times, CFSE histograms were obtained by flow cytometry and analysed with the FlowJo software. The total numbers of live lymphocytes, N^s, and the distribution of lymphocytes with respect to the number of divisions they have undergone, N_i^s, $i = 0, 1, \ldots, 7$, were followed during 96 h from the start of the experiment at the indicated times t_s, $s = 0, 1, \ldots, 5$

Time hours t_s	Total number of live cells N^s	Numbers of cells w.r.t. the number of divisions (i) they undergone N_i^s							
		0	1	2	3	4	5	6	7
0	55827	55827	0	0	0	0	0	0	0
36	30492	18283	12197	0	0	0	0	0	0
48	52272	11678	15635	19550	5395	0	0	0	0
60	63180	6722	8315	13609	18493	12598	5332	0	0
72	50787	5104	3316	6282	11046	14622	8974	1432	0
96	20849	715	915	1418	2815	3628	4545	4629	2773

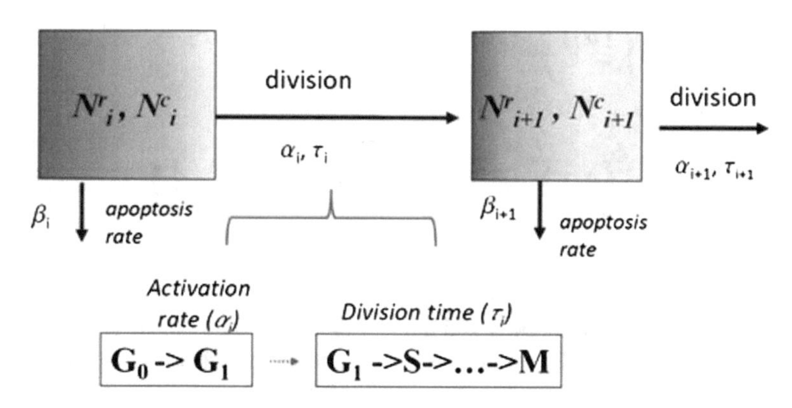

Fig. 3.13 The scheme of the division-structured model with a time lag for T cell proliferation. Resting and cycling cell compartments of generations i and $i + 1$ are shown. The parameters of the model are explained in the text

The two terms on the right-hand side of the equations for $N_i^c(t)$ consider the rates at which cells enter the i-th round of division and complete the cell cycle after time τ_i, respectively. The experimental data on cell numbers correspond to the sum of the resting and cycling cell populations. Therefore, although the cycling cells do not appear on the right-hand side of the DDE for the resting cells, their number needs to be followed for correctly representing the total cell number for data fitting. In general, the per capita proliferation and death rates of T lymphocytes, α_i, respectively, β_i, depend on the number of divisions the lymphocytes have undergone. Model (3.3.44) is used for the description of the dynamics of cell generations and to estimate the rates of cell division (α_i) and death (β_i) from CFSE data in Table 3.6.

The model (3.3.44), formulated as a system of linear delay differential equations, allows an analytical solution, which is given by the following

Lemma 3.1 *Let* $c_i := \alpha_i + \beta_i$, $c_i \neq c_k$ *for* $i \neq k$, $i = 0, 1, \ldots, i_r$, *and* $T_i := \sum_{j=0}^{i} \tau_j$, $i = 0, 1, \ldots, i_r - 1$. *The solution of model (3.3.44) is*

$$
\begin{aligned}
N_0^r(t) &= N^0 \exp^{-c_0 t}, \quad t \geq 0, \\
N_i^r(t) &= 0, \quad t \in [0, T_{i-1}), \quad i = 1, \ldots, i_r, \\
N_i^r(t) &= 2^i N^0 \Big(\prod_{j=0}^{i-1} \alpha_j\Big) \sum_{j=0}^{i} \exp^{c_j(T_{i-1}-t)} \prod_{k=0,k\neq j}^{i} (c_k - c_j)^{-1}, \\
&\qquad t \geq T_{i-1}, \quad i = 1, \ldots, i_r, \\
N_0^c(t) &= N^0 \alpha_0 (1 - \exp^{-c_0 t}) c_0^{-1}, \quad t \in [0, T_0), \\
N_0^c(t) &= N^0 \alpha_0 (\exp^{c_0(T_0-t)} - \exp^{-c_0 t}) c_0^{-1}, \quad t \geq T_0, \\
N_i^c(t) &= 0, \quad t \in [0, T_{i-1}), \quad i = 1, \ldots, i_r - 1, \\
N_i^c(t) &= 2^i N^0 \Big(\prod_{j=0}^{i} \alpha_j\Big) \sum_{j=0}^{i} (1 - \exp^{c_j(T_{i-1}-t)}) c_j^{-1} \prod_{k=0,k\neq j}^{i} (c_k - c_j)^{-1}, \\
&\qquad t \in [T_{i-1}, T_i), \quad i = 1, \ldots, i_r - 1. \\
N_i^c(t) &= 2^i N^0 \Big(\prod_{j=0}^{i} \alpha_j\Big) \sum_{j=0}^{i} (\exp^{c_j(T_i-t)} - \exp^{c_j(T_{i-1}-t)}) c_j^{-1} \prod_{k=0,k\neq j}^{i} (c_k - c_j)^{-1}, \\
&\qquad t \geq T_i, \quad i = 1, \ldots, i_r - 1.
\end{aligned}
\tag{3.3.45}
$$

The proof of this lemma can be found in [7]. Lemma 3.1 gives an analytical solution for the linear DDE model with proliferation, time lag and death parameters which are considered to be different for every cell generation. Notice that similar results were obtained for an ODE model in the fully heterogeneous parameters case [5] and partly heterogeneous ODE and DDE models [90].

3.3.6 Asymmetric Division and Label-Structured Delay hPDE Model

In this section, we introduce a novel mathematical model for the dynamics of CFSE-labelled lymphocyte populations which considers asymmetry and time lag in cell division. We consider (as before [4]) a population of cells which are structured according to a single continuous variable x that characterizes the CFSE amount (unit of intensity, UI) in the cells with the cell distribution function $n(t, x)$ (cell/UI). Similarly to [85–87, 89], we split the overall CFSE-labelled cell population into generations of cells which differ in terms of the completed divisions. Therefore, the

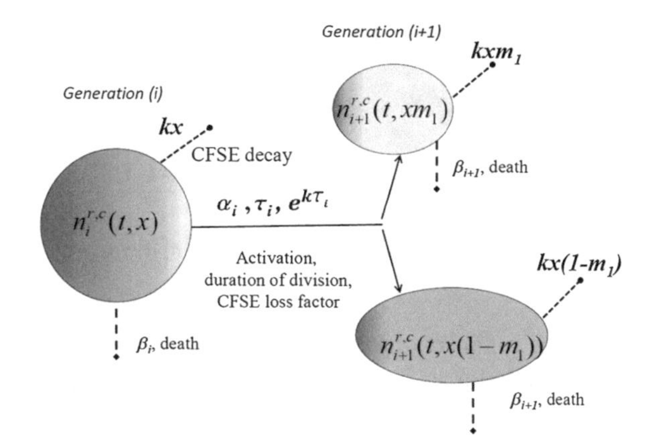

Fig. 3.14 The scheme of the division- and label-structured model with an asymmetry and a time lag for T cell proliferation. Resting and cycling cell compartments of generations i and $i+1$ are shown. The parameters of the model are explained in the text

state of the population in the i-th generation of cells at time t is described by the distribution (density) function $n_i(t,x)$, so that the number of cells with the CFSE intensity between x_1 and x_2 is given by $\int_{x_1}^{x_2} n_i(t,x)dx$. As in the division-structured DDE model (3.3.44) each generation of cells is further subdivided into the resting and cycling subsets as follows: $n_i(t,x) = n_i^r(t,x) + n_i^c(t,x)$, $i = 0, 1, \ldots, i_r - 1$, $n_{i_r}(t,x) = n_{i_r}^r(t,x)$.

Now, we relax the fundamental assumption in the analysis of the CFSE proliferation assay, namely, that as a cell divides the fluorescently tagged cellular proteins are allocated equally to each daughter cell [91]. The asymmetric T cell division with respect to partitioning of proteins has been clearly shown to take place in the adaptive immune responses in a number of recent studies (e.g. [92]). To take into account an unequal partitioning of fluorescent proteins between daughter cells, we introduce the label dilution parameters m_1 and m_2 such that when a mother cell has CFSE label intensity (equivalently, the fluorescently tagged proteins) x, then upon division one daughter cell has the fluorescence intensity m_1x, while another daughter cell has the fluorescence intensity m_2x. We assume that cells do not lose label during division, i.e. $m_1 + m_2 = 1$. The schematic illustration of the mathematical model is given in Fig. 3.14.

The evolution of the generation-structured cell distributions for resting and cycling subsets, $n_i^r(t,x)$ and $n_i^c(t,x)$, respectively, is modelled by the following label-structured cell population balance equations formulated as a system of delay hyperbolic PDEs:

$$\frac{\partial n_0^r(t,x)}{\partial t} - k\frac{\partial(xn_0^r(t,x))}{\partial x} = -(\alpha_0 + \beta_0)n_0^r(t,x),$$

$$\frac{\partial n_i^r(t,x)}{\partial t} - k\frac{\partial(xn_i^r(t,x))}{\partial x} = -(\alpha_i + \beta_i)n_i^r(t,x) + \alpha_{i-1}exp^{k\tau_{i-1}}\Big(\frac{1}{m_1}n_{i-1}^r(t-\tau_{i-1}, \exp^{k\tau_{i-1}}\frac{x}{m_1})+$$

$$\frac{1}{m_2}n_{i-1}^r(t-\tau_{i-1}, \exp^{k\tau_{i-1}}\frac{x}{m_2})\Big), \quad i = 1, 2\ldots, i_r,$$

$$\frac{\partial n_i^c(t,x)}{\partial t} - k\frac{\partial(xn_i^c(t,x))}{\partial x} = \alpha_i(n_i^r(t,x) - \exp^{k\tau_i} n_i^r(t-\tau_i, \exp^{k\tau_i} x)), \quad i = 0, \ldots, i_r - 1, \tag{3.3.46}$$

with the initial conditions: $n_0^r(\xi, x) = 0$ for $\xi \in [-\tau_0, 0)$; $n_0^r(0, x)$ is the initial cell density determined by the given experimental histogram at $t = 0$ as described below; $n_i^r(\xi, x) = 0$ for $\xi \in [-\tau_i, 0]$, $i = 1, \ldots, i_r$ and $n_i^c(\xi, x) = 0$ for $\xi \in [-\tau_i, 0]$, $i = 0, \ldots, i_r - 1$. The solution of this model, as Theorem 3.1 below shows, is uniquely determined by the initial conditions so that boundary conditions are not required. This model takes into account the time lags in cell division and asymmetry in protein partition in the daughter cells.

The model equations consist of the following terms:

$k\partial(xn_i^r(t,x))/\partial x$, the advection term, describes the natural decay of the CFSE fluorescence intensity of the labelled cells with the rate kx;

$(\alpha_i + \beta_i)n_i^r(t,x)$ describes the local disappearance of cells with the CFSE intensity x due to the division and the death with α_i and β_i being the proliferation and death rates, respectively;

$\alpha_{i-1}\exp^{k\tau_{i-1}}\Big(\frac{1}{m_1}n_{i-1}^r(t-\tau_{i-1}, \exp^{k\tau_{i-1}}\frac{x}{m_1}) + \frac{1}{m_2}n_{i-1}^r(t-\tau_{i-1}, \exp^{k\tau_{i-1}}\frac{x}{m_2})\Big)$. It describes the birth of two cells, appearing in the interval $[x + dx]$ at time t, after division, which started at time $t - \tau_{i-1}$, of one cell from the interval $\eta_1 = \exp^{k\tau_{i-1}}[x + dx]/m_1$ and another one from the interval $\eta_2 = \exp^{k\tau_{i-1}}[x + dx]/m_2$. Since the length of the intervals η_1 and η_2 differs by the factor $\exp^{k\tau_{i-1}}/m_1$, respectively, $\exp^{k\tau_{i-1}}/m_2$ from the length of the interval $[x + dx]$, the factors in the source terms $\exp^{k\tau_{i-1}}/m_1$ and $\exp^{k\tau_{i-1}}/m_2$ appear before $n_{i-1}^r(t-\tau_{i-1}, \exp^{k\tau_{i-1}}\frac{x}{m_j})$, $j = 1, 2$. The terms $\exp^{k\tau} x/m_j$, $j = 1, 2$, in the definition of the intervals η_1 and η_2 have the following meaning: if the label intensity of a cell at time t is $x(t)/m_j$, then at time $t - \tau$ it is $\exp^{k\tau} x(t)/m_j$. This follows from the equation for the rate of the label dilution,

$$\frac{dx(t)}{dt} = -kx(t), \tag{3.3.47}$$

which implies that $x(t) = \exp^{-kt} x(0)$ and $x(t-\tau) = \exp^{-k(t-\tau)} x(0) = \exp^{k\tau} x(t)$. Note that if $m_1 = m_2 = 1/2$ and $\tau_i = 0$, then the last term in the equation for $n_i^r(t,x)$ takes the standard form specific for symmetric division $4\alpha_{i-1}n_{i-1}(t, 2x)$.

$\alpha_i(n_i^r(t,x) - \exp^{k\tau_i} n_i^r(t-\tau_i, \exp^{k\tau_i} x))$ describes the density of cells with label intensity x which are in the division cycle. The factor $\exp^{k\tau_i}$ in the disappearance term with $n_i^r(t-\tau_i, \exp^{k\tau_i} x)$ arises due to the following reason: setting $m_1 + m_2 = 1$, we require that the total amount of label

$$\exp^{k\tau_i}\left(\int_0^\infty \frac{x}{m_1} n_i^r(t-\tau_i, \exp^{k\tau_i}\frac{x}{m_1})dx + \int_0^\infty \frac{x}{m_2} n_i^r(t-\tau_i, \exp^{k\tau_i}\frac{x}{m_2})dx\right) =$$

$$\frac{m_1}{\exp^{k\tau_i}}\int_0^\infty y n_i^r(t-\tau_i, y)dy + \frac{m_2}{\exp^{k\tau_i}}\int_0^\infty y n_i^r(t-\tau_i, y)dy = \frac{1}{\exp^{k\tau_i}}\int_0^\infty y n_i^r(t-\tau_i, y)dy, \tag{3.3.48}$$

is conserved during the cell division. Therefore, the following equality should be fulfilled for the last term in the equation for $n_i^c(t,x)$:

$$\omega\int_0^\infty x n_i^r(t-\tau_i, \exp^{k\tau_i}x)dx = \frac{\omega}{\exp^{2k\tau_i}}\int_0^\infty y n_i^r(t-\tau_i, y)dy = \frac{1}{\exp^{k\tau_i}}\int_0^\infty y n_i^r(t-\tau_i, y)dy. \tag{3.3.49}$$

Hence, we arrive at the expression for the scaling factor $\omega = \exp^{k\tau_i}$.

The parameter k characterizes the exponential decay of the CFSE fluorescence intensity of cells. This is a result of complex processes including the turnover of intracellular proteins to which the fluorescent conjugate binds and the outflow of CFSE from the cells [93]. Both processes depend on the activation status of the cells with the corresponding differences in the cellular metabolism, internal biochemistry and the cell membrane properties. The current knowledge of these processes is rather limited (see further discussion in [93]). Therefore, we used as simplifying assumption the same constant value of parameter k for resting and cycling cells. The assumption can be relaxed and different values of k for resting and activated cells can be readily considered in the model. However, the corresponding increase of the model complexity should be considered in the context of the data sets available for parameter estimation.

Clearly, the state variables of models (3.3.44) and (3.3.46) are related as

$$N_i^r(t) = \int_0^\infty n_i^r(t,x)dx, \qquad N_i^c(t) = \int_0^\infty n_i^c(t,x)dx \tag{3.3.50}$$

and the overall population size at time t is

$$\begin{aligned} N(t) &= \sum_{i=0}^{i_r} N_i(t) = \sum_{i=0}^{i_r-1}(N_i^r(t) + N_i^c(t)) + N_{i_r}^r(t) = \\ &\int_0^\infty (\sum_{i=0}^{i_r-1}(n_i^r(t,x) + n_i^c(t,x)) + n_{i_r}^r(t,x))dx = \int_0^\infty n(t,x)dx, \end{aligned} \tag{3.3.51}$$

where $n(t,x)$ is the cell density in the overall population at time t.

Model (3.3.46) is a system of linear hyperbolic PDEs with time delays. Recently, an efficient approach was proposed in [85–87] to treat analytically similar type of hyperbolic PDEs, however without time delays (i.e. only equations for aggregated cell population $n_i(t,x)$ are considered) and under the assumption of equal partition of the marker intensity x between daughter cells upon division ($m_1 = m_2 = 0.5$). The approach is based upon decomposition of a system of coupled hyperbolic PDEs into a system of ODEs and a set of decoupled hyperbolic PDEs which can be solved analytically. We extend this method to a more complex system (3.3.46) of delay hyperbolic PDEs by decomposing (3.3.46) into the system of DDEs (3.3.44) and a set of decoupled hPDEs. Since the background necessary for this method is detailed in [87], we only present one issue specific for our dhPDE model (3.3.46) with the two states for cells (i.e. resting and cycling ones). Similar to [85–87], for cells which have divided i times we introduce the probability density for a single cell to have a certain label intensity x as

$$p_i(t,x) = \frac{n_i^r(t,x)}{N_i^r(t)} \tag{3.3.52}$$

for $N_i^r(t) > 0$. Note that

$$\frac{n_i^r(t,x)}{N_i^r(t)} = \frac{n_i^c(t,x)}{N_i^c(t)}, \tag{3.3.53}$$

since cells of the resting and cell cycle progressing populations in the i-th generation have the same distribution of the label intensity. The set of model equations (3.3.46) has an explicit solution defined by the following theorem.

Theorem 3.1 *The solution of model (3.3.46) is*

$$\begin{aligned} n_i^r(t,x) &= N_i^r(t)p_i(t,x),\ i = 0,1,\ldots,i_r, \\ n_i^c(t,x) &= N_i^c(t)p_i(t,x),\ i = 0,1,\ldots,i_r-1, \end{aligned} \tag{3.3.54}$$

where

(i) $N_i^r(t)$ *and* $N_i^c(t)$ *are the solutions of the system of DDEs (3.3.44),*
(ii) $p_i(t,x)$ *is the solution of the PDE:*

$$\frac{\partial p_i(t,x)}{\partial t} - k\frac{\partial(xp_i(t,x))}{\partial x} = 0, \quad i = 0,1,2\ldots,i_r, \tag{3.3.55}$$

with initial conditions:

$$\begin{aligned} p_0(0,x) &= \tfrac{n_0^r(0,x)}{N^0}, \\ p_{i+1}(0,x) &= \tfrac{1}{2}\Big(\tfrac{1}{m_1}p_i(0,\tfrac{x}{m_1}) + \tfrac{1}{m_2}p_i(0,\tfrac{x}{m_2})\Big), \quad i = 0,1,\ldots,i_r-1. \end{aligned} \tag{3.3.56}$$

The proof of Theorem 3.1 can be found in [7].

Note that, according to (3.3.52), $p_i(t,x)$ is not defined in case $N_i^r(t) = 0$. Since in this case $n_i^r(t,x) = 0$, Eq. (3.3.54) simplifies to $0 = 0 \cdot p_i(t,x)$, allowing, as suggested in [85–87], any specification for $p_i(t,x)$. Therefore, in such cases we use $p_i(t,x)$ defined by (3.3.55)–(3.3.56).

Corollary 3.1 *The solutions* $n_i^r(t,x)$ *and* $n_i^c(t,x)$ *of model (3.3.46) are*

$$n_0^r(t,x) = N_0^r(t)\exp^{kt}\frac{n_0^r(0,\exp^{kt}x)}{N^0}, \quad n_0^c(t,x) = N_0^c(t)\exp^{kt}\frac{n_0^r(0,\exp^{kt}x)}{N^0},$$

$$n_i^r(t,x) = N_i^r(t)C_{i-1}(t,x), \quad i = 1,\dots,i_r,$$

$$n_i^c(t,x) = N_i^c(t)C_{i-1}(t,x), \quad i = 1,\dots,i_r-1,$$

$$C_i(t,x) = \frac{\exp^{kt}}{2}\left(\frac{1}{m_1}p_i(0,\frac{\exp^{kt}x}{m_1}) + \frac{1}{m_2}p_i(0,\frac{\exp^{kt}x}{m_2})\right), \quad i = 0,1,\dots,i_r-1, \tag{3.3.57}$$

where $N_i^r(t)$ and $N_i^c(t)$ are the solutions of the DDE model (3.3.44).

The proof of this corollary follows from the proof of Theorem 3.1.

The division stage- and label-structured population balance model (3.3.46) with asymmetry and time lag in cell division is developed for the description of the evolution of CFSE histograms and to estimate per each cell generation the average rates of cell division (α_i) and death (β_i), the duration of the cell cycle (τ_i), the rate of label decay (k) and the division asymmetry coefficient (m_1). Note that the dhPDE model (3.3.46) is formulated using a linear scale for the structure variable x, while the initial cell density n_j^0, $j = 1,\dots,d$, is determined in the z-coordinate by

$$n_j^s = \frac{c_{s,j}N^s}{C_s\Delta_z}, \quad s = 0,1,\dots,M, \quad j = 1,\dots,d, \tag{3.3.58}$$

where N^s is the given total number of cells at time t_s (cf. Table 3.6), $c_{s,j} = c(t_s, z_j)$ and $C_s = \sum_{j=1}^{d} c_{s,j}$ is the total number of cell counts at time t_s; $M = 5$ and $d = 4096$ for the data set used. Note that the values n_j^s are the experimental data to estimate parameters of the label-structured dhPDE model. Therefore, we translate n_j^0 in the x-coordinate as

$$n_0^r(0,x_j) = \frac{n_j^0}{\ln(10)10^{z_j}}, \quad j = 1,\dots,d, \tag{3.3.59}$$

where the factor $\ln(10)10^{z_j}$ is necessary to preserve the fluorescence intensity after the change of variables.

3.3.7 dhPDE Model

The division- and label-structured mathematical model of the CFSE proliferation assay formulated with a system of delay hyperbolic PDEs has the same parameters as the division-structured DDE model plus the two ones which characterize the rate of label loss and the fraction of the label which is acquired by the daughter cells from the mother cell upon division. Therefore, the task is to estimate the following parameters of the dhPDE model (3.3.46):

$$\boldsymbol{\theta} = (k, m_1, \boldsymbol{\alpha}, \boldsymbol{\beta}, c, \tau_0, \tau_1) \in R^{19}, \tag{3.3.60}$$

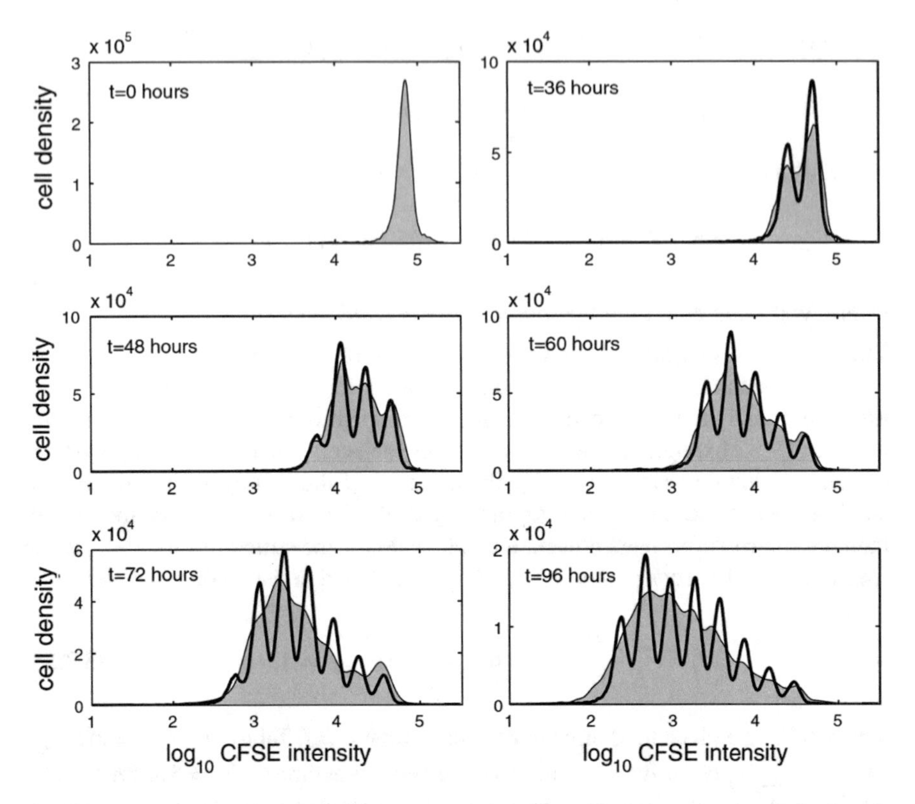

Fig. 3.15 Symmetric cell division. Experimental data (grey filled histograms) and the best-fit solution of the dhPDE model (3.3.46) with fixed $m_1 = 0.5$ (solid curves)

assuming that $\tau_i = \tau_1,\ i = 2, \ldots, 7$ and using the information on the total number of live T cells at the given times (columns 1 and 2 of Table 3.6) and the corresponding histograms of the CFSE distribution (Fig. 3.15). No decomposition of the histograms into generational clusters is needed. The experimental histograms of the cell population density, to which the model solution is fitted are characterized by poorly resolved peaks (and therefore, poorly identifiable generations) after time $t = 36$. The absence of clear peaks of the population density, corresponding to different generations of cells, cannot be explained by cellular autofluorescence. Indeed, the fluorescence of the labelled proliferating cells at the final time $t = 96$ h is larger than the determined background fluorescence (the median fluorescence of unstained control sample), which is below 100 UI.

One can readily observe from Fig. 3.15 that as time increases, the pattern of the CFSE distribution of the cell population becomes smoother suggesting that the fluorescent dye is divided unequally between the daughter cells. The computed best-fit solution for model (3.3.46) under the assumption of an equal division of the CFSE label between the two daughter cells (i.e. with $m_1 = 0.5$ fixed) is characterized by clearly pronounced peaks in histograms representing the different cell generations,

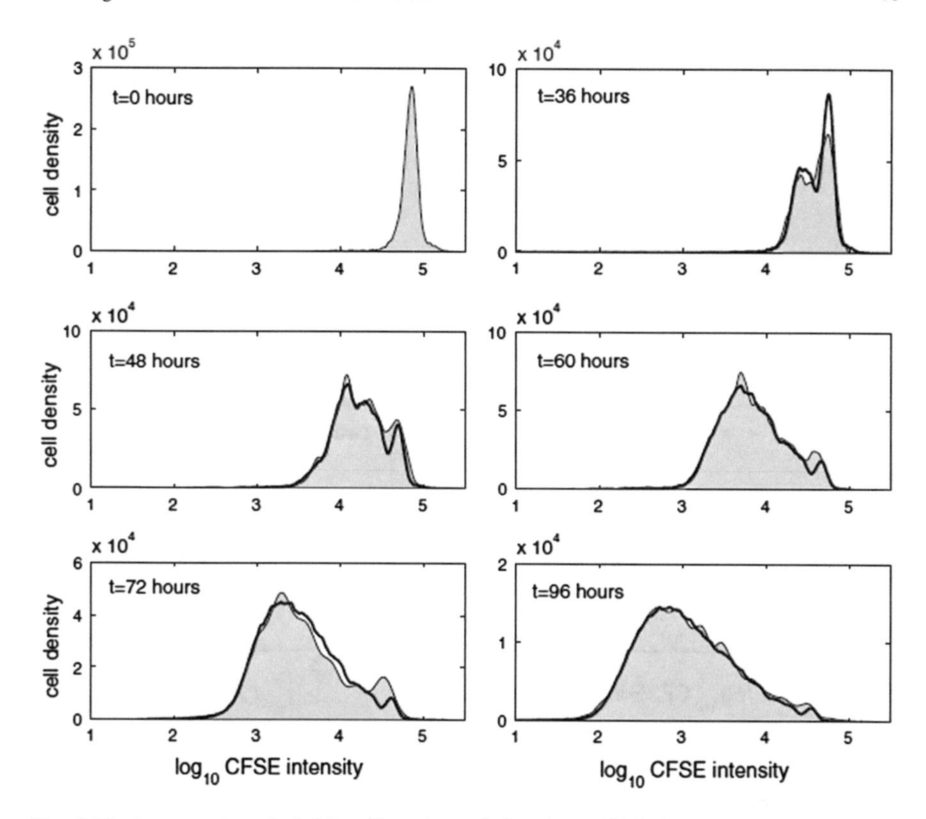

Fig. 3.16 Asymmetric cell division. Experimental data (grey filled histograms) and the best-fit solution of the dhPDE model (3.3.46) with estimated m_1 (solid curves)

cf. Figure 3.15. The parameter estimation for model (3.3.46) with a released asymmetry parameter m_1 leads to the best-fit solution shown in Fig. 3.16 with the estimated parameter values presented in Table 3.7. The model solution with released asymmetry provides an improved fit to the histogram data as well as the data on the clonal expansion of T lymphocytes.

We showed in a higher resolution Fig. 3.17 (upper row) that the symmetric division leads to a best-fit solution which is characterized by a systematic bias from the data as compared to an asymmetric consideration (left versus right figures, respectively). A direct consequence of the asymmetry in the lymphocyte division is the broadening of the CFSE range associated with a particular generation with the increase in the generation number as one can see at the middle row of Fig. 3.17. Therefore, the use of the label partition parameter $m_1 = 0.5$, corresponding to the symmetric division scenario, does not result in a qualitatively and quantitatively consistent approximation of the given set of experimental histograms.

We note that, starting with different initial points, the minimization procedure either converges to the best-fit minimum indicated in Table 3.7, or much poorer

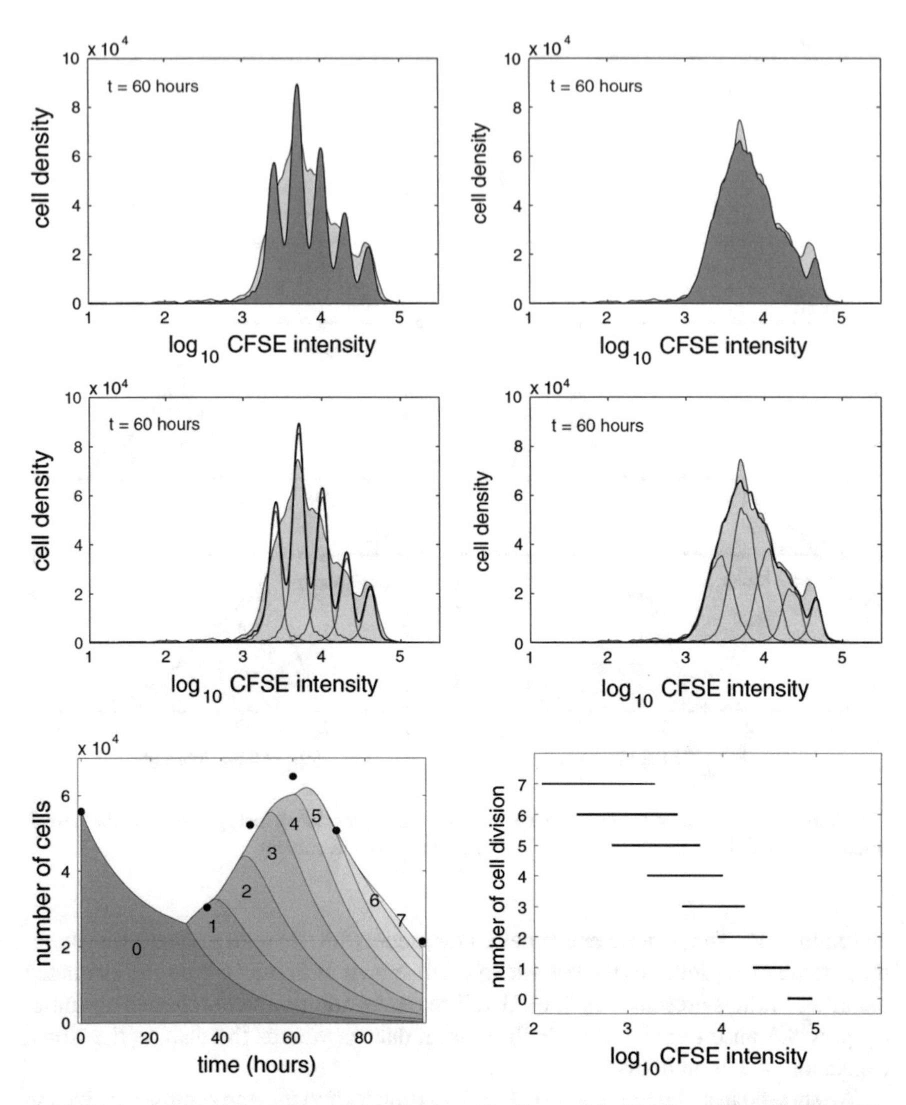

Fig. 3.17 Best-fit model solution. First two rows: Consistency with the data of the versions of the dhPDE model (3.3.46) based upon symmetric (left) versus asymmetric (right) division with the support of the individual generations shown in the middle row. Last row: Predicted structure of cell generations in the asymmetric case (left); the generation numbers are indicated; bullets indicate the data on the total cell numbers (cf. Table 3.6). Evolution of the initial CFSE support, the interval $[4 \times 10^4, 9 \times 10^4]$ of CFSE intensity of nondivided cells, with the cell divisions (right)

local minima are found. The estimated value $m_1 \approx 0.423$ clearly suggests that an unequal partition of the fluorescent dye during cell division takes places. The extent of the asymmetry is about 16% when compared to $m_1 = 0.5$.

The presence of asymmetry leads to the following important consideration for the generational resolution of CFSE histograms. Let an undivided cell have the

Table 3.7 Best-fit parameter estimates for the dhPDE model (3.3.46), the estimates of their 95% CIs and the estimates of the 95% CIs of the best-fit values of the ratios β_i/α_i. The best-fit solution is characterized by $\Phi(\boldsymbol{\theta}) \approx 1.32 \times 10^{11}$

Parameter	Best-fit value	Estimate of CIs	Fraction of parameters	Best-fit value	Estimate of CIs
k	7.40×10^{-3}	$[7.06,\ 7.73] \times 10^{-3}$	β_0/α_0	1.65	[1.56, 1.74]
m_1	4.23×10^{-1}	$[4.19,\ 4.27] \times 10^{-1}$	β_1/α_1	2.7×10^{-5}	[0, 0.03]
α_0	2.48×10^{-2}	$[2.34,\ 2.63] \times 10^{-2}$	β_2/α_2	0.19	[0.12, 0.28]
α_1	2.73×10^{-1}	$[2.40,\ 3.16] \times 10^{-1}$	β_3/α_3	0.81	[0.69, 1.0]
α_2	489	$[21,\ > 10^5]$	β_4/α_4	2.5	[1.2, 6.4]
α_3	36.0	$[4.0,\ > 10^5]$	β_5/α_5	4.9	[0, 31]
α_4	3.07×10^{-1}	$[0.065,\ > 10^5]$	β_6/α_6	0.10	[0, 22]
α_5	5.75×10^{-2}	$[0.026,\ > 10^5]$			
α_6	3.82	$[0.05,\ > 10^5]$			
β_0	4.08×10^{-2}	$[3.9,\ 4.3] \times 10^{-2}$			
β_1	7.41×10^{-6}	$[0,\ 7 \times 10^{-3}]$			
β_2	94.9	$[0.6,\ > 10^5]$			
β_3	29.2	$[1.5,\ > 10^5]$			
β_4	7.52×10^{-1}	$[0.55,\ > 10^5]$			
β_5	2.84×10^{-1}	$[0.18,\ > 10^5]$			
β_6	3.83×10^{-1}	$[0,\ > 10^5]$			
c	5.08×10^{-1}	[0, 1.1]			
τ_0	30.0	[29.5, 30.5]			
τ_1	7.48	[7.25, 7.71]			

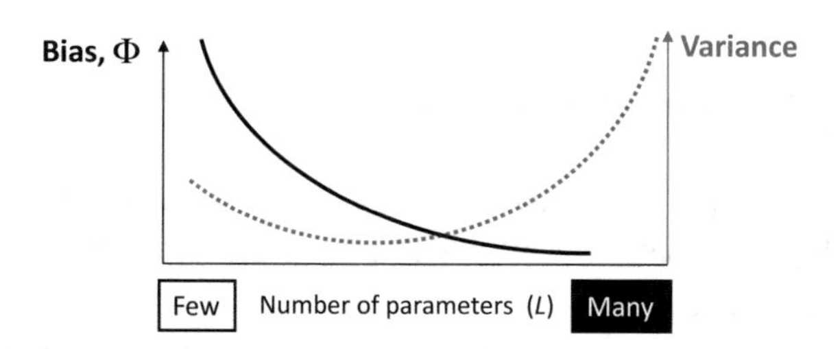

Fig. 3.18 Model performance characterization: accuracy versus parsimony. An increase in the complexity of the model associated with the number of the model parameters, leads to a better description of the data (bias reduction) but eventually results in a poorer estimation of the parameters (larger variance). A proper balance between under-fitting and over-fitting needs to be achieved for the mathematical model of optimal complexity

marker intensity x. Then, assuming loss of the label only due to cell division, its daughter cells, appearing after its i-th division, have the marker intensity in the interval $[m_1^i x,\ m_2^i x]$ ($m_2 = 1 - m_1$). If $m_2^{i+1} x > m_1^i x$, then the interval $[m_1^i x,\ m_2^{i+1} x]$ will contain cells which divided i and $i + 1$ times, i.e. the generations overlap. This imposes a limit on our ability to resolve the individual generations using a conventional decomposition approach with Gaussian functions of equal variance. Figure 3.17 (lower right panel) shows how the initial interval of the CFSE label intensity $[5 \times 10^4,\ 9 \times 10^4]$, to which most undivided cells belong to at time $t = 0$, moves to the left and broadens in size after each division of cells. As a result, an overlap of the support (CFSE range) of sequential generations becomes more and more prominent. For instance, the marker intensity 2.5×10^4 can be a characteristic of cells which have divided one or two times, whereas the marker intensity 10^3 can be observed in cells divided five, six and seven times. Hence, in case of nonequal partition of the marker at cell division, it is practically impossible to correctly predict the number of cells in later generations from the initial distribution of the marker intensity without using an appropriate analytical framework. The best-fit parameters of the division- and label-structured dhPDE model (3.3.46) allows one to predict the generational structure of the CFSE-labelled population at any given time via the solution of the related division-structured DDE model. The predicted number of cells $N_i(t)$ in each generation i using the parameters from Table 3.7 is shown in Fig. 3.17(left).

Overall, we quantified the uncertainty of the best-fit estimates of the model parameters using the profile-likelihood-based approach. The computed 95% confidence intervals are shown in Table 3.7. They indicate that the parameters $k, m_1, \alpha_0, \alpha_1, \beta_0, \beta_1, c, \tau_0$ and τ_1 ($\tau_i = \tau_1,\ i \geq 2$) have a much lower uncertainty than the others. The ratios β_i/α_i, $i = 0, 1, \ldots, 6$, are estimated with a high degree of confidence as indicated by their small CIs and can therefore be used as robust descriptors of cell proliferation performance.

In completion of the modelling aspects related to fine details of the CFSE-based proliferation analysis, we note the studies of the effects of autofluorescence and label decay presented in [93–95].

3.4 Model Ranking and Selection[4]

A priori immunological and mathematical knowledge enter into the models in the form of simplifying assumptions. Potentially, the interaction between a virus infec-

[4]Material of sect. 3.4 uses the results of the studies from Journal of Computational and Applied Mathematics, Vol. 184, C.T.H Baker et al., Computational approaches to parameter estimation and model selection in immunology, Pages 50–76, Copyright © 2005; Applied Numerical Mathematics, Vol. 53. C.T.H Baker et al., Computational modelling with functional differential equations: Identification, selection, and sensitivity, Pages 107–129, Copyright © 2005; Journal of Computational and Applied Mathematics, Vol. 205, S. Andrew et al., Rival approaches to mathematical modelling in immunology, Pages 669–686, Copyright © 2007, with permission from Elsevier.

tion and the immune system can be described by multiple mechanisms and considering various sets of differential equations. One may thus argue that different mechanisms and their functional forms might equally well describe the data set and the goodness of fit (i.e. the maximized likelihood function) is not sufficient to judge whether the model is correct. It has been observed elsewhere that the maximum likelihood principle leads to choosing the models with higher possible complexity (corresponding to more parameters) [96].

3.4.1 Accuracy and Parsimony

Given a set of related models, a fundamental question is how to rank them. One possibility is to consider the goodness of fit, i.e. the size of the minimized least-squares (or maximized likelihood) function. However, a consistent approach is based on the parsimony principle, i.e. a proper balance between underfitting and overfitting. Indeed, as shown in Fig. 3.18, a mathematical model that is based more closely on the biology of the system, would require to consider many parameterized processes. This would result in a multi-parameter estimation characterized by a better fit of the model solutions to the data. However, in practice, the empirical data sets (in R.A. Fisher's terms, the relevant information supplied by the sample) are limited as well as the information content per estimated parameter and eventually starts to decrease with an increase of dimensionality of the model parameter space. As a consequence, an uncertainty in the estimated parameter values will tend to become larger after some level of model complexity. Therefore, one needs a quantitative measure to implement the general principle that a model should be 'as simple as possible yet as complex as necessary' [97] with respect to the included variables, model structure and the number of parameters for adequate representation of the data.

3.4.2 Information-Theoretic Basis for Model Selection

The information-theoretic approach to model building has been presented systematically in [98]. The ranking methodology is that associated estimation of the minimum information loss for the model in hand. The model-specific information loss criterion is based on the Kullback–Leibler information-theoretic measure of the 'distance between' two probabilistic models and it characterizes the information lost when the model is used to approximate the reality or 'full truth' as outlined below.

The researcher aiming to find a true quantitative description of the complex system has a set of 'multiple working hypotheses' about the phenomenon at hand. The science of matter, experience and expertise are used to define an a priori set of candidate models g_i, $i = 1, \ldots, M$, representing each hypothesis. Then, one seeks to find a candidate model g^* that minimizes the 'information' $I(f, g_i)$ loss when the model g_i is used to approximate full reality or truth f, over the candidate models g_i. However, $I(f, g_i)$ cannot be used directly because neither the full truth model f nor the parameters in the approximating models g_i are known.

The problem of discrimination between statistical models describing the underlying data was solved by Kullback and Leibler who suggested a measure of the information distance of divergence between statistical populations represented by models f and g [99]. The Kullback–Leibler information is defined as the following multidimensional integral:

$$I(f, g) = \int_{\Omega_{\mathbf{x}}} f(\mathbf{x}) \log \left(\frac{f(\mathbf{x})}{g(\mathbf{x}, \mathbf{p})} \right) d\mathbf{x}, \tag{3.4.61}$$

where $I(f, g)$ denotes the information loss when g is used to approximate f. $\Omega_{\mathbf{x}}$ is the state space domain, and $\mathbf{p} \in \Omega_{\mathbf{p}}$ is the parameter space vector to be estimated. In fact, the Kullback–Leibler discrepancy is a directed or oriented distance from the various candidate models g_i to f. The Kullback–Leibler measure provides a basis for deriving 'information-theoretic' criteria, such as the Akaike, Schwarz and Takeuchi indices. One problem is that the full truth model is not known. Akaike in 1973 found a rigorous way to estimate the Kullback–Leibler directed distance of the candidate model g to the truth f based on the empirical log-likelihood function $\mathscr{L}(\mathbf{p})$ at its maximum point $\mathbf{p}^*$, $\mathscr{L}(\mathbf{p}^*)$ [100]. Given a family of mathematical models and a data set, the Akaike index uses maximized likelihood estimation to quantify the Kullback–Leibler information loss for each model. Therefore, the Akaike's information criterion (AIC) makes use of an estimate of the expected, relative distance between the fitted model and the unknown true mechanism that actually generated the observed data set with some cardinality (n).

3.4.3 Akaike Criteria

Given a hierarchy of models (each one with a best-fit set of parameters), the question is how to rank them by giving each a score. The goodness of fit associated with parameter estimates $\widetilde{\mathbf{p}}$ can be characterized when one has confidence in the form of the model by the size of an objective function $\Phi_\star(\widetilde{\mathbf{p}})$. This is the data-fitting approach, and here $\widetilde{\mathbf{p}}$ may be an approximation (however obtained) to $\widehat{\mathbf{p}}$ such that $\Phi_\star(\widehat{\mathbf{p}}) = \min_{\mathbf{p}} \Phi_\star(\mathbf{p})$. Thus, one criterion by which to judge a model may be the size of $\Phi_\star(\widetilde{\mathbf{p}})$ (see [101]). However, if there are a *number* of candidate models, our task is not simply to identify the one with the smallest objective function but to incorporate other criteria for discriminating between models of differing complexity. There are (information-theoretic) criteria, such as the Akaike, Schwarz and Takeuchi information criteria[5] and generalizations related to informational complexity of models, which depend not only upon the maximum likelihood estimation bias [96, 98, 100] but take into account the *number of parameters* and the *number of observations* in a quantitative evaluation of different models. Burnham and Anderson [98] review both the concept

[5]The Akaike criterion is based upon Kullback–Leibler notion of information or distance between two probabilistic models (information loss) [99] approximated using the maximum likelihood estimation [98, 100].

of Kullback–Leibler information and maximum likelihood as a natural basis for model selection.

For the Akaike and the corrected Akaike criteria, the indicators are the size of the measures μ_{AIC} and μ_{cAIC} given by

$$\mu_{AIC} = -2\,\ell n\,\mathscr{L}(\widehat{\mathbf{p}}) + 2(L+1), \tag{3.4.62a}$$

$$\mu_{cAIC} = -2\,\ell n\,\mathcal{L}(\widehat{\mathbf{p}}) + 2(L+1) + \frac{2(L+1)(L+2)}{n-L-2}, \text{ with } n = NM, \tag{3.4.62b}$$

respectively, see [98]. These indicators are expressed in terms of the MLE $\mathscr{L}(\widehat{\mathbf{p}})$. There are $L+1$ parameters being estimated, comprising $p_1, p_2, \ldots, p_L$ and σ, since we currently assume that a single value σ, which we also estimate, characterizes all the variances. The advice of Burnham and Anderson in [98] is that (3.4.62a) is satisfactory if $n > 40(L+1)$; otherwise, (3.4.62b) is preferred by these authors. We note that as $n \to \infty$, $\mu_{cAIC} \to \mu_{AIC}$.

Our interest is in the relative size of the indicators; thus (omitting technical details) it is convenient to discard extraneous terms and to employ the revised indicators

$$\breve{\mu}_{AIC} = NM\ell n(\Phi_{\Omega LS}(\widehat{\mathbf{p}})) + 2(L+1), \tag{3.4.63a}$$

$$\breve{\mu}_{cAIC} = \breve{\mu}_{AIC} + \frac{2(L+1)(L+2)}{NM-L-2}. \tag{3.4.63b}$$

3.4.3.1 Experimental Data on Viral Load—CTL Dynamics

The infection of a mouse with lymphocytic choriomeningitis virus (LCMV) provides a basic experimental system used in immunology to address fundamental issues of virus–host interactions [102]. The infection results in the activation of immune responses and clonal burst [103] of virus-specific cytotoxic T lymphocytes (CTL). We note that Ehl et al. [104] observed that

> The use of a well-characterized murine infectious disease, which has been shown to be almost exclusively controlled by CTL-mediated perforin-dependent cytotoxicity, provides an exceptionally solid basis for the formulation of [models].

At discrete times, it is possible to measure, experimentally, (i) the amount of the virus, measured in plaque forming units (*pfu*), and (ii) the virus-specific CTL (measured in the number of cells found per spleen[6]).

In general, it is possible that data comes from a single experiment, or that the data arises from several experiments or a series of observations. Our mathematical models rely upon data being of a certain type: we assume that the mean values of data are, at each time, normally or log-normally distributed, and independent.

The experimental data is provided in Table 3.8 and is shown in Fig. 3.19. It was obtained as follows. A batch of genetically identical C57BL/6 mice were infected with 200 *pfu* (plaque forming units) of LCMV (WE strain), delivered intravenously.

[6]Some modellers introduce as a variable the amount of virus-specific memory CTL, a subset of (ii) that is harder to quantify reliably.

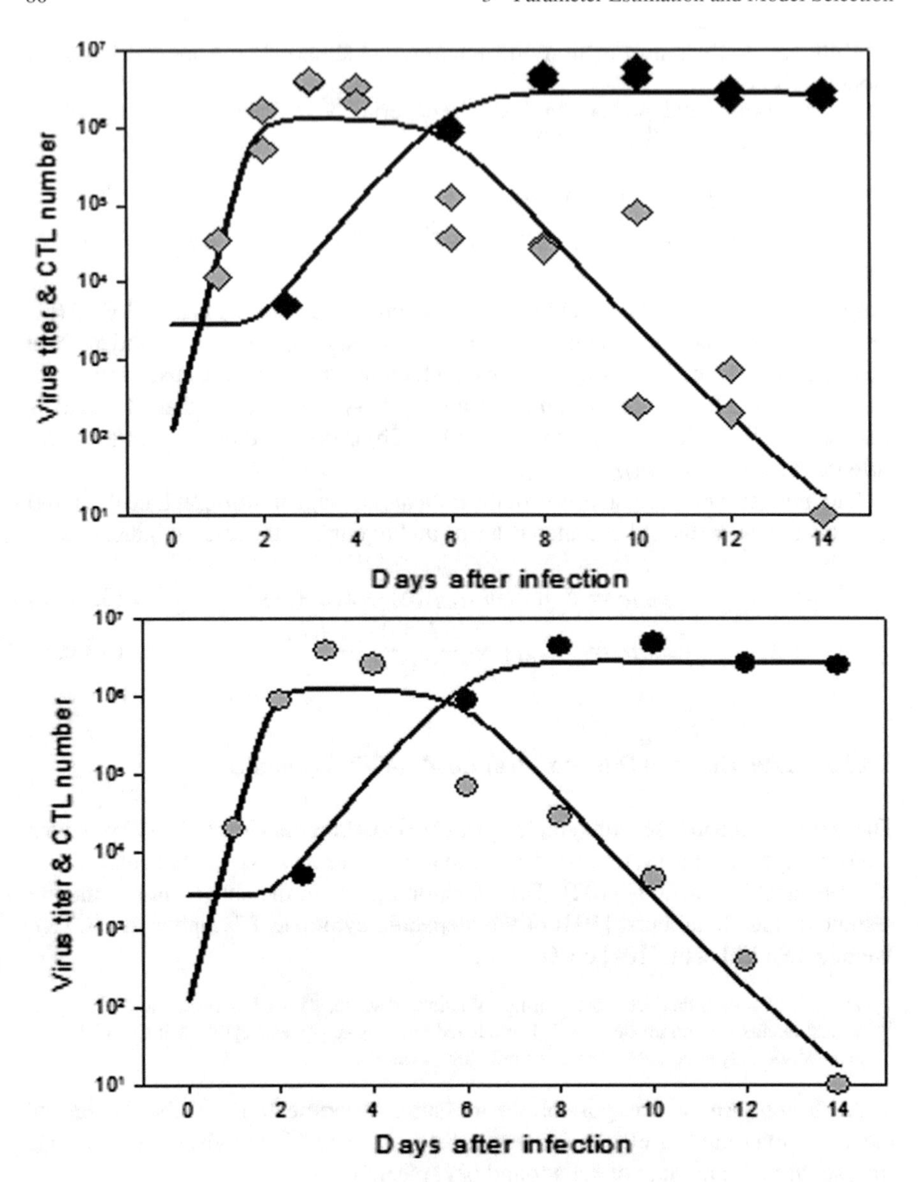

Fig. 3.19 The infection of a mouse with lymphocytic choriomeningitis virus (LCMV). At discrete times, the following characteristics of infection were determined experimentally, (i) the amount of the virus, measured in plaque forming units (pfu), and (ii) the virus-specific CTL (measured in the number of cells found per spleen). Upper: individual observations. Lower: averaged data. Reprinted from Journal of Computational and Applied Mathematics, Vol. 184, C.T.H Baker et al., Computational approaches to parameter estimation and model selection in immunology, Pages 50-76, Copyright © 2005, with permission from Elsevier

Viral titers in spleens were determined at 1, 2, 3, 4, 6, 8, 10, 12 and 14 days post infection and the clonal expansion of CTL cells specific for the gp33 epitope in spleens

Table 3.8 Data set for the virus and cytotoxic T lymphocyte kinetics in the spleen after systemic infection with 200 *pfu* of LCMV-WE ('b.d.l.' *means 'below the detection limit'*). Reprinted from Journal of Computational and Applied Mathematics, Vol. 184, C.T.H Baker et al., Computational approaches to parameter estimation and model selection in immunology, Pages 50-76, Copyright © 2005, with permission from Elsevier

	$V(t)$		$E(t)$	
Time (days)	Set 1—Virus population (*pfu*)	Set 2—Virus population (*pfu*)	Set 1—CTLs population (cells)	Set 2—CTL population (cells)
1	3.55×10^4	1.20×10^4	b.d.l.	b.d.l.
2	5.0×10^5	1.6×10^6	b.d.l.	b.d.l.
3	3.8×10^6	3.9×10^6	b.d.l.	b.d.l.
4	3.2×10^6	2.1×10^6	b.d.l.	b.d.l.
6	3.7×10^4	1.25×10^5	8.33×10^5	9.85×10^5
8	3.1×10^4	2.6×10^4	4.75×10^6	4.03×10^6
10	2.5×10^2	8.0×10^4	4.16×10^6	5.8×10^6
12	2.0×10^2	7.5×10^2	3.07×10^6	2.25×10^6
14	b.d.l.	b.d.l.	2.22×10^6	2.89×10^6

were assessed using tetramer analysis (see below). The techniques are standard; see, for example, [105, 106]. At the indicated time points after infection, blood samples were taken from two mice and single cell suspensions were prepared of spleen, prior to the determination of absolute cell counts using FACS and Neubauer equipment.

An important feature of this experiment is that the mice were genetically identical, produced by inbreeding. Inbred strains reduce experimental variation; their immune responses can be studied in the absence of variables introduced by individual genetic differences. If the mice are genetically identical, it is argued that large numbers of mice are not required and the mean obtained represents the mean of a larger set of data. This assertion merits closer examination and testing, but we proceed on the basis that it is correct.

For a reliable parameter estimation, it is useful to have an idea of the CTL kinetics at times earlier than 6 days post infection—before the virus population starts to decrease. The quantity of virus-specific CTL below 5000 cells/spleen cannot be detected using the tetramer technique. Our experience (arising from numerous studies with the LCMV system) suggests that after injection of 200 *pfu* of LCMV the proliferating CTLs should reach the detection threshold in about two and a half days. This evidence was considered in the parameter estimation by supplementing Table 3.8 with a CTL number at day 2.5 representing the least possible detection level.

The detection threshold for LCMV in the spleen is about 100 *pfu*. LCMV-WE dropped below the detection threshold by day 14; however, it is believed that the virus still persists below the detection level for some time. To ensure that the LCMV number in the model remains below the detection threshold between days 12 to 14, we supplement the data with an assumption that the virus quantity on day 14 was 10 *pfu*/spleen.

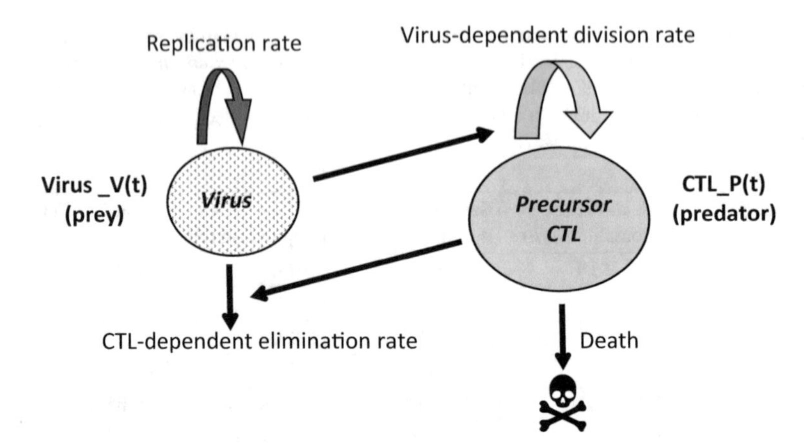

Fig. 3.20 Biological scheme of the virus–host organism interaction to model the dynamics of lymphocytic choriomeningitis virus infection in mice

3.4.4 Rival Models for Virus-CTL Dynamics

The simplest scheme of the interaction between the virus infection and the antiviral CTL response is presented in Fig. 3.20. It is equivalent to a predator–prey view of the above system which is valid for a low dose infection resulting in acute infection followed by CTL-mediated virus elimination.

The mathematical models for the virus-CTL interaction in LCMV infection can be introduced as a system of two or three ODEs or DDEs for the evolution of the virus, *V(t)*, and virus-specific CTL (activated and memory cells—$E(t)$, $E_m(t)$) population dynamics. As there are a number of candidate models, our task is not simply to identify the one with the smallest objective function but to consider the principle of parsimony in model evaluation, and the maximum use of information presented implicitly in the data. We consider and analyse a hierarchy of mathematical models that were distilled from the existing literature.

The equation for the rate of change of the virus population is the same for all the models. It is based upon a Verhulst–Pearl logistic growth term and second-order elimination kinetics. The models we consider here differ in the way the immune response is described—an issue of some controversy in today's mathematical immunology. Specifically, the models differ with respect to the following building blocks:

1. virus-dependent CTL proliferation (basic predator–prey versus the Holling type II response);
2. whether a time lag in the division of CTL (cell division time) is included;
3. consideration of homeostasis for naive CTL precursors;
4. whether a separate equation for the memory CTL is used.

The death rate of CTL is assumed constant. Overall, we consider here the following five models that have their prototypes in the literature (see Table 3.9 for biological definitions of the parameters included in the model).

Model 1: simplest predator–prey consideration of the CTL dynamics

$$\frac{d}{dt}V(t) = \beta \cdot V(t) \cdot \left(1 - \frac{V(t)}{K}\right) - \gamma \cdot V(t) \cdot E(t), \tag{3.4.64}$$

$$\frac{d}{dt}E(t) = b_1 \cdot V(t) \cdot E(t) - \alpha_E \cdot E(t). \tag{3.4.65}$$

Model 2: virus-dependent CTL proliferation with saturation

$$\frac{d}{dt}V(t) = \beta \cdot V(t) \cdot \left(1 - \frac{V(t)}{K}\right) - \gamma \cdot V(t) \cdot E(t), \tag{3.4.66}$$

$$\frac{d}{dt}E(t) = \underbrace{b_2 \cdot V(t) \cdot E(t)/(\theta_{Sat} + V(t))}_{\text{A modification of model 1}} - \alpha_E \cdot E(t). \tag{3.4.67}$$

Model 3: virus-dependent CTL proliferation with saturation and with time lag

$$\frac{d}{dt}V(t) = \beta \cdot V(t) \cdot \left(1 - \frac{V(t)}{K}\right) - \gamma \cdot V(t) \cdot E(t), \tag{3.4.68}$$

$$\frac{d}{dt}E(t) = \underbrace{b_3 \cdot V(t-\tau) \cdot E(t-\tau)/(\theta_{Sat} + V(t))}_{\text{As in model 2 but incorporating delay}} - \alpha_E \cdot E(t). \tag{3.4.69}$$

Model 4: primary CTL homeostasis

$$\frac{d}{dt}V(t) = \beta \cdot V(t) \cdot \left(1 - \frac{V(t)}{K}\right) - \gamma \cdot V(t) \cdot E(t), \tag{3.4.70}$$

$$\frac{d}{dt}E(t) = b_4 \cdot V(t-\tau) \cdot E(t-\tau)/(\theta_{Sat} + V(t)) - \underbrace{\alpha_E \cdot E(t) + T^*}_{\text{includes additive term}}. \tag{3.4.71}$$

Model 5: Additional equation for the population of memory CTL

$$\frac{d}{dt}V(t) = \beta \cdot V(t) \cdot \left(1 - \frac{V(t)}{K}\right) - \gamma \cdot V(t) \cdot E(t), \tag{3.4.72}$$

$$\frac{d}{dt}E(t) = b_5 \cdot V(t-\tau) \cdot E(t-\tau)/(\theta_{Sat} + V(t)) - \alpha_E \cdot E(t) - r_m \cdot E(t) + T^*, \tag{3.4.73}$$

Table 3.9 Biological definition of the model parameters for virus-CTL dynamics in the spleen during primary LCMV infection. The spleen volume is estimated to be about 0.1 (millilitres). Reprinted from Journal of Computational and Applied Mathematics, Vol. 184, C.T.H Baker et al., Computational approaches to parameter estimation and model selection in immunology, Pages 50-76, Copyright © 2005, with permission from Elsevier

Parameter (units) The units are d (days), *pfu* (plaque forming units)	Notation
Virus exponential growth rate (d^{-1})	β
Carrying capacity for the virus ($copies/$spleen)	K
Virus elimination rate ($1/copy/d$)	γ
CTL stimulation rate ($1/copy/d$, M_1; d^{-1}, M_2 to M_5)	b_i
CTL division time (d)	τ
Viral load for half-maximal CTL stimulation ($copy/$spleen)	θ_{Sat}
Death rate of CTL (d^{-1})	α_E
Specific precursor CTL export from thymus ($cell/$spleen$/d$)	T^*
Reversion activated CTL into the memory state (d^{-1})	r_m
Death rate of memory CTL (d^{-1})	α_m

$$\frac{d}{dt}E_m(t) = r_m \cdot E(t) - \alpha_m \cdot E_m(t). \tag{3.4.74}$$

Model 1 cannot be regarded as a special case of Model 2 if some of the Model 2 parameters are set to zero. However, Models 2 to 5 have a common subset of parameters, i.e. they are nested.

The general initial data are

$$V(t) = 0, t \in [-\tau, 0), V(0) = V_0; \;\; E(t) = E_0, t \in [-\tau, 0]; \;\; E_m(0) = 0.$$

We set the initial values for the dose of infection and for the number of naive CTL as follows: $V_0 = 200\ pfu$ and $E_0 = 265\ cells$. These parameters are considered to be fixed.

3.4.5 Information-Theoretic Model Evaluation

Parameter estimation results obtained using an ordinary least-squares approach for Models 1 to 5 are summarized in Table 3.10. The numerical treatment of the parameter estimation problem is based on the following software:

- the code `Archi`, at www.ma.man.ac.uk/~chris/reports/rep283.pdf, for FORTRAN (and the related codes `Archi-L`, `Archi-N`) [107, 108] for the initial value problem;
- the code `LMDIF1` at http://www.netlib.org/minpack/lmdif1.f to solve nonlinear least-squares problems (with an approximate Jacobian), `Archi-L`;
- `Archi-N` invokes a NAg constrained minimization routine `E04UNF` [109].

Table 3.10 Best-fit parameter estimates for ordinary least squares and the corrected Akaike indicator. Calculations are based on `LMDIF` and the values `eps` $= 10^{-6}$, `ftol` $= 10^{-6}$, `xtol` $= 10^{-6}$, `epsfcn` $= 0$. Reprinted from Journal of Computational and Applied Mathematics, Vol. 184, C.T.H Baker et al., Computational approaches to parameter estimation and model selection in immunology, Pages 50-76, Copyright © 2005, with permission from Elsevier

Parameter	M_1	M_2	M_3	M_4	M_5
β	4.44×10^0	4.36×10^0	4.52×10^0	4.52×10^0	4.50×10^0
K	3.99×10^6	3.23×10^6	3.17×10^6	3.17×10^6	3.19×10^6
γ	3.02×10^{-6}	3.48×10^{-6}	3.45×10^{-6}	3.48×10^{-6}	3.63×10^{-6}
b_i	1.23×10^{-6}	1.92×10^0	2.52×10^0	2.41×10^0	2.40×10^0
θ_{Sat}	–	2.46×10^4	1.34×10^5	1.31×10^5	1.15×10^5
τ	–	–	7.17×10^{-2}	8.98×10^{-2}	9.54×10^{-2}
α_E	0.0	9.14×10^{-2}	8.62×10^{-2}	9.1×10^{-2}	9.31×10^{-2}
T^*	–	–	–	1.24×10^2	1.40×10^2
r_m	–	–	–	–	5.17×10^{-3}
α_m	–	–	–	–	2.55×10^{-1}
Φ_{OLS}	1.05×10^{13}	4.49×10^{12}	4.04×10^{12}	3.91×10^{12}	3.78×10^{12}
$\check{\mu}_{cAIC}$	472.2	467.0	475.4	488.9	544.4

In optimization procedures the tolerances specified by the user govern the successful conclusion of an iterative process for the determination of the minimum of an objective function: (i) `ftol` governs the relative change in the estimated minimum value of the objective function, (ii) `xtol`, governs the relative change in the argument at which the estimate of the minimum is attained. In `Archi`, `eps` specifies the error-per-step tolerance in the ODE or DDE solver, and `epsfcn` sets the relative errors in the objective functions.

An increase in the number of model parameters provides a better description of the data in terms of the minimized value of the objective function. However, the increasing values of the corrected Akaike index indicate a gradual information loss for the given data set as the complexity of models increases. The variation of the best-fit parameter estimates between the models is within $\pm 10\%$, except for the estimate of θ_{Sat}. Further, the data set does not provide a biologically correct estimate of the time lag of cell division τ. Rather, the delay estimate obtained via ordinary least squares corresponds to a realistic duration of some stage of the cell cycle. Visual inspection of graphs of $V(t)$ and $E(t)$ in Fig. 3.21 (upper raw) suggests that Model 1 nicely approximates the viral load data, but rather poorly approximates the CTL data. The other models (see lower raw in Fig. 3.21 corresponding to M_2) describe much better the CTL kinetics at the expense of a somewhat poorer agreement with the virus data ($V(t)$).

The calculations represented in Table 3.10 were checked by refining the parameters that govern the accuracy. The original values `eps` $= 10^{-6}$, `ftol` $= 10^{-6}$, `xtol` $= 10^{-6}$, `epsfcn` $= 0$ were replaced by `eps` $= 10^{-15}$, `ftol` $= 10^{-12}$, `xtol` $= 10^{-12}$, `epsfcn` $= 10^{-15}$, and the computed values of interest are presented in Table 3.11. The refined tolerances have a noticeable effect on the parameter val-

Table 3.11 Best-fit parameter estimates for ordinary least squares and the corrected Akaike indicator when the higher accuracy numerical solution is used: Calculations were based on `Archi-N` with `E04UNF`, and `eps` = 10^{-15}, `ftol` = 10^{-12}, `xtol` = 10^{-12}, `epsfcn` = 10^{-15}. Reprinted from Journal of Computational and Applied Mathematics, Vol. 184, C.T.H Baker et al., Computational approaches to parameter estimation and model selection in immunology, Pages 50-76, Copyright © 2005, with permission from Elsevier

Parameter	M_1	M_2	M_3	M_4	M_5
β	4.61×10^0	4.51×10^0	4.62×10^0	4.61×10^0	4.61×10^0
K	2.70×10^6	4.69×10^6	5.01×10^6	4.98×10^6	5.07×10^6
γ	1.39×10^{-6}	8.04×10^{-5}	3.29×10^{-4}	2.96×10^{-4}	2.45×10^{-4}
b_i	9.22×10^{-7}	1.42×10^0	1.14×10^0	1.16×10^0	1.22×10^0
θ_{Sat}	–	$0\ (3.23 \times 10^{-176})$	8.79×10^{-6}	4.59×10^{-6}	2.45×10^{-4}
τ	–	–	4.38×10^{-2}	4.15×10^{-2}	4.38×10^{-2}
α_E	9.29×10^{-2}	2.01×10^{-1}	1.02×10^{-1}	1.02×10^{-1}	1.03×10^{-14}
T^*	–	–	–	1.09×10^0	134×10^0
r_m	–	–	–	–	2.12×10^{-1}
α_m	–	–	–	–	2.20×10^{-1}
Φ_{OLS}	6.54×10^{12}	7.82×10^{11}	1.60×10^{12}	1.60×10^{12}	1.37×10^{12}
$\breve{\mu}_{cAIC}$	465.1	440.8	461.5	475.5	529.2

Table 3.12 Best-fit parameter estimates for weighted least-squares and the corrected Akaike indicator when the higher accuracy numerical solution is used: Calculations were based on `Archi-N` with `E04UNF`, and `eps` = 10^{-15}, `ftol` = 10^{-12}, `xtol` = 10^{-12}, `epsfcn` = 10^{-15}. Reprinted from Journal of Computational and Applied Mathematics, Vol. 184, C.T.H Baker et al., Computational approaches to parameter estimation and model selection in immunology, Pages 50-76, Copyright © 2005, with permission from Elsevier

Parameter	Model 1 (M_1)	Model 2 (M_2)	Model 3 (M_3)	Model 4 (M_4)	Model 5 (M_5)
β	5.14×10^0	4.56×10^0	4.57×10^0	4.60×10^0	4.60×10^0
K	1.23×10^5	4.42×10^6	4.46×10^6	3.27×10^6	3.26×10^6
γ	1.90×10^{-6}	5.47×10^{-5}	7.93×10^{-5}	2.14×10^{-5}	2.23×10^{-5}
b_i	1.37×10^{-5}	1.41×10^0	1.41×10^0	2.34×10^0	2.41×10^0
θ_{Sat}	–	$0\ (9.19 \times 10^{-77})$	3.49×10^{-13}	4.854×10^4	5.39×10^4
τ	–	–	1.76×10^{-2}	5.18×10^{-1}	5.04×10^{-1}
α_E	2.09×10^{-2}	1.09×10^{-1}	1.09×10^{-1}	1.09×10^{-1}	2.52×10^{-10}
T^*	–	–	–	1.663×10^3	1.508×10^3
r_m	–	–	–	–	1.32×10^{-1}
α_m	–	–	–	–	5.10×10^{-1}
$\Phi_{\Omega LS}$	5.23×10^0	5.36×10^0	5.15×10^0	4.18×10^0	4.18×10^0
$\breve{\mu}_{cAIC}$	47.3	55.2	64.6	75.5	131.5

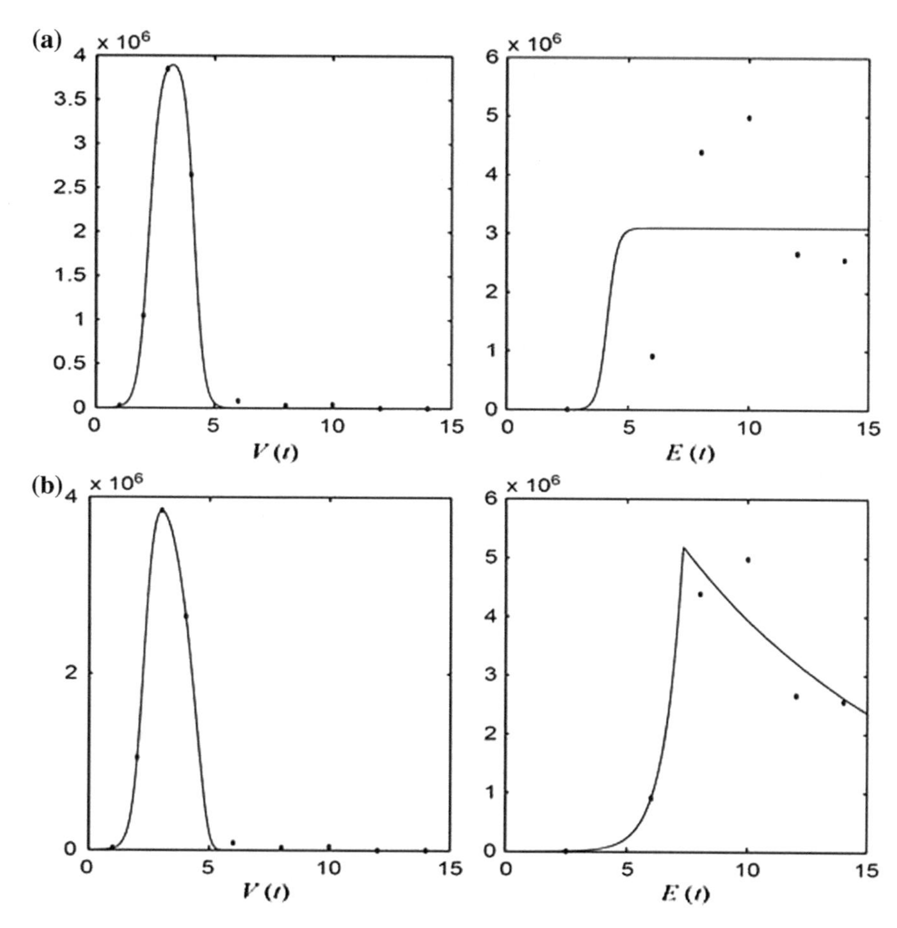

Fig. 3.21 Ordinary least squares. **a** Model 1 computed using low accuracy (Table 4); **b** Model 2 computed using high accuracy (Table 7). Shown are the best-fit predictions (solid lines) for the viral load, $V(t)$ and for the number of CTLs, $E(t)$, and the mean values (* symbols) for the original data. Reprinted from Journal of Computational and Applied Mathematics, Vol. 184, C.T.H Baker et al., Computational approaches to parameter estimation and model selection in immunology, Pages 50-76, Copyright © 2005, with permission from Elsevier

ues and, as a consequence, on the ranking of the parameterized models. Models with parameters computed with lower tolerance are ranked with respect to the information loss as follows:

$$M2 \text{ (best)} - M1 - M3 - M4 - M5;$$

with parameters computed to the higher tolerance, we obtain the ranking

$$M2 \text{ (best)} - M3 - M1 - M4 - M5.$$

In both cases, Model 2 has the least Akaike index. The parameter θ_{Sat} (which represents the viral load for half-maximal CTL stimulation) does not occur in Model 1;

Table 3.13 Estimates of 95% confidence intervals, for the best-fit high-accuracy parameter estimates for Models 1 to 3 using ordinary least squares. Estimates were computed using `Archi-N`, with `E04UNF`, `eps` $= 10^{-15}$, `ftol` $= 10^{-12}$, `xtol` $= 10^{-12}$, `epsfcn` $= 10^{-15}$ as in Table 3.11. Reprinted from Journal of Computational and Applied Mathematics, Vol. 184, C.T.H Baker et al., Computational approaches to parameter estimation and model selection in immunology, Pages 50-76, Copyright © 2005, with permission from Elsevier

Parameter	M_1	M_2	M_3
β	4.61×10^0	4.51×10^0	4.62×10^0
95% CI:	$[4.00 \times 10^0, 5.23 \times 10^0]$	$[4.23 \times 10^0, 4.76 \times 10^0]$	$[4.43 \times 10^0, 4.85 \times 10^0]$
K	2.70×10^6	4.69×10^6	5.01×10^6
95% CI:	$[2.28 \times 10^6, 3.00 \times 10^6]$	$[4.20 \times 10^6, 5.20 \times 10^6]$	$[4.62 \times 10^6, 5.45 \times 10^6]$
γ	1.39×10^{-6}	8.04×10^{-5}	3.29×10^{-4}
95% CI:	$[1.17 \times 10^{-6}, 1.71 \times 10^{-6}]$	$[7.54 \times 10^{-5}, 8.58 \times 10^{-5}]$	$[3.20 \times 10^{-4}, 3.31 \times 10^{-4}]$
b_i	9.22×10^{-7}	1.42×10^0	1.141×10^0
95% CI:	$[7.96 \times 10^{-7}, 1.01 \times 10^{-6}]$	$[1.40 \times 10^0, 1.43 \times 10^0]$	$[1.134 \times 10^0, 1.143 \times 10^0]$
θ_{Sat}	–	$0\ (3.23 \times 10^{-176})$	8.79×10^{-6}
95% CI:	–	$[0, 1.2 \times 10^{-164}]$	$[5.25 \times 10^{-6}, \theta^{max}]$
		If $\theta_{Sat} = 0$, $bE(t)$ growth term results.	where $\theta^{max} \geq 8 \times 10^{-5}$.
τ	–	–	4.38×10^{-2}
95% CI:	–	–	$[4.24 \times 10^{-2}, 4.43 \times 10^{-2}]$
α_E	9.29×10^{-2}	2.01×10^{-1}	1.02×10^{-1}
95% CI:	$[4.84 \times 10^{-2}, 1.73 \times 10^{-1}]$	$[1.19 \times 10^{-1}, 2.14 \times 10^{-1}]$	$[1.01 \times 10^{-1}, 1.15 \times 10^{-1}]$
Φ_{OLS}	6.54×10^{12}	7.82×10^{11}	1.6×10^{12}
$\breve{\mu}_{cAIC}$	465.1	440.8	461.5

in the high-accuracy figures for Model 2 it is close to zero (see Tables 3.11, 3.12). Such a small value represents an effective immediate response to the infection (what is considered to be a ‘programmed’ response, in [110]), irrespective of the viral load. If we use Model 5, the data do not support a biologically correct estimate of the memory cell lifespan α_m. We note that in a similar parameter estimation study [110], the value of parameter α_m was assigned rather than estimated.

Table 3.13 summarizes the analysis of the confidence in the best-fit parameter estimates for Models 1 to 3 using high-accuracy solutions and following OLS. The ranking of the models according to the Akaike criteria suggests that the least information loss is the feature of Model 2 as compared to Models 1 or 3. The reader may compare the widths of the confidence intervals for the same parameter in differing models displayed in Table 3.13. Other things being equal, a narrower interval is to be preferred. An unduly large confidence interval indicates that the parameter is unidentifiable in practice (for the given data set).

Note that the structure of the parsimonious model M_2 is such that the following conclusions can be drawn: the CTL response to a low dose LCMV-WE infection is regulated by virus antigen load in an on and off way [110].

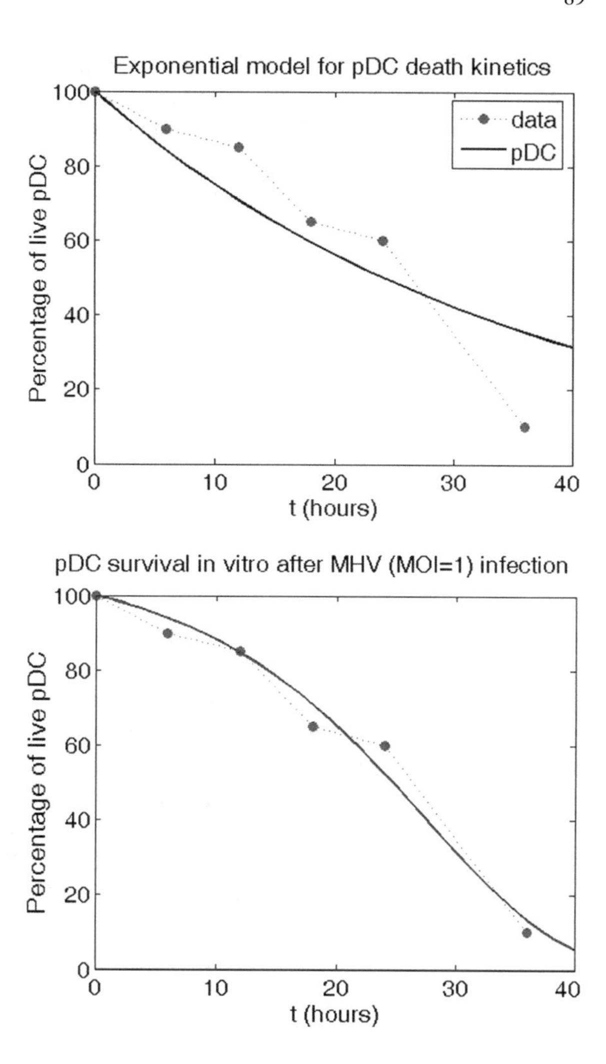

Fig. 3.22 Data on persistence of plasmacytoid dendritic cells: comparison of the exponential decay and Gompertz models

3.4.6 *Minimum Description Length*

The minimum description length (MDL) provides a selection method that is sensitive to models functional form and favours the model that permits the greatest compression of data in its description. The use of MDL for model selection with reduced complexity is reviewed in [112]. The quantitative measure of the model description length for specific data which needs to be computed is given by the following expression:

$$\mu_{MDL} = -\ln(L(\mathbf{p}^*)) + 0.5\, n_p \,\ln(n_d/(2\pi)) + \ln \int_{\Omega_p} \sqrt{det(I(\mathbf{p}))d\mathbf{p}}, \quad (3.4.75)$$

where n_d is the total number of scalar observations, n_p is the number of optimized parameters, $L(\mathbf{p}^*)$ is the maximized likelihood function, $I(\mathbf{p})$ is the Fisher information matrix, and Ω_p is the domain of the parameter space in which the model is defined.

To illustrate its practical use, we examined whether the persistence of plasmacytoid dendritic cells (pDC) in vitro shown in Fig. 3.22 is consistent with the exponential decay (E-model), or the Gompertz (G-model) (see Chap. 2). To test whether the increased complexity of the G- versus E-model of decay is justified by the pDCs data in hand, we evaluated the Akaike criterion of the information loss μ_{AIC} and the μ_{MDL} for the models. Both criteria of the model parsimony turned out to be smaller for the Gompertz model: the μ_{AIC} is 30 versus 38 and the μ_{MDL} value is 19.5 versus 22.1 [111]. For a comprehensive coverage of the MDL methodology, we refer to [113].

3.4.7 Summary

We presented a computational methodology for developing mathematical models of different complexities formulated with various types of differential equations, e.g. ODEs, DDEs, hPDEs and delay hPDES. The models in immunology have no a priori claim of validity. Therefore, their consistency with available data and knowledge has to be checked. Maximum likelihood approach to data fitting and information-theoretic criteria of model ranking are shown to provide necessary tools for identification of the most parsimonious model from a family of plausible ones. The efficiency of approach was illustrated by considering data on CFSE-labelled cell proliferation and antiviral immune response for a low dose LCMV infection.

References

1. Baker, C.T.H., Bocharov, G.A., Paul, C.A.H., Rihan, F.A., Computational modelling with functional differential equations: identification, selection and sensitivity, *Appl. Numer. Math.*, **53** (2005) 107–129.
2. Baker, C.T.H., Bocharov, G.A., Ford, J.M., Lumb, P.M., Norton, S.J., Paul, C.A.H., Junt, T., Krebs, P., Ludewig, B., Computational approach to parameter estimation and model selection in immunology, *J. Comput. Appl. Math.*, **184** (2005) 50–76.
3. Andrew, S.M., Baker, C.T.H., Bocharov, G.A. Rival approaches to mathematical modelling in immunology, *J. Comput. Appl. Math.*, **205** (2007) 669–686.
4. Luzyanina, T., Roose, D., Bocharov, G.: Distributed parameter identification for a label-structured cell population dynamics model using CFSE histogram time-series data. J. Math. Biol. **59**(5): 581–603 (2009)
5. Luzyanina, T., Mrusek, S., Edwards, J.T., Roose, D., Ehl, S., Bocharov, G.: Computational analysis of CFSE proliferation assay. J. Math. Biol. **54**(1) 57–89 (2007)
6. Luzyanina, T., Roose, D., Schinkel, T., Sester, M., Ehl, S., Meyerhans, A., Bocharov, G.: Numerical modelling of label-structured cell population growth using CFSE distribution data. Theor. Biol. Math. Model. **4** 1–26 (2007)

7. T. Luzyanina, J. Cupovic, B. Ludewig, G. Bocharov. (2014) Mathematical models for CFSE labelled lymphocyte dynamics: asymmetry and time-lag in division. Journal of Mathematical Biology. 69(6–7):1547–83
8. Antia, R., Ganusov, V.V., Ahmed, R., The role of models in understanding CD8+ T-cell memory, *Nat. Rev. Immunol.*, **5** (2005) 101–111.
9. Goldstein, B., Faeder, J.R., Hlavacek, W.S., Mathematical and computational models of immune-receptor signalling. *Nat Rev Immunol.*, **4** (2004) 445–456.
10. Mohler, R.R., Bruni, C., Gandolfi, A., A systems approach to immunology, *Proc. IEEE*, **68** (1980) 964–990.
11. Morel, P.A., Mathematical modeling of immunological reactions, *Front Biosci.*, **16** (1998) d338–347.
12. Perelson A.S., Modelling viral and immune system dynamics, *Nat Rev Immunol.*, **2** (2002) 28–36.
13. Perelson, A.S., Ribeiro, R.M., Hepatitis B virus kinetics and mathematical modeling, *Semin Liver Dis.*, **24** (2004) Suppl 1, 11–16.
14. Perelson, A.S., Nelson, P.W., Mathematical analysis of HIV-1 dynamics in vivo, *SIAM Review*, **41** (1999) 3–44.
15. Petrovsky, N., Brusic, V., Computational immunology: The coming of age, *Immunol. Cell Biol.*, **80** (2002) 248–254.
16. Ribeiro, R.M., Lo, A., Perelson, A.S., Dynamics of hepatitis B virus infection, *Microbes Infect.*, **4** (2002) 829–835.
17. Wodarz, D., Mathematical models of HIV and the immune system, *Novartis Found Symp.*, **254** (2003) 193–207.
18. Yates, A., Chan, C.C., Callard, R.E., George, A.J., Stark, J., An approach to modelling in immunology, *Brief Bioinform.*, **2** (2001) 245–257.
19. Asquith B, Borghans JA, Ganusov VV, Macallan DC. Lymphocyte kinetics in health and disease. Trends Immunol. (2009); 30(4):182–9.
20. Germain RN, Meier-Schellersheim M, Nita-Lazar A, Fraser ID. Systems biology in immunology: a computational modeling perspective. Annu Rev Immunol. (2011); 29:527–85. Review.
21. Kirschner DE, Linderman JJ. Mathematical and computational approaches can complement experimental studies of host-pathogen interactions. Cell Microbiol. (2009); 11(4):531–9.
22. Klauschen F, Angermann BR, Meier-Schellersheim M. Understanding diseases by mouse click: the promise and potential of computational approaches in Systems Biology. Clin Exp Immunol. (2007); 149(3):424–9.
23. Wodarz D. Ecological and evolutionary principles in immunology. Ecol Lett. (2006); 9(6):694–705.
24. Yan Q. Immunoinformatics and systems biology methods for personalized medicine. Methods Mol Biol. (2010); 662:203–20.
25. van den Berg HA, Rand DA. Quantitative theories of T-cell responsiveness. Immunol Rev. (2007); 216:81–92.
26. Mirsky HP, Miller MJ, Linderman JJ, Kirschner DE. Systems biology approaches for understanding cellular mechanisms of immunity in lymph nodes during infection. J Theor Biol. (2011); 287:160–70.
27. Narang V, Decraene J, Wong SY, Aiswarya BS, Wasem AR, Leong SR, Gouaillard A. Systems immunology: a survey of modeling formalisms, applications and simulation tools. Immunol Res. (2012); 53(1–3):251–65.
28. Ganusov VV, Pilyugin SS, de Boer RJ, Murali-Krishna K, Ahmed R, Antia R. Quantifying cell turnover using CFSE data. J Immunol Methods. 2005; 298(1-2):183-200. Erratum in: J Immunol Methods. (2006); 317(1–2):186–7.
29. De Boer RJ, Ganusov VV, Milutinovi D, Hodgkin PD, Perelson AS. Estimating lymphocyte division and death rates from CFSE data. Bull Math Biol. (2006); 68(5):1011–31.
30. Hawkins ED, Hommel M, Turner ML, Battye FL, Markham JF, Hodgkin PD. Measuring lymphocyte proliferation, survival and differentiation using CFSE time-series data. Nat Protoc. (2007); 2(9):2057–67.

31. Len K, Faro J, Carneiro J. A general mathematical framework to model generation structure in a population of asynchronously dividing cells. J Theor Biol. (2004); 229(4):455–76.
32. Asquith B, Debacq C, Florins A, Gillet N, Sanchez-Alcaraz T, Mosley A, Willems L. Quantifying lymphocyte kinetics in vivo using carboxyfluorescein diacetate succinimidyl ester (CFSE). Proc Biol Sci. (2006); 273(1590):1165–71.
33. Yates A, Chan C, Strid J, Moon S, Callard R, George AJ, Stark J. Reconstruction of cell population dynamics using CFSE. BMC Bioinformatics. (2007); 8:196.
34. Perelson AS, Ribeiro RM. Modeling the within-host dynamics of HIV infection. BMC Biol. (2013); 11:96.
35. Canini L, Perelson AS. Viral kinetic modeling: state of the art. J Pharmacokinet Pharmacodyn. (2014); 41(5):431–43.
36. Eftimie R, Gillard JJ, Cantrell DA. Mathematical Models for Immunology: Current State of the Art and Future Research Directions. Bull Math Biol. (2016); 78(10):2091–2134.
37. Ganusov VV. Strong Inference in Mathematical Modeling: A Method for Robust Science in the Twenty-First Century. Front Microbiol. (2016); 7:1131
38. Castro M, Lythe G, Molina-Pars C, Ribeiro RM. Mathematics in modern immunology. Interface Focus. (2016); 6(2):20150093
39. Deem MW, Hejazi P. Theoretical aspects of immunity. Annu Rev Chem Biomol Eng. (2010); 1:247-76.
40. Rapin N, Lund O, Bernaschi M, Castiglione F. Computational immunology meets bioinformatics: the use of prediction tools for molecular binding in the simulation of the immune system. PLoS One. (2010); 5(4):e9862.
41. Belfiore M, Pennisi M, Aric G, Ronsisvalle S, Pappalardo F. In silico modeling of the immune system: cellular and molecular scale approaches. Biomed Res Int. (2014); 2014:371809.
42. Thakar J, Poss M, Albert R, Long GH, Zhang R. Dynamic models of immune responses: what is the ideal level of detail? Theor Biol Med Model. (2010); 7:35.
43. Lundegaard C, Lund O, Kesmir C, Brunak S, Nielsen M. Modeling the adaptive immune system: predictions and simulations. Bioinformatics. (2007); 23(24):3265–75.
44. Arazi A, Pendergraft WF 3rd, Ribeiro RM, Perelson AS, Hacohen N. Human systems immunology: hypothesis-based modeling and unbiased data-driven approaches. Semin Immunol. (2013); 25(3):193–200.
45. Kidd BA, Peters LA, Schadt EE, Dudley JT. Unifying immunology with informatics and multiscale biology. Nat Immunol. (2014); 15(2):118–27
46. Proserpio V, Mahata B. Single-cell technologies to study the immune system. Immunology. (2016); 147(2):133–40.
47. Tang J, van Panhuys N, Kastenmller W, Germain RN. The future of immunoimaging–deeper, bigger, more precise, and definitively more colorful. Eur J Immunol. (2013); 43(6):1407–12.
48. Bocharov G, Argilaguet J, Meyerhans A. Understanding Experimental LCMV Infection of Mice: The Role of Mathematical Models. J Immunol Res. (2015); 2015:739706.
49. Stephen P. Ellner, John Guckenheimer. Dynamic Models in Biology. Princeton University Press. (2006). 330 pp. ISBN: 9780691125893.
50. Bell G, Perelson AS, Pimbley G (eds): Theoretical Immunology. New York, Marcer Dekker, (1978). 646 pp.
51. Polderman, J.W., Willems, J.C., *Introduction to Mathematical Systems Theory. A behavioral approach*, Texts in Applied Mathematics, **26**, Springer-Verlag, New York, 1998.
52. Chakraborty, A.K., Dustin, M.L., Shaw, A.S. *In silico* models for cellular and molecular immunology: successes, promises and challenges, *Nature Immunology*, **4** (2003) 933–936.
53. Baker, C.T.H., Bocharov, G.A., Paul, C.A.H., Rihan, F.A., Modelling and analysis of time-lags in some basic patterns of cell proliferation, *J. Math. Biol.*, **37** (1998) 341–371.
54. Armitage, P., Berry G., Matthews, J.N.S., *Statistical Methods in Medical Research.* (Fourth Edition) Blackwell Science, Oxford (2001).
55. Gershenfeld, N.A., *The Nature of Mathematical Modelling*, Cambridge University Press, Cambridge, (2000).
56. Bard, Y., *Nonlinear Parameter Estimation* (Academic Press, 1974).

57. Myung, I.J. Tutorial on maximum likelihood estimation. *J. Mathematical Physiology*, **47** (2003) 90–100.
58. Pascual, M.A., Kareiva, P. Predicting the outcome of competition using experimental data: maximum likelihood and Bayesian approaches. *Ecology*, **77** (1996) 337–349.
59. Gingerich, P.D. Arithmetic or geometric normality of biological variation: an empirical test of theory. *J. Theor. Biology***204** (2000) 201–221.
60. Venzon, D.J., Moolgavkar, S.H.: A method for computing profile-likelihood-based confidence intervals. Appl. Statist. **37**(1) 87–94 (1988)
61. B. Efron and R. Tibshirani. Bootstrap methods for standard errors, confidence intervals, and other measures of statistical accuracy. Stat. Sci., 1(1):54–77, (1986).
62. B. Efron and R. Tibshirani. Introduction to the bootstrap. Chapman and Hall, New York, (1993).
63. Rubinov S.I. Cell kinetics. In: *Mathematical models in molecular and cellular biology* Segel L.A. (Ed) Cambridge University Press, Cambridge (1980), pp 502–522.
64. Pilyugin S.S., Ganusov V.V., Murali-Krishna K., Ahmed R. and Antia R. The rescaling method for quantifying the turnover of cell populations. *J. Theor. Biol.* (2003) **225**: 275–283.
65. Ganusov V.V., Pilyugin S.S., de Boer R.J., Murali-Krishna K., Ahmed R. and Antia R. Quantifying cell turnover using CFSE data. *J. Immunol. Methods.* (2005) **298**: 183–200.
66. De Boer R.J. and Perelson A.S. Estimating division and death rates from CFSE data. *J. Comput. Appl. Math.* (2005) **184**: 140–164.
67. Hadamard, J.: Le probléme de Cauchy et les équations aux dérivées partielles linéaires hyperboliques. Paris, Hermann (1932)
68. Tikhonov, A.N., Arsenin, V.Y.: Solutions of ill-posed problems. Washington, V. H. Winston & Sons (1977)
69. Hasanov, A., DuChateau, P., Pektas, B.: An adjoint problem approach and coarse-fine mesh method for identification of the diffusion coefficient in a linear parabolic equation. J. Inv. Ill-Posed Problems **14**(5) 435–463 (2006)
70. Bitterlich, S., Knabner, P.: An efficient method for solving an inverse problem for the Richards equation. J. Comput. Appl. Math. **147** 153–173 (2002)
71. Tikhonov, A.N.: Solution of incorrectly formulated problems and the regularization method. Sov. Math. Dokl. **4** 1035–1038 (1963)
72. Engl, H.W., Rundell, W., Scherzer, O.: A regularization scheme for an inverse problem in age-structured populations. J. Math. Anal. Appl. **182** 658–679 (1994)
73. Grebennikov, A.: Local regularization algorithms of solving coefficient inverse problems for some differential equations. Inverse Probl. Eng. **11**(3) 201–213 (2003)
74. Navon, I.M.: Practical and theoretical aspects of adjoint parameter estimation and identifiability in meteorology and oceanography. Dynam. Atmos. Oceans **27**(1) 55–79 (1997)
75. DuChateau, P., Thelwell, R., Butters, G.: Analysis of an adjoint problem approach to the identification of an unknown diffusion coefficient. Inverse Problems **20** 601–625 (2004)
76. Tautenhahn, U., Jin, Q.: Tikhonov regularization and a posteriori rules for solving nonlinear ill posed problems. Inverse Problems **19** 1–21 (2003)
77. Perthame, B., Zubelli, J.P.: On the inverse problem for a size-structured population model. Inverse Problems **23** 1037–1052 (2007)
78. Miao, H., Jin, X., Perelson, A.S., Wu, H.: Evaluation of multitype mathematical models for CFSE-labeling experiment data. Bull. Math. Biol. **74**(2) 300–326 (2012)
79. Banks, H.T., Thompson, W.C.: Mathematical models of dividing cell populations: Application to CFSE data. Math. Model. Nat. Phenom. **7**(5) 24–52 (2012)
80. De Boer RJ, Perelson AS. Quantifying T lymphocyte turnover. J Theor Biol. (2013) 21;327:45–87.
81. Hross S, Hasenauer J. Analysis of CFSE time-series data using division-, age- and label-structured population models. Bioinformatics. (2016); 32(15):2321–9
82. Ackleh, A.S., Banks, H.T., Deng, K., Hu, S.: Parameter estimation in a coupled system of nonlinear size-structured populations. Math. Biosci. Engin. **2**(2) 289–315 (2005)

83. Morozov, V.A.: On the solution of functional equations by the method of regularization. Sov. Math. Dokl. **7** 414–417 (1966)
84. Morozov, V.A.: Methods for solving incorrectly posed problems. New York, Springer-Verlag (1984)
85. Schittler, D., Hasenauer, J., Allgöwer, F.: A generalized model for cell proliferation: Integrating division numbers and label dynamics. Proc. Eight International Workshop on Computational Systems Biology (WCSB 2011), Zurich, Switzerland, 165–168 (2011)
86. Hasenauer, J., Schittler, D., Allgöwer, F.: A computational model for proliferation dynamics of division- and label-structured populations. arXiv:1202.4923v1 [q-bio.PE] (2012)
87. Hasenauer, J., Schittler, D., Allgöwer, F.: Analysis and simulation of division- and label-structured population models: a new tool to analyze proliferation assays. Bull. Math. Biol. **74**(11) 2692–2732 (2012)
88. Sabrina Hross, Jan Hasenauer; Analysis of CFSE time-series data using division-, age- and label-structured population models, Bioinformatics, Volume 32, Issue 15, 1 August 2016, Pages 23212329, https://doi.org/10.1093/bioinformatics/btw131
89. Banks, H.T., Thompson, W.C., Peligero, C., Giest, S., Argilaguet, J., Meyerhans, A.: A division-dependent compartmental model for computing cell numbers in CFSE-based lymphocyte proliferation assay. CRSC-TR12-03, North Carolina State University (2012)
90. De Boer, R.J., Perelson, A.S.: Estimating division and death rates from CFSE data. J Comput. Appl. Math. **184** 140–164 (2005)
91. Roederer, M.: Interpretation of cellular proliferation data: avoid the panglossian. Cytometry A **79**(2) 95–101 (2011)
92. Chang, J.T., Palanivel, V.R., Kinjyo, I., Schambach, F., Intlekofer, A.M., Banerjee, A., Longworth, S.A., Vinup, K.E., Mrass, P., Oliaro, J., Killeen, N., Orange, J.S., Russell, S.M., Weninger, W., Reiner, S.L.: Asymmetric T lymphocyte division in the initiation of adaptive immune responses. Science **315** (5819) 1687–1691 (2007)
93. Banks, H.T., Choi, A., Huffman, T., Nardini, J., Poag, L., Thompson, W.C.: Quantifying CFSE label decay in flow cytometry data. Appl. Math. Lett. **26**(5) 571–577 (2013)
94. Banks, H.T., Sutton, K.L., Thompson, W.C., Bocharov, G., Roose, D., Schenkel, T., Meyerhans, A.: Estimation of cell proliferation dynamics using CFSE data. Bull. Math. Biol. **70** 116–150 (2011)
95. Banks, H.T., Sutton, K.L., Thompson, W.C., Bocharov, G., Doumic, M., Schenkel, T., Argilaguet, J., Giest, S., Peligero, C., Meyerhans, A.: A new model for the estimation of cell proliferation dynamics using CFSE data. J. Immunol. Methods **373** 143–160 (2011)
96. Schwarz, G. Estimating the dimension of a model. *The Annals of Statistics*, **6** (1978) 461–464.
97. Garny A, Noble D, Kohl P. Dimensionality in cardiac modelling. Prog Biophys Mol Biol. (2005); 87(1):47–66.
98. Burnham, K.P., Anderson, D.R., *Model selection and inference - a practical information-theoretic approach* (Springer, New York, 1998).
99. Kullback, S., Leibler, R.A. On information and sufficiency. *Ann. Math. Stat.*, **22** (1951) 79–86.
100. Akaike H., A new look at the statistical model identification, *IEEE Transactions on Automatic control*, **19** (1974) 716–723.
101. Borghans, J.A., Taams, L.S., Wauben, M.H.M., De Boer, R.J., Competition for antigenic sites during T cell proliferation: a mathematical interpretation of in vitro data, *Proc. Natl. Acad. Sci. USA.*, **96** (1999) 10782–10787.
102. Zinkernagel RM: Lymphocytic choriomeningitis virus and immunology. Curr Top Microbiol Immunol (2002), 263:1–5.
103. Burnet, F.M. *The Clonal Selection Theory of Acquired Immunity* (Cambridge University Press, 1959).
104. Ehl, S., Klenerman, P., Zinkernagel, R.M., Bocharov, G. The impact of variation in the number of $CD8^+$ T-cell precursors on the outcome of virus infection. *Cellular Immunology*, **189** (1998) 67–73.
105. Altman, J.D., Moss, P.A.H., Goulder, P.J.R., Barouch, D.H., McHeyzer-Williams, M.G., Bell, J.I., McMichael, A.J., Davis, M.M. Phenotypic analysis of antigen-specific T lymphocytes *Science*, **274** (1996) 94–96.

106. Battegay, M., Cooper, S., Althage,A., Banziger, H., Hengartner, H., Zinkernagel, R.M. Quantification of lymphocytic choriomeningitis virus with an immunological focus assay in 24- or 96-well plates *J. Virol. Methods*, **33** (1991) 191–198.
107. Paul, C.A.H., A User Guide to Archi, MCCM Rep. 283, University of Manchester. http://www.maths.man.ac.uk/~chris/reports/rep283.pdf
108. Paul, C.A.H., Archi FORTRAN listing. http://www.maths.man.ac.uk/~chris/software/ University of Manchester.
109. Numerical Algorithms Group The NAg FORTRAN Library http://www.nag.co.uk/numeric/Fortran_Libraries.asp.
110. De Boer, R.J., Oprea, M., Antia, R., Murali-Krishna, K., Ahmed, R., Perelson, A.S. Recruitment times, proliferation, and apoptosis rates during the $CD8^+$ T-cell response to lymphocytic choriomeningitis virus. *J. Virology*, **75** (2001) 10663–10669.
111. Bocharov G, Züst R, Cervantes-Barragan L, Luzyanina T, Chiglintsev E, Chereshnev VA, Thiel V, Ludewig B. A systems immunology approach to plasmacytoid dendritic cell function in cytopathic virus infections. PLoS Pathog. (2010); 6(7):e1001017.
112. Pitt, M.A., Myung, I.J. When a good fit can be bad. *Trends Cogn Sci.* 2002, 6(10): 421–425.
113. Grünwald, P.D., Myung, J.I., Pitt N.A. (Editors) Advances in Minimum Description Length: Theory and Applications (MIT Press, 2007)

Chapter 4
Modelling of Experimental Infections

This chapter aims to give a clear idea of how mathematical analysis for experimental systems could help in the process of data assimilation, parameter estimation and hypothesis testing. In particular, we illustrate the potential of a question-, and data-driven mathematical modelling in the

- estimation of model parameters for the 'virus–host' system,
- understanding kinetic regulation of virus infection dynamics,
- prediction of various phenotypes of virus infections and antigen-specific immune responses,
- testing specific hypothesis about the feedback regulation of T-cell responses.

The material of this chapter is based on our previous work published in [1–4, 9, 51, 84].

4.1 Why Experimental Infections?

Experimental systems of various types are used in fundamental immunology to unravel the complex cellular interactions of the immune responses. In vivo systems, which involve the whole animal provides the most natural experimental conditions. However, the in vivo systems have many unknown and uncontrollable interactions that add ambiguity to the interpretation of empirical data. The study of the immune system in vertebrates requires a suitable animal model. For most basic research in immunology, mice have been the experimental animal of choice. To control experimental variation caused by differences in the genetic background of experimental animals, immunologists work with inbred or knock-out or knock-in strains that are genetically identical animals produced by inbreeding. Hundreds of different strains of mice are available these days, e.g. CBA, BALB, C57BL/6, etc.

In this chapter, we present examples of mathematical models developed for experimental virus infection systems to answer specific questions concerning the kinetic

G. Bocharov et al., *Mathematical Immunology of Virus Infections*,
https://doi.org/10.1007/978-3-319-72317-4_4

regulation of virus infection dynamics, which are beyond the realm of purely empirical analysis techniques. The role of mathematical modelling in infection immunology can be summarized as follows:

- Descriptive
  - qualitative and quantitative characterization of process dynamics;
- Explanatory
  - interpretation of the experimental observations,
  - understanding the numbers game;
- Predictive
  - testable predictions; suggestion of new experiments,
  - sensitivity performance quantification,
  - hidden effects.

4.2 The LCMV System: Gold Standard for Infection Biology[1]

One of the best-studied model systems of viral infections is the lymphocytic choriomeningitis virus (LCMV) infection in mice.

4.2.1 *Immunobiology of LCMV*

LCMV is an RNA virus of the Arenaviridae family that is non-cytopathic in vivo, i.e. the virus itself does not cause direct damage to cells and tissues. This feature enables relating any damage that appears in the course of an infection to host immune responses against the virus. Another important feature of the LCMV model system is the existence of several well-characterized viral strains that differ in their replicative capacity, host range (cell tropism and mouse strain) and experimental routes of

[1]Material of subsections (4.2.2–4.2.4) uses the results of our studies from Bocharov, Modelling the dynamics of LCMV infection in mice: conventional and exhaustive CTL responses. J. Theor. Biol. 192, 283–308, Copyright © 1998; Ehl et al., The impact of variation in the number of CD8+T-cell precursors on the outcome of virus infection. Cell. Immunol. 189, 67–73, Copyright © 1998; Bocharov et al., Modelling the dynamics of LCMV infection in mice: II. Compartmental structure and immunopathology. J. Theor. Biol. 221, 349–78, Copyright © 2003; Luzyanina et al., Low level viral persistence after infection with LCMV: a quantitative insight through numerical bifurcation analysis. Math. Biosci. 173, 1–23, Copyright © 2001, with permission from Elsevier and the results of the studies from Proc. Natl. Acad. Sci. USA. (PNAS USA), Bocharov et al., Feedback regulation of proliferation vs. differentiation rates explains the dependence of CD4 T-cell expansion on precursor number, 108, 3318–23, Copyright © 2011 with permission from PNAS USA.

infection (intracranial versus intraperitoneal (i.p.) or intravenous (i.v.)) and thus show different infection outcomes. This enables directly linking easily measurable viral dynamic properties to pathogenic consequences and studying the kinetic mechanisms of chronic infections.

With the use of the LCMV infection model system, a large number of conceptual discoveries in immunology have been made, which are as follows:

- back in 1974/75, Zinkernagel and Doherty demonstrated that cytotoxic T-lymphocytes (CTLs) recognize foreign antigens only in the context of proteins of the major histocompatibility complex (MHC) [5, 6]. For this finding of MHC restriction, they were awarded the Nobel Prize in 1996.
- with the help of knockout mice, the mechanism of CTL-mediated destruction of LCMV-infected target cells in vivo was directly linked to perforin, a pore-forming protein contained in granules of this cell type [7, 8].
- fundamental properties of ‘memory’ of the adaptive immune response have been understood, in particular, the requirements for CTL memory to prevent the establishment of a persistent LCMV infection [9].
- NK cells of the innate immune response have been recognized as an important regulator of the helper T-cell support for antiviral CTL [10].
- a critical role of organized secondary lymphoid organs in the induction of naive T and B cells and subsequent virus control was established [11].
- the concept of immunopathology, that is the damage of tissues and organs due to the antiviral immune response rather than the infecting virus itself, was established. Mediators of immunopathology include CTL, macrophages, neutrophils and interferons [12–14].
- based on the amino acid similarities between viral antigens and host proteins, the so-called molecular mimicry, viral infections can trigger autoimmunity and influence the course of subsequent infections with other viral pathogens [15–17].
- important observations towards an acute versus a persistent infection outcomes were made as shown in Fig. 4.1 [18–21].

Which infection fate is followed depends on the infecting viral dose and the viral strain and thus can be easily directed experimentally. LCMV persistence is associated with CTL exhaustion, a reversible, non-functional state of CTL. CTL exhaustion is a physiological consequence of persistent antigen exposure and has been observed both in persistent human viral infections and in cancers, the LCMV system was instrumental to understand infection fate regulation in general terms. As CTL exhaustion can be reversed by antibodies against PD1 or PD-L1 that block the negative signalling pathway, novel immunotherapeutic modalities arose which show exciting promises as antiviral and anticancer therapies [22–24].

The LCMV infection model system offers sufficient experimental data to develop mathematical models in a problem-oriented manner. The mathematical model-driven studies of LCMV resulted in experimentally testable predictions concerning the mechanisms of the infection control, for example (i) threshold numbers of initial specific CTL precursors to protect from a chronic LCMV infection outcome, (ii) minimal number of antigen-presenting DCs in spleen for robust induction of CTL

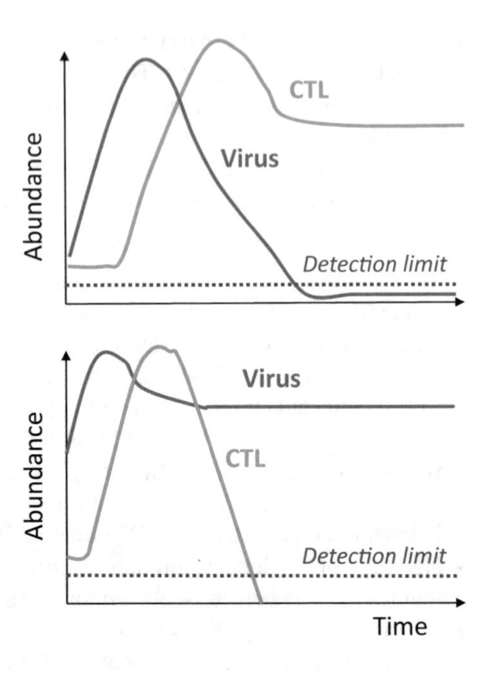

Fig. 4.1 Scheme of acute (top) and chronic (bottom) LCMV infection. Phenotypically different dynamic patterns of viral load and CTL activity are shown

responses, and (iii) the effect of virus growth rate on the magnitude of the clonal expansion of CTLs, to name just the major of them.

The basic biological features of LCMV, relevant to the mathematical models presented below can be summarized as follows:

- Family: Arenaviridae;
- Strains: Docile, Traub, WE, Aggressive, Armstrong, Clone 13;
- Host: mice, hamsters; humans: acute hemorrhagic fever;
- Target cells: macrophages and lymphocytes;
- Cytopathicity in vivo: non-cytopathic;
- CTL responses play a dominant role in virus clearance: appear early and are high;
- Neutralizing antibody responses: appear only late after infection;
- Immunopathology is a recovery fee: is observed in spleen, liver, central nervous system.

The spatial distribution (compartmental structure) of LCMV infection is presented in Fig. 4.2. It must be noted that spleen plays a central role in LCMV infection as it is a target organ for virus replication and the lymphoid organ in which the antiviral immune response takes place.

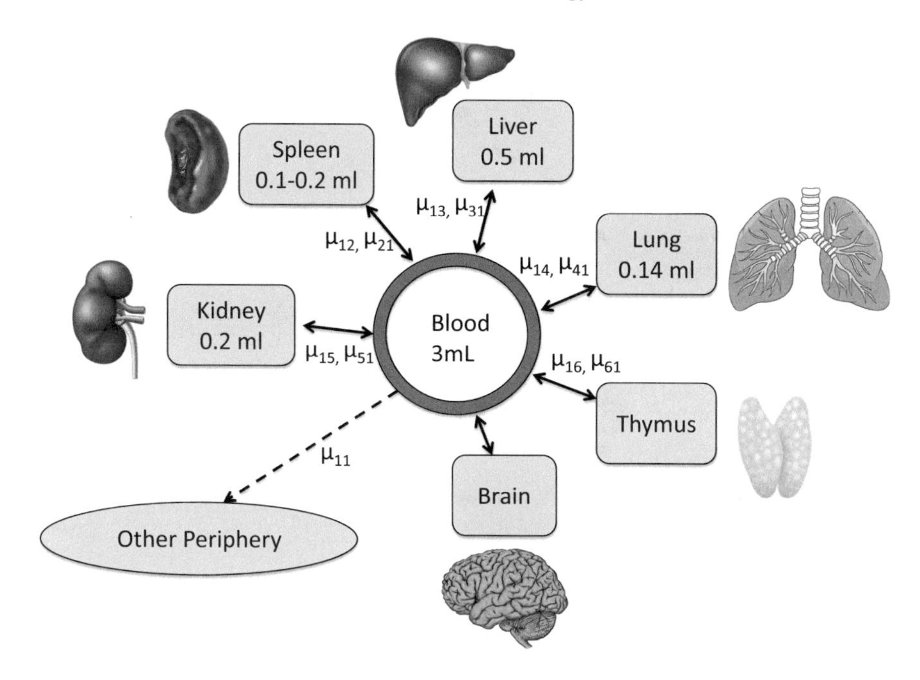

Fig. 4.2 Scheme of the compartmental structure of LCMV infection spreading in mice. The volumes of the organs, considered as well-mixed reactors, are indicated

4.2.2 *Basic Mathematical Model of LCMV Infection*

The biological scheme underlying the mathematical model of LCMV infection is presented in Fig. 4.3. The mathematical model of antiviral CTL response developed previously [1] is based upon assumptions reflecting general mechanisms of the virus–host interaction: (i) virus-specific CTLs are primarily responsible for control of infection with non-cytopathic viruses; (ii) the virus population stimulates clonal expansion and differentiation of the specific CTL precursors (CTLp) into effector cells; (iii) a high viral load leads eventually to inhibition of CTL responses via anergy and activation-induced cell death by apoptosis; (iv) in the absence of viral antigens the homeostasis of naive CTLs reflects a balance between the input of the precursor CTLs from thymus and their death at the periphery; (v) virus replication in the host exhibits a logistic-type growth, whereas the elimination follows a second-order kinetics. Only one organ in which both the virus infection and immune response take place (compartment), i.e. the spleen is considered in the basic model. The time-dependent variables of the model are as follows:

- $V(t)$ virus titer in spleen at time t (pfu/ml);
- $E_p(t)$ number of virus-specific precursor CTLs in spleen at time t (cell/ml);
- $E(t)$ number of virus-specific effector CTLs in spleen at time t (cell/ml);
- $W(t)$ cumulative virus antigen load in spleen at time t (pfu/ml).

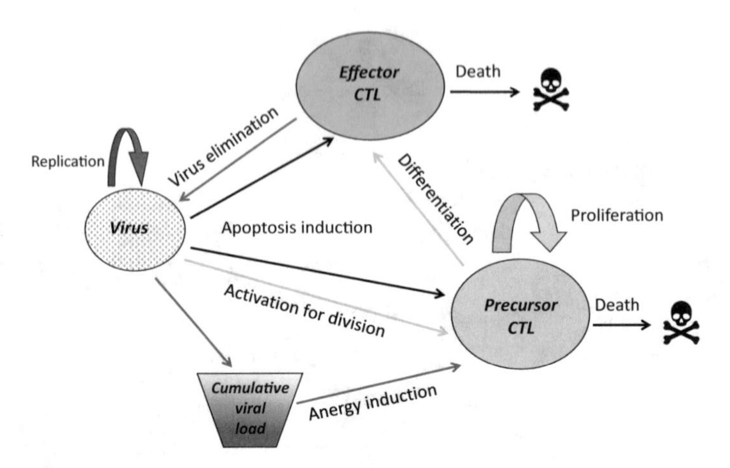

Fig. 4.3 Schematic representation of the variables and processes considered in the mathematical model of LCMV infection in mice

Their choice is guided by the availability of data for model identification [25] and a clearly elaborated understanding of the most relevant processes in the control of acute LCMV infection. The above-listed processes define the structure of a system of delay differential equations describing the rates of change of the population densities in the course of infection:

$$\frac{dV}{dt} = \beta V(t)(1 - V(t)/K) - \gamma_{VE} V(t)E(t), \tag{4.1}$$

$$\frac{dE_p}{dt} = \alpha_{E_p}\left(E_p^0 - E_p(t)\right) + b_p g_p(W)V(t-\tau)E_p(t-\tau) - \alpha_{AP} V(t-\tau_A)V(t)E_p(t), \tag{4.2}$$

$$\frac{dE_e}{dt} = b_e g_e(W)V(t-\tau)E_p(t-\tau) - \alpha_{AE} V(t-\tau_A)V(t)E_e(t) - \alpha_{E_e} E_e(t), \tag{4.3}$$

$$\frac{dW}{dt} = b_W V(t) - \alpha W(t), \tag{4.4}$$

where $g_p(W) = \frac{1}{(1+W/\theta_p)^2}$, $g_e(W) = \frac{1}{(1+W/\theta_e)^2}$. The equations are supplemented by initial data reflecting the low-, intermediate- and high-dose infections of C57BL/6 mice with LCMV-Docile i.v.:

- $V(t) = 0,\ t \in [-\tau^*, 0),\ V(0) = 10^2,\ 10^4,\ 10^7$ pfu/ml, $\tau^* = \max[\tau, \tau_A]$,
- $E_p(t) = 265$ cell/ml, $t \in [-\tau^*, 0]$,
- $E_e(0) = 0$ cell/ml,
- $W(0) = 0$ pfu/ml.

In the equation for $V(t)$, the first term on the right-hand side describes the virus growth with an upper limit K due to the limited amount of sensitive tissue cells supporting virus replication, and the second term takes into account the clearance of viruses due to lysis of virus-infected cells by effector CTLs.

In the equation for $E_p(t)$, the first term describes the maintenance of virus-specific precursor CTL at a certain level through their export from thymus and death in the periphery. The second term accounts for an increase in the number of CTL precursors resulting from virus-induced proliferation with the inhibitory effect of cumulative virus load on clonal expansion. The last term describes activation-induced cell death by apoptosis.

The dynamics of $E_e(t)$ is determined by the appearance of mature effector CTLs due to the division and differentiation of antigen-stimulated precursor CTLs with the downregulation of the differentiation process of CTLp due to high virus antigen load (the first term); the decrease in the number of effector CTLs as a consequence of lytic interactions with virus-infected target cells (the second term); the activation-induced cell death of effector CTLs and natural death of effector CTL due to their finite lifespan (two last terms).

In the equation for $W(t)$, the first term describes the increase in the total viral antigen load due to virus spread in the host and the second one accounts for the decrease of the inhibitory effect of high virus loads on the virus-specific CTLs as the virus is eliminated.

The model is based upon a fundamental assumption which reflects results of empirical analysis [25] that continuous exposure of virus-specific CTLs to LCMV induces a sequence of proliferation, anergy and activation-induced cell death by apoptosis. The balance between the above processes depends on the cumulative viral load and shifts towards the anergy and death phenotype in a high viral load infection.

This low-dimensional model is based on (i) a Verhulst logistic form for virus growth; (ii) second-order virus elimination kinetics by CTLs; (iii) the Holling type II response curve for CTLs expansion with a time lag representing cell division time and antigen-independent production/death of CTLs in the immune system (homeostasis).

The relevant information about the model parameters is summarized in Table 4.1.

The model parameters were estimated via a maximum likelihood approach using experimental data characterizing the virus-CTL dynamics after low-, intermediate- and high-dose i.v. infections of C57BL/6 mice [1, 25] and permit a good consistency of the model with the data, Fig. 4.4. The phenomenology of conventional and exhaustive CTL responses is quantitatively captured in the mathematical model.

The phenomenon of exhaustion in the model is defined as disappearance of CTL activity and the functional impairment of virus-specific CTLs. The exhaustion of antiviral CTL responses is modelled as a stepwise process observed in an overwhelming infection with LCMV-Docile. Following the initial activation, LCMV-specific T cells become anergic for 3–5 days and then disappear because of activation-induced cell death (apoptosis). (Of note, the observed lack of T-cell functionality was in time of the described experiments termed anergy; however, this functional state of T cells has been studied in more detail and shown to be a non-responsive state after continuous antigen exposure that is now termed exhausted; for a detailed discussion, see Wherry and Kurachi [26]).

Table 4.1 The LCMV infection model parameters and their best-fit estimates. We considered one pfu is one infectious $particle$

Parameter	Biological meaning	Units	Best-fit estimate
β	Replication rate constant of viruses	1/day	3.35
γ_{VE}	Rate constant of virus clearance due to effector CTLs	ml/(cell day)	1.34×10^{-6}
K	Virus carrying capacity of spleen	particles/ml	4.82×10^{7}
τ	Duration of CTL division cycle	day	0.6
b_p	Rate constant of precursor CTL stimulation	ml/(particle day)	7.73×10^{-5}
b_e	Rate constant of precursor CTL stimulation	ml/(particle day)	7.73×10^{-4}
θ_p	Cumulative viral load threshold for anergy induction in precursor CTL	particle/ml	3.25×10^{6}
θ_e	Cumulative viral load threshold for anergy induction in differentiation of CTL	particle/ml	3×10^{5}
α_{E_p}	Rate constant of precursor CTL natural death	1/day	0.542
α_{E_e}	Rate constant of effector CTL natural death	1/day	0.01
E_p^0	Homeostatic concentration of LCMV-specific precursor CTL in spleen of unprimed mice	cell/day	265
τ_{AP}	Duration of commitment of CTLs for apoptosis	day	5.6
α_{AP}	Rate constant of precursor CTL apoptosis	(ml/particle)^2/day	7.5×10^{-16}
α_{EP}	Rate constant of effector CTL apoptosis	(ml/particle)^2/day	4.36×10^{-14}
b_W	Rate constant of the cumulative viral load growth	1/day	1
α_W	Rate constant of the restoration from the inhibitory effect of cumulative viral load	1/day	0.11

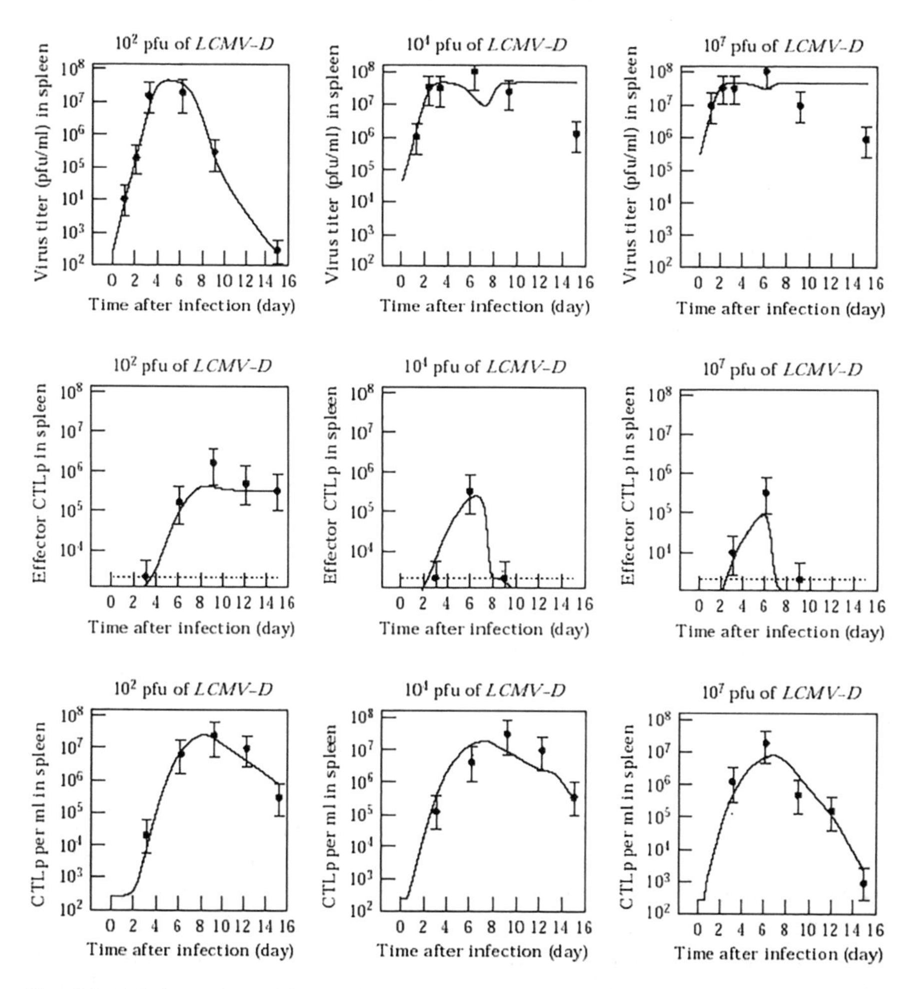

Fig. 4.4 Solutions of model (4.1–4.4) with the parameters estimated to best-fit the low-, intermediate- and high-dose i.v. infections (Table 4.1). Experimental data are denoted by o. Reprinted from Journal of Theoretical Biology, Vol. 192, Bocharov, Modelling the Dynamics of LCMV Infection in Mice: Conventional and Exhaustive CTL Responses, Pages 283–308, Copyright © 1998 with permission from Elsevier

The single characteristic that appeared to be sufficient to control conventional versus exhaustive responses of CTLs was the cumulative viral load (cvl) since the beginning of the infection. The increase of cvl above a certain threshold value in conjunction with the high viral load in the host for about 5 days results in the shift of the infection phenotype from an acute with recovery to a chronic infection.

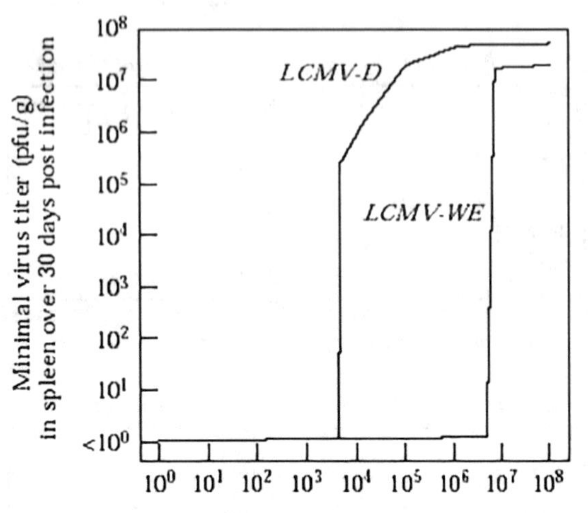

Fig. 4.5 Model prediction for the dose dependence of virus clearance from the spleen of C57BL/5 mice inoculated i.v. with the indicated doses of LCMV-Docile and LCMV-WE. Reprinted from Journal of Theoretical Biology, Vol. 192, Bocharov, Modelling the Dynamics of LCMV Infection in Mice: Conventional and Exhaustive CTL Responses, Pages 283-308, Copyright © 1998 with permission from Elsevier

4.2.3 *Viral Parameters: Impact on the Infection Phenotype*

The model predicts that the virus population reaching the spleen after i.v. infection depends on the inoculum size (IS) in a nonlinear way, as described by the following formula:

$$V_{Spleen}(day0) = \frac{0.37 \cdot IS}{1 + IS/0.84 \cdot 10^5}. \tag{4.5}$$

It suggests that the fraction of virus population reaching the spleen and establishing a productive infection decreases from 48% to 27% and 0.3% after infection with 10^2, 10^4 and 10^7 pfu, respectively. A continuous dose dependence of the extent of virus elimination from the spleen of infected mice is shown in Fig. 4.5 (left curve). The extent of virus elimination was assessed by the minimal value of the variable $V(t)$ over 30 days post-infection. The extent of virus (LCMV-Docile) clearance displays a threshold-type behaviour in relation to the dose of infection. The doses below 2×10^3 pfu are eliminated due to the CTL response, while the infections with higher doses lead to CTL exhaustion and virus persistence.

4.2.3.1 Why Does LCMV-WE Strain Fail to Cause Exhaustion of CTLs After i.v. Infection of C57BL/6 Mice?

It is known that some LCMV isolates (WE or Armstrong) do not induce viral persistence after high-dose i.v. infection. However, under certain conditions LCMV-WE can also establish persistent infection like in congenital LCMV-WE carrier C57BL/6 mice. The calibrated model can be used to examine the shape and position of the dose of infection-clearance curve for the LCMV-WE isolate. To this end, using additional data on the growth kinetics of LCMV-WE, one needs to quantify the exponential growth rate of the virus in spleen and the carrying capacity value. These appear to be smaller than in the cases of LCMV-D, i.e. $\beta = 2.57$ pfu ml^{-1} day^{-1} and $K = 0.18 \times 10^8$ pfu ml^{-1}, respectively. Neglecting the differences in CTL stimulation rate due to the variation in amino acids of the LCMV-GP epitope, the infection dose-dependent clearance curve for LCMV-WE can be computed as shown in Fig. 4.5. It suggests that the threshold dose of infection separating clearance and persistence phenotypes is around 7.0×10^6 pfu. This is an order of magnitude larger than the virus population reaching the spleen after i.v. infection (see 4.5). The difference provides an explanation of why LCMV-WE fails to cause chronic infection after i.v. injection of 10^7 pfu. Thus, minor variations between the distinct LCMV strains in the values of virus multiplication parameters might underline the about 10^3 increase in the virus dose threshold separating the two phenotypes of the virus–mouse interaction, i.e. virus clearance and persistent infection.

4.2.3.2 Can Underwhelming Infection Lead to Chronic Persistence?

According to the balance of growth and differentiation concept by Grossman and Paul [80–82], the immune system responds to a strong perturbation of the antigenic homeostasis. The implication is that a slower replicating virus could lead to a weaker immune response. The basic mathematical model of LCMV infection can be used to predict the impact of the virus replication kinetics on the magnitude of the CTL response in acute LCMV infection. To this end, we varied the exponential growth rate of the virus from 1 to 4.8 day^{-1}. This corresponds to an increase of the virus population per day by factors of 2.7 and 122. The predicted dependence of the maximum value of $E_p(t) + E_e(t)$, $t \in [0, 30]$ days is shown in Fig. 4.6. It appears to be bell shaped.

The experimental analysis of the clonal expansion of CTLs in C57BL/6 mice to LCMV strains (Armstrong, WE-Armstrong, WE, Traub and Docile) differing in their replication rate confirmed that there is a bell-shaped relationship between the LCMV growth rate and the peak CTL response (see Fig. 4.6). Both slow and fast replicating LCMV strains produce weaker CTL responses. Thus, a mechanism of virus persistence by sneaking surveillance due to slow replication kinetics can be hypothesized. The 'underwhelming' infection mechanism (supplementing the 'overwhelming' infection) fits the above-mentioned concept of the sensitivity of immune responses to perturbations.

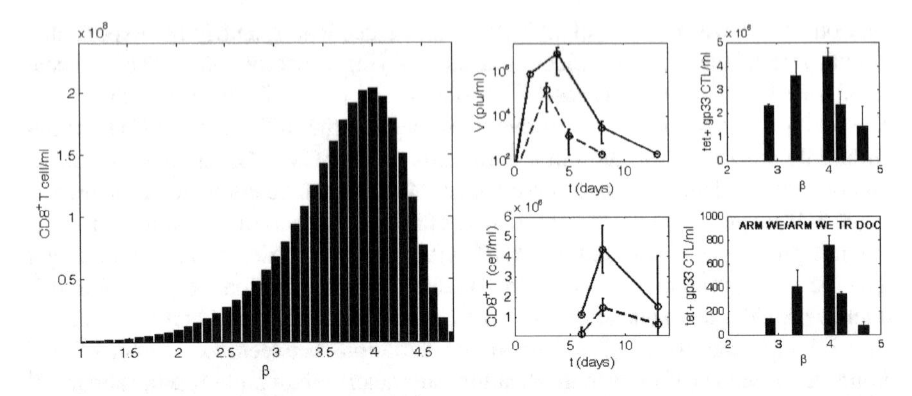

Fig. 4.6 Dependence of CTL expansion on virus growth rate. Left: Model prediction for i.v. infection with 200 pfu of LCMV stains differing in their replication rates. Center: Data on virus and CTL kinetics for the WE (solid line) and Armstrong (dashed line) strains. Right: Experimental data on the peak CTL responses in blood and spleen for 5 LCMV strains and C57BL/5 mice

4.2.3.3 Low-Level LCMV Persistence[2]

Acute infection with a low-dose LCMV is characterized by virus and CTL dynamics as shown in Fig. 4.7. The viral load drops below the level of detection by conventional assays. Experimental evidence indicates that (i) after acute infection with LCMV the virus might persist for some time in spleen cells at a frequency of 1 copy per 10^4–10^5 splenocytes, giving an estimate of about 500–5000 DNA copies per spleen; (ii) a difference in total LCMV RNA copies between the peak of infection ($10^8 - 10^9$ copies per spleen) and the memory phase (10^3 or fewer copies per spleen) has been observed; (iii) infectious LCMV may persist at no more than 100 pfu per spleen [27] and in some cases at the level of 1000 pfu per kidney [28].

How can the virus population persist in the face of CTL memory? There is a diverse array of biological mechanisms that are used by viruses to escape complete elimination by the immune system, ranging from those based on a limited growth, cell-to-cell passage without maturation, localization in an immunologically privileged site, integration into the host cell chromosome to those based on decreasing immune detection and destruction, e.g. via downregulation of MHC-restricted antigen presentation [29–31]. In terms of kinetics, the implication is that the replication rate and CTL-mediated elimination rate of LCMV (represented by β and γ_{VE}, respectively) might well be reduced during transition from the acute to the low-level persistence phase. Indeed, available data on the growth kinetics of LCMV after immune therapy of a persistent viral infection [32] or in $CD4^+$ T cell or B-cell-deficient mice [33] show a much lower rate of viral growth compared to the acute infection.

[2]The material of this subsection uses the results from Luzyanina et al., Low level viral persistence after infection with LCMV: a quantitative insight through numerical bifurcation analysis. Math. Biosci. 173, 1–23, Copyright © 2001, with permission from Elsevier.

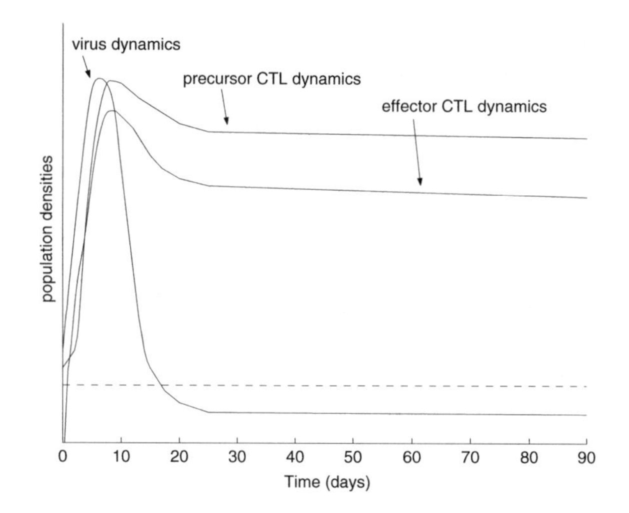

Fig. 4.7 Scheme of the within-host dynamics of virus and CTL populations characterized by expansion-, contraction- and memory phases. Reprinted from Mathematical Biosciences, Vol. 173, Luzyanina et al., Low level viral persistence after infection with LCMV: a quantitative insight through numerical bifurcation analysis, Pages 1–23, Copyright © 2001, with permission from Elsevier

One can investigate coexistence of viral and CTL populations in the memory phase through numerical bifurcation analysis of the virus–host interaction model.

$$\begin{cases} \frac{d}{dt}V(t) = \beta V(t)(1 - \frac{V(t)}{K}) - \gamma_{VE}E_e(t)V(t), \\ \frac{d}{dt}E_p(t) = \alpha_{E_p}(E_p^0 - E_p(t)) + \frac{b_p}{(1+W(t)/\theta_p)^2}V(t-\tau)E_p(t-\tau) - \alpha_{AP}V(t-\tau_A)V(t)E_p(t), \\ \frac{d}{dt}E_e(t) = \frac{b_d}{(1+W(t)/\theta_E)^2}V(t-\tau)E_p(t-\tau) - b_{EV}V(t)E_e(t) - \alpha_{AE}V(t-\tau_A)V(t)E_e(t) \\ \qquad -\alpha_{E_e}E_e(t), \\ \frac{d}{dt}W(t) = b_W V(t) - \alpha_W W(t). \end{cases} \tag{4.6}$$

In the context of dynamical system analysis, the coexistence of a low-level virus population and CTL memory corresponds to a stable steady-state solution (equilibrium) or to a stable oscillatory solution of the model (4.6) with V, respectively $V(t)$, below a (small) value, e.g. below the detection level of the virus in experiments. Concerning oscillatory solutions, we are interested in periodic solutions, i.e. solutions both existing in the long term and repeating themselves after a finite time.

For the model analysis, we use the software package DDE-BIFTOOL [34, 35]. DDE-BIFTOOL is a MATLAB package (The MathWorks, Inc.) for bifurcation analysis of systems of DDEs with several discrete delays. The package can be used to compute and analyse the stability of steady-state and periodic solutions of a given system as well as to study the dependence of these solutions on system parameters via continuation.

4.2.3.4 Steady-State Solutions

Introduce the notation $S := [V,\ E_p,\ E_e,\ W]^T$ for a vector of solutions of Eq. (4.6) and $F := F(S(t),\ S(t-\tau),\ S(t-\tau_A),\ p)$ for a vector defined by the right-hand sides of (4.6) with p a vector of parameters. A steady-state solution, S^*, of (4.6) is a solution of the following nonlinear algebraic system,

$$\begin{cases} \beta V(1-\frac{V}{K}) - \gamma_{VE} E_e V = 0, \\ \alpha_{E_p}(E_p^0 - E_p) + \frac{b_p}{(1+W/\theta_p)^2} V E_p - \alpha_{AP} V^2 E_p = 0, \\ \frac{b_d}{(1+W/\theta_E)^2} V E_p - b_{EV} V E_e - \alpha_{AE} V^2 E_e - \alpha_{E_e} E_e = 0, \\ b_W V - \alpha_W W = 0. \end{cases} \tag{4.7}$$

This system is solved by a Newton iteration starting from an initial guess for S^*.

The linearization of (4.6) around a solution trajectory $S^*(t)$ is the variational equation,

$$\frac{\mathrm{d}}{\mathrm{d}t} y(t) = A_0(t)y(t) + A_1(t)y(t-\tau) + A_2(t)y(t-\tau_A), \tag{4.8}$$

where A_i equals the derivative of F with respect to its $(i+1)$-th argument evaluated at $S^*(t)$.

For a steady-state solution, $S^*(t) \equiv S^*$, the matrices $A_i(t)$ are constant, $A_i(t) \equiv A_i$, and the variational equation (4.8) leads to a characteristic equation,

$$\det(\lambda I - A_0 - A_1 e^{-\lambda\tau} - A_2 e^{-\lambda\tau_A}) = 0, \tag{4.9}$$

with I the identity matrix. The characteristic roots, $\lambda \in \mathbb{C}$, determine the stability of the steady-state solution S^*. In general, (4.9) has an infinite number of roots. However, it is known that $\Re(\lambda_j) \to -\infty$ as $j \to \infty$ and that the number of roots in any right half- plane $\Re(\lambda) > \eta$, $\eta \in \mathbb{R}$, is finite. Hence, the stability is always determined by a finite number of roots. The rightmost (stability determining) characteristic roots are approximated using a linear multi-step method applied to variational equation (4.8), see [34–36] for details. A steplength heuristic is implemented to ensure accurate approximations of the roots with real part greater than a given constant. The approximations thus obtained are corrected using a Newton iteration on the characteristic equation.

Dependence of the steady-state solution S^* on a physical parameter (a component of p) can be studied by computing a branch of steady-state solutions as a function of the parameter using a continuation procedure [34]. The stability of the steady state can change during continuation whenever characteristic roots cross the imaginary axis. Generically a *fold* bifurcation (or *turning point*) occurs when a real characteristic root passes through zero and a *Hopf* bifurcation occurs when a pair of complex conjugate characteristic roots crosses the imaginary axis. Once a Hopf point is detected it can be followed in a two-parameter space using an appropriate determining system [34]. In this away, for instance one computes the stability region of the steady-state solution in the two-parameter space (if no other bifurcations occur in this region).

The relevant parameters of virus and CTL memory persistence are those characterizing virus replication and precursor-, effector CTL lifespans: β, α_{E_p}, and α_{E_e}.

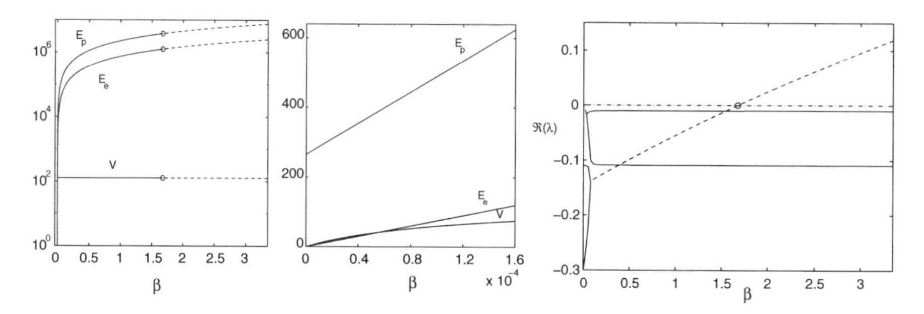

Fig. 4.8 Left, middle: Solutions V (pfu/ml), E_p (cell/ml) and E_e (cell/ml) along a branch of steady-state solutions of (4.6) versus parameter β for $\alpha_{E_e} = 0.3,\ \alpha_{E_p} = 0.01$. Middle figure is a blow up of the left figure. Stable and unstable parts of the branch are denoted by solid, respectively dashed lines. A logarithmic scale is used for the $Y-$axis (Left figure). Right: Real part of the rightmost roots (real (–) and complex (– –) roots) of the characteristic equation along the same branch. Hopf bifurcation (○) at $\beta \approx 1.675$

We choose the Hopf bifurcation point indicated in Fig. 4.8, as a starting point to continue a branch of Hopf bifurcation points in the (β, α_{E_p})-plane, see Fig. 4.9 (left). The corresponding Hopf curve bounds the stability region of the steady state corresponding to virus population–CTL memory coexistence because no other bifurcations were found in this region. Using a sequence of similar continuations, we computed branches of Hopf points in the (β, α_{E_p})-plane for different values of α_{E_e}, see Fig. 4.9 (right). Whenever $\alpha_{E_p} < 0.9\alpha_{E_e}$, it can be shown that the numerically established stability regions in the three- parameter space can be approximated by the formula,

$$\beta < 1.7 - 1.8\alpha_{E_p}/\alpha_{E_e}, \tag{4.10}$$

quantitatively describing the nature of the coupling between the parameters necessary to ensure a stable steady state with viral persistence and CTL memory. It indicates an opposite effect of parameters α_{E_p} and α_{E_e} on the value of β.

Some information about the numerical values of virus and CTL population densities for the steady states in the stability region shown in Fig. 4.9 (left) is given in Fig. 4.10. Figure 4.10 (left) presents the regions in the (β, α_{E_p})-plane where virus persists below the detection limit ($V < 1000$ pfu/ml) and below 100 pfu/ml. One can see that the value of V almost does not depend on β unless β gets close to 0 (see also Fig. 4.8) and virus can persist at a very low level if the death rate of CTLp (α_{E_p}) is small enough.

4.2.3.5 Periodic Solutions

A periodic solution $S^*(t)$ is a solution which repeats itself after a finite period T, i.e. $S^*(t+T) = S^*(t)$ for all $t > 0$. A discrete approximation to a periodic solution on a mesh in $[0, T]$ and its period are computed as solutions of the corresponding

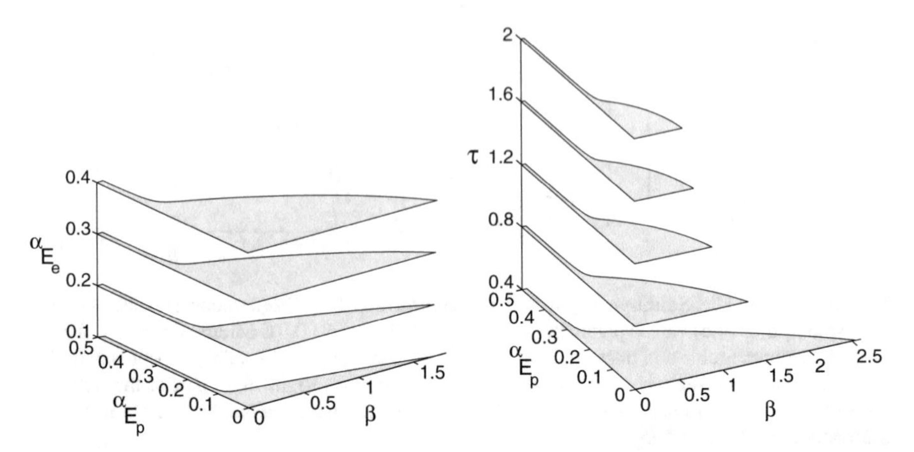

Fig. 4.9 Stability regions (depicted in grey) of the steady- state solution of (4.6) in the $(\beta, \alpha_{E_p}, \alpha_{E_e})$-space (left) and in the $(\beta, \alpha_{E_p}, \tau)$-space (right). Stability regions are visualized for Left: $\tau = 0.6$; Right: $\alpha_{E_e} = 0.3$. Solid lines correspond to curves of Hopf bifurcations. Reprinted from Mathematical Biosciences, Vol. 173, Luzyanina et al., Low level viral persistence after infection with LCMV: a quantitative insight through numerical bifurcation analysis, Pages 1–23, Copyright © 2001, with permission from Elsevier

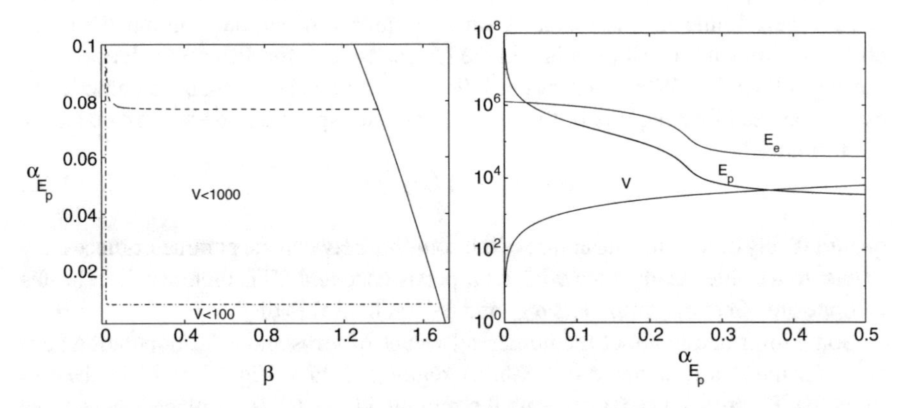

Fig. 4.10 Left: Regions in the (β, α_{E_p})-plane corresponding to solutions with $V < 1000$ pfu/ml and $V < 100$ pfu/ml. The solid line denotes a Hopf bifurcation curve. Right: Steady-state solutions V (pfu/ml), E_p (cell/ml), E_e (cell/ml) along the Hopf curve shown in Fig. 4.9 (left). Reprinted from Mathematical Biosciences, Vol. 173, Luzyanina et al., Low level viral persistence after infection with LCMV: a quantitative insight through numerical bifurcation analysis, Pages 1–23, Copyright © 2001, with permission from Elsevier

periodic boundary value problem using piecewise polynomial collocation [37, 38]. Adaptive mesh selection (the lengths of the mesh subintervals are adapted to the solution gradient) allows the computation of solutions with steep gradients.

Stability of a periodic solution is determined by the spectrum of the linear so-called Monodromy operator, which integrates the variational equation (4.8) around $S^*(t)$ from time $t = 0$ over the period T. Any nonzero eigenvalue μ of this operator is called a characteristic (Floquet) multiplier. Furthermore, $\mu = 1$ is always an eigenvalue and it is referred to as the trivial Floquet multiplier. A discrete approximation of this operator, a matrix M, is obtained using the collocation equations. The eigenvalues of M form approximations to the Floquet multipliers.

A branch of periodic solutions can be traced as a function of a system parameter using a continuation procedure [34]. The branch can be started from a Hopf point or from an initial guess (e.g. resulting from time integration). Bifurcations of periodic solutions occur when Floquet multipliers move into or out of the unit circle. Generically this is a *turning* point when a real multiplier crosses through 1, a *period doubling* point when a real multiplier crosses through -1 and a *torus* bifurcation when a complex pair of multipliers crosses the unit circle.

In the neighbourhood of a Hopf bifurcation point, solutions which belong to a branch of periodic solutions emanating from this point oscillate around the steady-state value corresponding to the Hopf point. Hence, Hopf points with low values of V can be sources of periodic solutions with oscillatory low-level viral persistence. We use the Hopf point shown in Fig. 4.8 as our 'basic Hopf point' and study the existence of oscillatory patterns in viral persistence by computing branches of periodic solutions emanating from this point as a function of the parameters listed in Table 4.1. Note that we depict periodic solutions on the time interval [0, 1], i.e. after time is scaled by the factor T^{-1} with T the period of the solution.

Influence of β. As β grows from its Hopf point value ($\beta \approx 1.675$), the amplitude of oscillations of $V(t)$ grows rapidly, see Fig. 4.11. The sensitivity of the dynamics to changes of β is also well characterized by the fact that a subtle change in β (from 1.675 to 2.06) leads to 'pulse' oscillations in virus population size, see Fig. 4.11 where solutions are shown for three values of β: close to the Hopf point (a), when $V_{\max} \approx 10^3$ pfu/ml (b) and when $V_{\max} \approx 2 \cdot 10^3$ pfu/ml (c).

We summarized the bifurcation analysis results for (β, α_{E_p}) in Fig. 4.12, where the curves of the turning points bound regions with different numbers of (stable and unstable) periodic solutions are shown. Note that left parts of the curves of turning points end at Hopf bifurcation points of steady-state solutions. The dynamic complexity of the system is well characterized by the fact that in region 3 steady-state solutions coexist with periodic solutions and in region 2 two stable periodic solutions coexist. However, the region of our interest, where periodic oscillations are such that $V_{\max} < 10^3$ pfu/ml, is quite small. Much smaller is the region with V varying in between 10 and 10^3 pfu/ml (or equivalently in between 1 and 100 pfu/spleen), see Fig. 4.12 (right). In this region, the period of oscillations varies from 10 to 20 days.

Influence of b_p. Larger values of b_p increase the region in the (β, α_{E_p})-plane where oscillatory solutions with $V_{\max} < 10^3$ pfu/ml exist, see region A in Fig. 4.13. However, due to a high sensitivity of the amplitude of oscillations to changes in

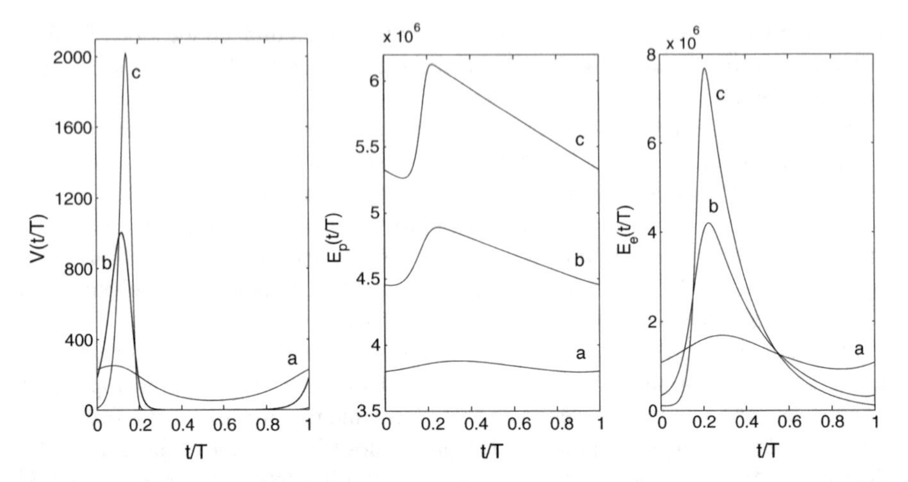

Fig. 4.11 Sensitivity of periodic solutions to changes in virus growth rate, β. Solutions V (pfu/ml), E_p, and E_e (cell/ml) corresponding to three points on the branch of periodic solutions: $\beta = 1.7$ (**a**), $\beta = 2.06$ (**b**), $\beta = 2.5$ (**c**). Values of the period T (days): $T \approx 9.5$ (**a**), $T \approx 13.2$ (**b**), $T \approx 18.2$ (**c**). Reprinted from Mathematical Biosciences, Vol. 173, Luzyanina et al., Low level viral persistence after infection with LCMV: a quantitative insight through numerical bifurcation analysis, Pages 1–23, Copyright © 2001, with permission from Elsevier

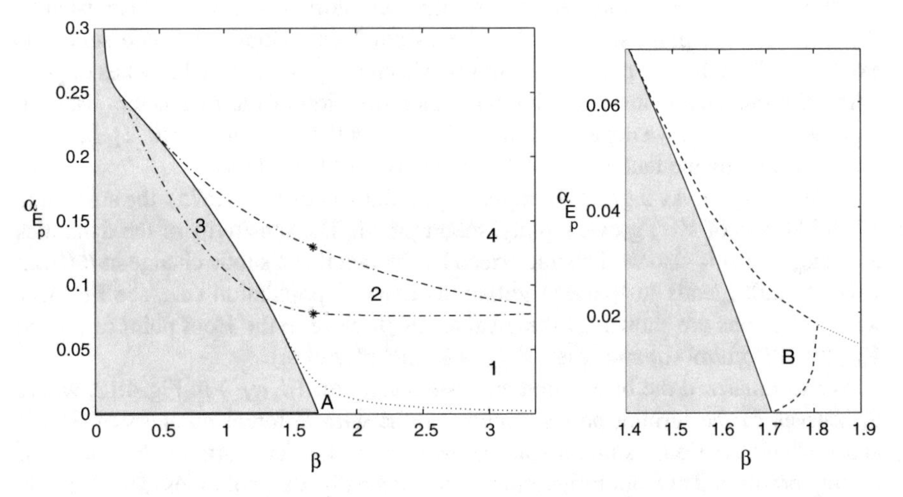

Fig. 4.12 Left: Stability region (depicted in grey) of the steady-state solution and Hopf curve (solid line) as in Fig. 4.9 (left). Two curves of turning points (−·) of periodic solutions. (∗)—two turning points corresponding to $\alpha_{E_p} \approx 0.130$, α_{E_p}, ≈ 0.078 (see [2] for further details). Regions of existence of periodic solutions: 1 stable solution (1, 4), 2 stable and 1 unstable solution (2), 1 stable and 1 unstable solutions (3). The dotted line bounds a region (A) of existence of periodic solutions with $V_{\max} < 1000$ pfu/ml. Right: A blow up of the left figure. The dashed line bounds a region (B) where $V_{\min} > 10$ pfu/ml, i.e. the region with V varying in between 1 and 100 pfu/spleen. $\alpha_{E_e} = 0.3$. Reprinted from Mathematical Biosciences, Vol. 173, Luzyanina et al., Low level viral persistence after infection with LCMV: a quantitative insight through numerical bifurcation analysis, Pages 1–23, Copyright © 2001, with permission from Elsevier

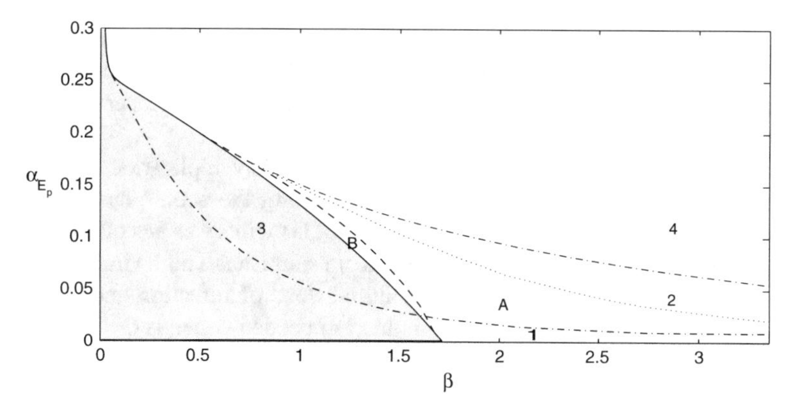

Fig. 4.13 Stability regions of steady-state- and periodic- solutions for $b_p = 10^{-3}$. All notations are analogous to Fig. 4.12. $\alpha_{E_e} = 0.3$. Reprinted from Mathematical Biosciences, Vol. 173, Luzyanina et al., Low level viral persistence after infection with LCMV: a quantitative insight through numerical bifurcation analysis, Pages 1–23, Copyright © 2001, with permission from Elsevier

values of β and α_{E_p} the region where $V_{\min} > 10$ pfu/ml (B) remains quite small. Although two stable periodic solutions coexist in a part of region A, one of them is not biologically realistic because of very large values of the amplitude and the period of oscillations. Note that the upper left part of region A is bounded by the curve of turning points (which ends at a Hopf point), i.e. for the corresponding values of β and α_{E_p} periodic solutions lose stability before $V_{\max}$ reaches 10^3 pfu/ml.

The effect of other parameters can be briefly summarized as follows.

Influence of α_{E_e}. As α_{E_e} decreases from 0.3 (Hopf point) to 0.1, the period of oscillations increases to 22 days and $V_{\max}$ increases to 950 pfu/ml. For $\alpha_{E_e} = 0.1$, Hopf bifurcation occurs at $\beta \approx 1.54$, which implies that for $\alpha_{E_e} = 0.1$ the value of $V_{\max}$ grows from 129 to 950 pfu/ml as β changes from 1.54 to 1.675. Hence, for $\alpha_{E_e} \in [0.1, 0.4]$ the size of the region in (β, α_{E_p})-plane where $V_{\max} < 10^3$ pfu/ml is also quite small, and the location of this region with respect to the corresponding Hopf curve is similar to the one shown in Fig. 4.12.

Influence of τ. As τ increases from 0.6 (Hopf point), the amplitude of oscillations grows rapidly and $V_{\max}$ reaches 10^3 pfu/ml at $\tau \approx 0.8$. At this point the period of oscillations is about 15 days. For $\tau = 0.8$ Hopf bifurcation occurs at $\beta \approx 1.3$, which implies that for $\tau = 0.8$ the value of $V_{\max}$ grows from 129 to 10^3 pfu/ml as β changes from 1.3 to 1.675. Hence for $\tau = 0.8$ the size of the region in (β, α_{E_p})-plane where $V_{\max} < 10^3$ pfu/ml is also quite small and the location of this region with respect to the corresponding Hopf curve (see Fig. 4.9) is similar to the one shown in Fig. 4.12.

Variations of parameters b_d and γ_{VE} within some admissible ranges (see Table 2 in [2] for details) have much lesser impact on the amplitude of oscillations compared to variations of β, α_{E_p} and τ and do not change it significantly.

Overall, we found that the periodic solutions with V varying in between 10 and 10^3 pfu/ml exist in quite narrow intervals of β and α_{E_p} values and the amplitude of oscillations grows rapidly as parameters β, α_{E_p} and τ increase. So the model predicts that oscillatory patterns in low level viral persistence (with virus population varying in between 1 and 100 pfu/spleen) are possible for quite 'special' combinations of the rates of virus growth and precursor CTLs death because of a high sensitivity of the amplitude of oscillations to changes in the above parameters.

The main result of our analysis is that unless LCMV replication rate does not reduce to smaller values, as compared to that during the acute phase of primary infection, a low-level persistence in the face of CTL memory as an equilibrium state is not possible: the virus will either be cleared or establishes a high viral load chronic infection, both outcomes depend on the initial dose of infection and the relative kinetics of viral growth. The extent of reduction needed depends on the responder status of the host, in particular, the lifespan of CTL memory subsets, duration of CTL division cycle, activation thresholds of CTL for proliferation, and differentiation. Since the virus remains the same during acute and persistence phase (it should not acquire attenuating mutations) we propose that the reduction in LCMV replication rate resulting in the low-level persistence could be either due to changes in the host cells, e.g. mediated by type I interferons, or intrinsic features of the virus replicatory cycle [39], which slow down the virus growth. This mechanism seems to be in agreement with virus reappearance after in $CD4^+$ T cell help deficient mice, since the deficiency primarily impairs the LCMV-specific $IFN\gamma$ production by CTLs and $CD4^+$ T cells.

4.2.4 Role of $CD8^+$ T Cells: Protection, Exhaustion, Immunopathology

LCMV infection of mice is a highly dynamic process with high sensitivity to variation in both host and virus parameters: virus control and functional CTL memory versus virus persistence and complete exhaustion of virus-specific CTL precursors reflect the two extreme ends of this spectrum. While both of these outcomes are of limited pathological consequences for the host, extensive T-cell-mediated immunopathology represents an unfortunate intermediate in the balance of virus–host interactions. Important host and virus parameters that determine the outcome of infection include those controlled by MHC and non-MHC genes, presumably affecting T-cell precursor frequencies and T-cell responsiveness, and virus strain, the route and dose of infection affecting the kinetics of initial virus multiplication and virus distribution. Thus, the susceptibility to the establishment of a virus carrier state is increased with lower CTL responses (low responder status) and slower CTL expansion on the one hand and the ability of the virus to spread rapidly and widely on the other hand.

4.2.4.1 How Many Precursor T Cell are Needed to Protect Against Chronic Infection?[3]

The calibrated mathematical model allows the examination of the effects of variations in virus dose and initial CTLp number on the phenotype of the LCMV infection. Two basic outcomes of the infection can be assessed: virus clearance, i.e. virus titer on day 20 (V_{min}) less than the detection limit of 30 pfu per gram of spleen associated with an elevated number of CTLp versus virus persistence ($V_{min} \geq 30$ pfu/g of spleen) and exhaustion of virus-specific CTL.

The impact of variations in the initial number of virus-specific CTLp on control of early virus spread is shown in Fig. 4.14 (left). The effect can also be studied experimentally. To this end C57BL/6 mice were adoptively transfused with 10^7 spleen virus antigen-specific CTLp from TCR-P14 mice (closed symbols) or left unmanipulated (open symbols). One day later, mice were infected with 500 pfu LCMV-Docile and splenic virus titers were determined daily thereafter in two to three mice per group (Fig. 4.14 centre). The results show that a 1000-fold elevation in the number of CTLp reduces the time until virus clearance was achieved by about 2–3 days. The peak virus concentration reached during the course of the infection is also reduced about 2–3 orders of magnitude. The model and data consistently predict that while further increases in CTLp have an only limited effect, any decrease in the number of CTLp below that of a C57BL/6 mouse results in significant increase in the time until virus elimination is achieved (Fig. 4.14 right).

The model can be further used to predict the impact of variation in the number of virus-specific CTLp on the prevention of virus persistence. The results are sum-

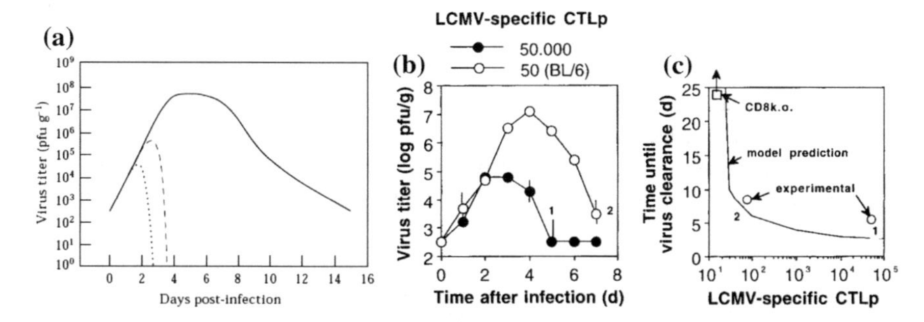

Fig. 4.14 Impact of variation in the number of virus-specific CTLp on the kinetics of virus clearance. Left: Mathematical model prediction. Centre: Experimental assessment. Right: The relation of the number of initial naive LCMV-specific CTL precursors and the time needed to clear virus from the spleen as predicted by the model and validated experimentally. Reprinted from Cellular Immunology, Vol. 189, Ehl et al., The Impact of Variation in the Number of CD8+T-Cell Precursors on the Outcome of Virus Infection, Pages 67–73, Copyright © 1998, with permission from Elsevier

[3]Material of this subsection uses the results of our studies from Ehl et al., The impact of variation in the number of CD8+T-cell precursors on the outcome of virus infection. Cell. Immunol. 189, 67–73, Copyright © 1998, with permission from Elsevier.

marized in Fig. 4.15, right. In particular, the model predicts that a minimal threshold number of about 25 splenic virus-specific CTLp is needed to prevent virus persistence after infection with 1 pfu. Some of the model-generated predictions were tested experimentally as described below. To address the question of whether and how efficiently an increase in the initial number of virus-specific naive CTL precursors can protect against the establishment of virus persistence, C57BL/6 mice were adoptively transfused with different numbers of spleen cells from TCR-P14 mice such that the splenic CTLp number was varied in the range from 50 to 50×10^3 cells. One day later, the recipient mice were infected with varying doses of LCMV-Docile i.v. and 20 days after infection, virus titers were determined in the spleen and LCMV-specific CTL activity was assessed after restimulation in vitro. Overall, the following conclusions can be made (see Fig. 4.15, left):

- a minimal threshold number of about 2550 naive LCMV-specific CTL precursors are necessary for control of infections in the range of $1 - 10^4$ pfu;
- with a tenfold higher dose, a 100-fold increase is required to restore virus control;
- in high-dose infection (above 10^6 pfu), elevations in CTLp were found to be detrimental as they changed the outcome of infection from harmless virus persistence to lethal immunopathology.

In the range where the model predictions could be tested, they were in good agreement with observational data and supported the conclusion that above a certain threshold increases in the number of naive CTLp must be enormous in order to improve virus control. However, the limiting parameter for the efficacy of CTL-mediated virus control is not only the achievement of a critical CTL number in relation to the number of virus-infected cells. Of equal importance is the time required for CTL to mature to be antivirally protective, i.e. the earliest time point when the CTL can efficiently eliminate a population of infected target cells.

The opportunity to compare the model predictions and experimental allows one to define the limitations of the model as a predictive tool related to the fact that it neglected virus spread outside the spleen. While this assumption is presumably justified for low-dose infection, it is responsible for the fact that the model does not account for the significant immunopathology observed after infections with higher doses.

Since the model neglects spread of virus to extralymphatic organs, it is not suited to predict the extent of immunopathology associated with virus clearance from these tissues. The model requires organ-oriented extention to be relevant for examination of the balance between protection and immunopathology by effector memory versus naive precursor CTLs against intravenous or peripheral infections.

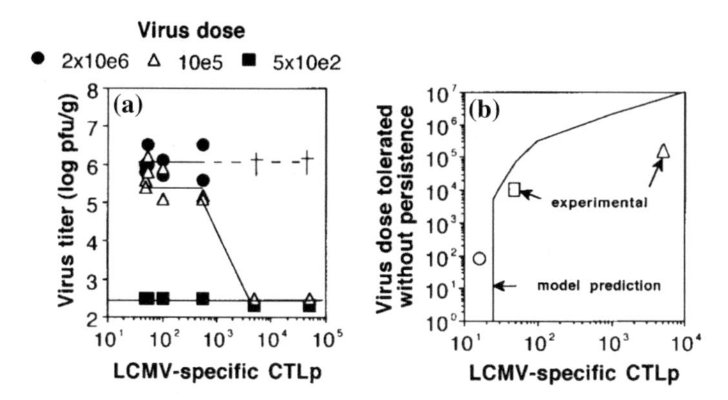

Fig. 4.15 Impact of variation in the number of virus-specific CTLp on the prevention of virus persistence. Left: Experimental data on splenic virus titers determined in surviving mice 20 days after infection for the indicated doses and initial CTLp number. Right: model prediction on the relation of the number of initial naive LCMV-specific CTL precursors and the maximum dose of virus that can still be eliminated from the spleen within 20 days after infection. Reprinted from Cellular Immunology, Vol. 189, Ehl et al., The Impact of Variation in the Number of CD8+T-Cell Precursors on the Outcome of Virus Infection, Pages 67–73, Copyright © 1998, with permission from Elsevier

4.2.4.2 Modelling LCMV-Associated Liver Disease[4]

Infection of mice LCMV represents an example of a systemic infectious process, where the localization, dose and time of availability of virus antigens are important parameters determining the outcome of infection by affecting the antiviral immune response and pathological consequences of the cytotoxic T lymphocyte- (CTL) mediated destruction of virus-infected cells [40, 41]. It provides an experimental model system for studying diseases mediated by cytotoxic activity of effector CTL against cells expressing virus antigen such as diabetes [42, 43], aplastic anemia [44], choriomeningitis [45], liver disease [46], to name just few of them. A classical example is the LCMV-WE-induced liver hepatitis in mice [47].

The problem closely related to systemic virus spread is CTL-mediated immunopathology. This depends on the extent of virus distribution in peripheral tissues as well as the relative kinetics of the CTL response and is an important determinant of the outcome of infection. Virus-induced CTL responses represent heterogeneous populations of cells in different activation and differentiation states: activated cycling, cytolytic effector and quiescent memory cells. These subsets differ essentially in their function and ability to migrate to peripheral sites of infection. Because LCMV is non-cytopathic, virus clearance from the host through CTL-mediated perforin-dependent destruction of infected cells is always associated with a varying degree of immunopathology. Under small infectious doses virus replication

[4]Material of this subsection uses the results of our studies from Bocharov et al., Modelling the dynamics of LCMV infection in mice: II. Compartmental structure and immunopathology. J. Theor. Biol. 221, 349–78, Copyright © 2003, with permission from Elsevier.

is localized mainly to the spleen so that virus-triggered CTL-dependent destruction of infected antigen-presenting cells (macrophages and dendritic cells) is manifested as acquired immune suppression [48, 49]. Large doses of systemic or any peripheral infections lead to a wider spread of the virus in the host and can induce other types of immunopathology, depending on the tissue damage involved (e.g. choriomeningitis). The type of pathology in natural and experimental systems depends on

- the route of infection (intravenous, intracerebral, intrahepatic, etc.),
- tissue tropism of the virus,
- the dose of infection and
- immune status of the host.

All these factors interact nonlinearly to produce various infection outcomes ranging from virus elimination to lifelong persistence. A quantitative characterization and prediction of the outcome of infection in murine LCMV system requires consideration of the three aspects of infection:

1. systemic virus spread,
2. lymphocyte migration during immune responses to tissue sites outside the spleen and
3. the pathological consequences of virus elimination via perforin-dependent CTL-mediated destruction of infected cells.

In this section, we formulate a mathematical model to investigate the demands to CTL memory for protection against LCMV infection with minimal immunopathology. To address the immunopathology question, the basic model of LCMV infection in spleen was extended to consider additional organs, i.e. blood and liver. Such extension should allow to examine the severity of LCMV-associated CTL-induced hepatitis.

Formulation of a multi-compartmental mathematical model integrating the kinetics of LCMV spread in various tissues of mice with effector CTL activation and trafficking allow one to specify the parameters which have to be achieved for CTL vaccination/immunization to ensure virus elimination with minimal immunopathology versus vaccination for disease. To keep the mathematical model in accord with what is experimentally controlled [47], one can consider the dynamics of two enzymes signalling liver cells destruction, $AST(t)$ and $ALT(t)$ as disease characteristics.

The mathematical model for CTL-mediated hepatitis in LCMV infection considers the population dynamics of infection and immune response in three organs (compartments), i.e. the blood, spleen and liver, as shown in Fig. 4.2. The compartmental structure of the model is formulated as a linear mamillary compartmental system [50]:

$$\frac{d}{dt}\mathbf{y}(t) = \mathbf{M} \times \mathbf{y}(t), \tag{4.11}$$

where $\mathbf{y}(t)$ is a state vector of spatially distributed species.

The corresponding set of differential equations for LCMV-induced hepatitis considers the population dynamics of

- virus titer in spleen, blood and liver: $V_{Spleen}(t),\ V_{Blood}(t),\ V_{Liver}(t)$;

- precursor CTLs in spleen: $E_p(t)$;
- recirculating effector CTLs in spleen, blood and liver: $E_{e,Spleen}(t)$, $E_{e,Blood}(t)$, $E_{e,Liver}(t)$;
- cumulative viral load in spleen: $W(t)$;
- liver enzymes levels in blood: $AST(t)$, $ALT(t)$.

The basic model of LCMV infection in spleen (developed in Sect. 4.2) has to be modified and extended to take into account the virus transfer between blood and spleen as well as the recirculation of effector CTLs between blood and spleen. The procedure is outlined in detail in [51]. Implementing a building block approach, it can be described in the following structured form. *Splenic LCMV infection module*:

$$
\begin{cases}
\frac{d}{dt} V_{Spleen}(t) = \mu_{SB} V_{Blood}(t) - \mu_{BS} V_{Spleen}(t) \\
\qquad + \beta V_{Spleen}(t)(1 - \frac{V_{Spleen}(t)}{V_{mvc,Spleen}}) - \gamma_{VE} E_{e,Spleen}(t) V_{Spleen}(t)/(1 + E_{e,Spleen}(t)/\theta_{VE,S}), \\
\frac{d}{dt} E_p(t) = \alpha_{E_p}(E_p^0 - E_p(t)) + \frac{b_p}{(1+W(t)/\theta_p)^2(1+(E_p(t)+E_{e,Spleen}(t))/E_{PE}^{Sat})} V_{Spleen}(t-\tau) E_p(t-\tau) \\
\qquad - \alpha_{AP} V_{Spleen}(t-\tau_A) V_{Spleen}(t) E_p(t), \\
\frac{d}{dt} E_{e,Spleen}(t) = \eta_{SB} E_{e,Blood}(t) - \eta_{BS} E_{e,Spleen}(t) \\
\qquad + \frac{b_d}{(1+W(t)/\theta_E)^2(1+(E_p(t)+E_{e,Spleen}(t))/E_{PE}^{Sat})} V_{Spleen}(t-\tau) E_p(t-\tau) - b_{EV} V_{Spleen}(t) E_{e,Spleen}(t) \\
\qquad - \alpha_{AE} V_{Spleen}(t-\tau_A) V(t) E_{e,Spleen}(t) - \alpha_{E_e} E_{e,Spleen}(t) \\
\qquad - \alpha_{PCD}/(1 + V_{Spleen}(t)/\theta_{PCD}) E_{e,Spleen}(t), \\
\frac{d}{dt} W(t) = b_W V_{Spleen}(t) - \alpha_W W(t).
\end{cases}
\tag{4.12}
$$

Note that the equation for the effector CTLs in spleen has additional terms which describe saturation of CTL expansion rate at high population densities and passive effector cell death under condition of limiting antigen in the spleen. We assume $b_{EV} = 0$.

Additional equations for virus dynamics in blood and liver are as follows:

$$
\begin{cases}
\frac{d}{dt} V_{Blood}(t) = \mu_{BS} V_{Spleen}(t) + \mu_{BL} V_{Liver}(t) - (\mu_{SB} + \mu_{LB} + \varepsilon_B) V_{Blood}(t) \\
\frac{d}{dt} V_{Liver}(t) = \mu_{LB} V_{Blood}(t) - \mu_{BL} V_{Liver}(t) \\
\qquad + \beta V_{Liver}(t)(1 - \frac{V_{Liver}(t)}{V_{mvc,Liver}}) - \gamma_{VE} E_{e,Liver}(t) V_{Liver}(t)/(1 + E_{e,Liver}(t)/\theta_{VE,L})
\end{cases}
\tag{4.13}
$$

The module describing the recirculation of effector CTLs between blood and liver is

$$
\begin{cases}
\frac{d}{dt} E_{e,Blood}(t) = \eta_{BS} E_{e,Spleen}(t) + \eta_{BL} E_{e,Liver}(t) - (\eta_{LB} + \eta_{SB} + \delta_B) E_{e,Blood}(t) \\
\frac{d}{dt} E_{e,Liver}(t) = \eta_{LB} E_{e,Blood}(t) - (\eta_{BL} E_{e,Liver}(t) + \delta_L) E_{e,Liver}(t).
\end{cases}
\tag{4.14}
$$

The equations for enzymes dynamics in blood are as follows:

$$
\begin{cases}
\frac{d}{dt} AST(t) = \rho_{AST} V_{Liver}(t)) E_{e,Liver}(t)/(1 + E_{e,Liver}(t)/\theta_{VE,L}) \\
\qquad - \alpha_{AST} AST(t) \\
\frac{d}{dt} ALT(t) = \rho_{ALT} V_{Liver}(t)) E_{e,Liver}(t)/(1 + E_{e,Liver}(t)/\theta_{VE,L}) \\
\qquad - \alpha_{ALT} ALT(t).
\end{cases}
\tag{4.15}
$$

The model was calibrated using diverse sets of published data as described in details in [51]. The parameters of the model are listed in Table 4.2.

Table 4.2 List of the model parameters estimated for systemic LCMV-WE infection in CB57BL/6 mice

Parameter	Biological meaning	Units	Best-fit estimate
β	Replication rate constant of viruses in spleen	1/day	4.7
V_{mvc}	Maximal virus concentration in spleen	pfu/ml	6.5×10^8
β	Replication rate constant of viruses in liver	1/day	2.1
V_{mvc}	Maximal virus concentration in spleen	pfu/ml	3.0×10^8
γ_{VE}	Rate constant of virus clearance due to effector CTLs	ml/(cell day)	2.5×10^{-5}
θ_{VE}	CTL number of half-maximal virus clearance rate	cell/ml	2.6×10^5
E_p^0	Homeostatic concentration of LCMV-specific CTLs in spleen of unprimed mouse	cell/ml	1100
α_{E_p}	Rate constant of natural death for precursor CTLs	1/day	0.068
b_p	Rate constant of CTL stimulation	ml/(pfu day)	2×10^{-3}
τ	Duration of CTL division cycles	day	1.0
b_d	Rate constant of CTL differentiation	ml/(pfu day)	2×10^{-2}
θ_p	Cumulative viral load threshold for anergy induction in precursor CTLs (proliferation process)	pfu/ml	1×10^6
θ_E	Cumulative viral load threshold for anergy induction in effector CTLs (differentiation process)	pfu/ml	5.5×10^5
α_{PCD}	Rate constant of effector CTL death after virus clearance below a threshold	l/day	0.3

(continued)

Table 4.2 (continued)

Parameter	Biological meaning	Units	Best-fit estimate
θ_{PCD}	Extent of virus elimination at which the passive cell death is in effect	l/day	1.0
α_{E_e}	Rate constant of natural death for effector CTLs	pfu/ml	0.068
τ_A	Duration of commitment of CTLs for apoptosis	day	9.1
α_{AP}	Rate constant of apoptosis for precursor CTLs	$(ml/pfu)^2$/day	1.0×10^{-13}
α_{AE}	Rate constant of apoptosis for effector CTLs	$(ml/pfu)^2$/day	3×10^{-14}
b_W	Rate constant of viral load increase	1/day	1.7
α_W	Rate constant of restoration from the inhibitory effect of virus load	1/day	0.4
E_{EP}^{Sat}	Saturation rate constant for CTL expansion	cell/ml	1.0×10^7
ρ_{AST}	Rate constant of AST release into blood from CTL destroyed infected liver cell	U/l/(pfu cell day) ml^2	1.0×10^{-9}
ρ_{ALT}	Rate constant of ALT release into blood from CTL destroyed infected liver cell	U/l/(pfu cell day) ml^2	0.7×10^{-9}
α_{AST}	Decay rate of AST in blood	1/day	0.5
α_{ALT}	Decay rate of ALT in blood	1/day	0.5

Table 4.3 Transfer rates (hr^{-1}) of LCMV between Blood-, Spleen- and Liver compartments

Organ	Blood	Spleen	Liver
Blood	−0.74	0.33×10^{-3}	0.27×10^{-4}
Spleen	0.5×10^{-3}	-0.33×10^{-3}	0
Liver	0.74×10^{-2}	0	-0.27×10^{-4}

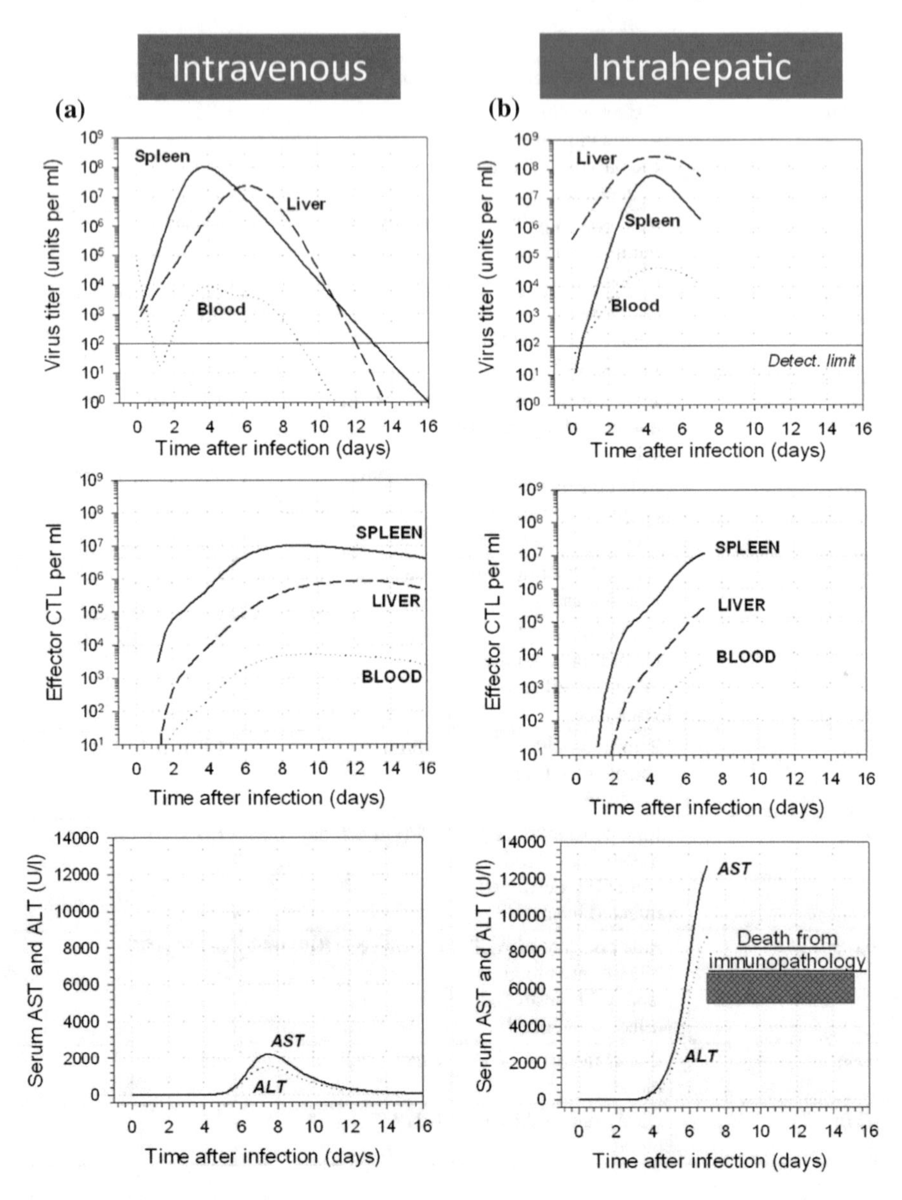

Fig. 4.16 Model prediction for the dynamics of liver disease associated with LCMV infection of C57BL/6 mice. Two qualitatively different routes of application of 2×10^5 pfu of LCMV-WE are considered: **a** systemic infection leads to acute infection with modest immunopathology and elimination of the virus; **b** peripheral route of infection via liver leads to severe immunopathology and death of the animal by day 6 post-infection. Reprinted from Journal of Theoretical Biology, Vol. 221, Bocharov et al., Modelling the Dynamics of LCMV Infection in Mice: II. Compartmental Structure and Immunopathology, Pages 349–378, Copyright © 2003, with permission from Elsevier

Table 4.4 Trafficking rates (day^{-1}) of effector CTLs between Blood-, Spleen- and Liver compartments

Organ	Blood	Spleen	Liver
Blood	−20	0.25	0.2
Spleen	10	−0.25	0
Liver	10	0	−0.2

The transfer rates, i.e. the elements of compartmental matrix **M**, for virus and effector CTL are specified in Tables 4.3 and 4.4, respectively.

The example of compartmental dynamics of LCMV infection predicted by the model is shown in Fig. 4.16a. It presents the simulation of the CTL-mediated liver disease after i.v. infection of C57BL/6 mice with 2×10^5 pfu of LCMV-WE. The model predicts that: (i) virus growth in the liver proceeds at a slower rate than in spleen and the viremia lasts for about 1 week; (ii) CTL response in spleen eliminates the splenic virus in about 10 days starting from day 4; (iii) it takes 3 days more to overcome virus replication in the liver and this time lag is needed for effector CTL to accumulate in liver above the threshold number, estimated to be 1.24×10^5 cells per ml of liver, for which the basic reproductive ratio of the virus in liver becomes less than one; (iv) the serum enzyme levels start to rise at high rate by day 5 after infection.

The dynamics and outcome of LCMV infection after peripheral route of infection is quite different. In Fig. 4.16b, the simulation of a direct injection of 2×10^5 pfu of LCMV-WE into the liver of C57BL/6 mouse is shown. The model predicts that virus extensively replicates in liver reaching the maximum possible titer of 3×10^8 pfu/ml by day 5, which implies that all target cells get infected. Virus growth in the spleen is decreased and delayed by about 1 day as compared to the i.v. infection, and therefore, the splenic CTL response starts later. By day 5, when effector CTLs accumulate in the liver in large number, the destruction of all the infected hepatocytes results in a fulminant immunopathology as is manifested in the model by the enormous elevation of AST and ALT levels. Therefore, this particular combination of viral and host parameters leads to a lethal outcome.

Adoptive transfer experiments demonstrated that virus-specific CTLs are crucial in production of LCMV-associated hepatitis. We examined the 'dose-effect' relationship between the number of effector CTLs injected into blood from one side and the peak serum AST levels and the time until virus in spleen declines below detection limit of 100 pfu/ml on the other side. The scenario of experimental i.v. infection of a naive C57BL/6 mouse with 2×10^5 pfu of LCMV-WE accompanied with adoptive transfer of effector $CD8^+$ T cells at day 0 was mathematically modelled to determine the maximum serum AST level. The predicted effect of the number of transferred effector CTL and peak AST is shown in Fig. 4.17 Left, (b). It suggests that a higher number of injected effectors decreases the severity of clinical disease, and injection of about 10^3 cells is enough to reduce the AST level below 500 U/l. The time required to eliminate virus below detection limit displays a non-monotone pattern, it declines from 14 to 7 days as the number of transferred effector CTL increases from 0 to 10^5. Further increase of transferred CTL above 4×10^5 cells leads to a rapid virus elimination within 1 day with no signs of disease. Note, that the narrow suboptimal range of transferred CTL represents the situation when the basic reproductive ration of virus infection is close to 1.

The effect of increase in number of virus-specific precursor CTL in spleen (the responder immune status) on the severity of LCMV-WE- induced liver disease and the time of virus elimination is summarized in Fig. 4.17 Left, (c). A lifelong virus persistence and CTL exhaustion are predicted by the model as an outcome of systemic infection with 2×10^5 pfu of mice with less than 20 precursor CTLs in spleen. The minimal number of precursor CTL to clear an infection is about 100 per spleen. The time needed for virus elimination decreases with the increase in the number of precursor CTLs but does not go below 4 days, in contrast to the case of effector CTLs. For initial numbers of precursor CTLs in spleen ranging from 30 to 100 cells the outcome of the high-dose infection would be a severe or fatal hepatitis, reflecting an unfavourable combination of viral and host parameters.

The validation of the model was conducted by comparing its predictions on the virus dose dependence of serum enzyme concentration with experimental observations. The results and data available for C57BL/6 and ICR mice, the last one known as being more susceptible to LCMV-WE-induced hepatitis than C57BL/6 mice, are shown in Fig. 4.17a. The data shown are the averages for 2–4 mice bled at the times indicated. The mathematical model based upon data for LCMV-WE infection of C57BL/6 mice predicts a dose-effect curve which is situated below the data for ICR mice. However, it is consistent with three available data points for C57BL/6 mice representing the severity of infection with 2×10^5 pfu of LCMV-WE. Overall, the model predicts the following functional relationship between the peak AST level in blood and the dose (ranging from 0 to 10^6 pfu) of i.v. infection:

$$AST_{max} \sim \sqrt{V_{blood}(0)} \tag{4.16}$$

that is the severity of the hepatitis increases as a squared root of the infection dose.

The mathematical model can be further used to examine the infection outcome/severity of the liver disease after peripheral LCMV-WE infection with 2×10^5 pfu via liver, i.e. intrahepatic infection. The liver infection of naive C57BL/6 mice would result in severe hepatitis for doses ranging in between 10 and 10^6 pfu, see Fig. 4.17 Right, (a). One might try to prevent this unfavourable outcome by adoptively transferring LCMV-specific effector or precursor CTLs. The impact of effector CTL is presented in Fig. 4.17 Right, (b). First, there exists a threshold number of effector CTL $\sim 2 \times 10^5$ cells conferring an immediate type of protection against virus spread and severe disease. With the CTL number above the threshold, the LCMV population is eliminated in less than 1 day, with no signs of the disease. If the number of transferred lytic CTL is below threshold then protection against the liver disease is not conferred, although the virus is likely to be eliminated.

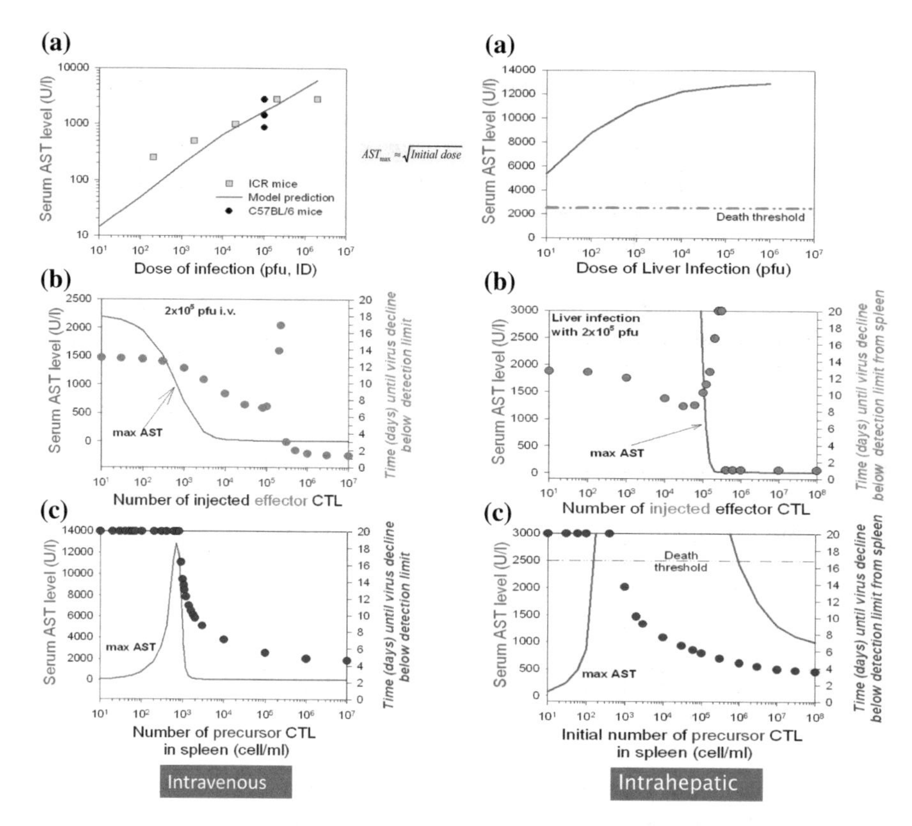

Fig. 4.17 Analysis of the impact of viral and immune status parameters on the essential characteristics of LCMV-WE infection associated liver disease. Left: Intravenous infection with 2×10^5 pfu is considered. Left: I.v. infection. **a** severity of the disease in terms of peak AST level as a function of initial virus dose; **b** effect of the initial number of virus-specific cytolytic effector $CD8^+$ T cells injected into blood at day 0 of infection on the serum AST level and the time until virus is eliminated below detection level of 100 pfu/ml; **c** effect of the initial number of precursor $CD8^+$ T cells present in spleen at day 0 of infection on the serum AST level and the time until virus is eliminated below detection level of 100 pfu/ml. Right: Peripheral infection. **a** severity of the disease in terms of peak AST level as a function of initial virus dose; **b** effect of the initial number of virus-specific cytolytic effector $CD8^+$ T cells injected into blood at day 0 of infection on the serum AST level and the time until virus is eliminated below detection level of 100 pfu/ml; **c** effect of the initial number of precursor $CD8^+$ T cells present in spleen at day 0 of infection on the serum AST level and the time until virus is eliminated below detection level of 100 pfu/ml. Reprinted from Journal of Theoretical Biology, Vol. 221, Bocharov et al., Modelling the Dynamics of LCMV Infection in Mice: II. Compartmental Structure and Immunopathology, Pages 349–378, Copyright © 2003, with permission from Elsevier

Figure 4.17 right, (c) predicts the effect of the LCMV-specific precursor CTLs present in the spleen at the moment of infection on the peak AST level and virus elimination time. Low numbers (less than 20 cells per spleen) of CTLp are associated with virus persistence, CTL exhaustion and no symptoms of hepatitis. For the initial number of CTLp in between 20 and 10^5 cells a severe or fatal hepatitis would be an outcome of the intrahepatic infection with 2×10^5 pfu. Only the population of splenic precursor CTLs larger than 10^5 cells would provide protection against virus persistence, but at the expense of a marked damage of the infected liver. Even with 10^7 LCMV-specific precursor CTLs in the spleen, the mouse would need at least 4 days to eliminate the virus and the associated immunopathology would still be above 1000 U/l. This time lag is required for them to get activated and generate sufficient number of effector CTLs.

Overall, a 'Complete' characterization of the outcome of virus–host organism interaction with a mathematical model requires consideration of not only the immune response and viral dynamics, but also some characteristics of tissue damage. A new 'spatial' dimension can be introduced into the model via compartmental analysis.

The extended model quantitatively predicts that there is a range for the initial number of precursor CTLs in spleen for which an elevation in the clonal size is accompanied by an increase of disease immunopathology. Thus, it overcomes the predictive limitations of a single-compartmental model as discussed in Sect. 4.2 and reflects what was described experimentally as 'vaccination for disease' [16].

4.3 Parameters Defining a Robust DC-Induced CTL Expansion[5]

Successful vaccination depends on the availability of specific antigens, efficient delivery of these antigens, and their optimal presentation to T cells within secondary lymphoid organs. The growing knowledge of the molecular identity of tumour-specific antigens has opened new avenues for effective cancer vaccines [52]. Immunotherapeutic approaches based on adoptive transfer of dendritic cells (DC) expressing relevant antigens may be used for active mobilization of cellular immune responses (CTLs, T-helper cells and NK cells) against tumours. DC-based immunotherapeutic approaches appear particularly promising because DC migrate to the T-cell zones of secondary lymphoid organs where they efficiently initiate both Th and CTL responses [53, 54]. The extraordinary efficacy of DC to prime immune responses is shown by the fact that only $10^2 - 10^3$ antigen-presenting DCs in the spleen are sufficient to achieve protective levels of CTL activation in mice [55]. A series of preclinical experimental studies in mice demonstrated that anti-tumour immunity can be induced using DC [56–59]. This preclinical experience has been translated into the performance of a variety of clinical trials, which have shown that application of DC is safe and that clinical efficacy of this treatment strategy can be obtained [60–62].

The efficacy of this active immunization depends on the complex biology of the DC life cycle and their interaction with T cells. The kinetics of this interaction and its sensitivity to relevant parameters are still incompletely understood. These parameters include antigen loading, DC maturation stage, frequency and route of DC injection,

[5] Material of this section uses the results from Ludewig et al., Eur J Immunol. 34 (2004), 2407–18.

frequency and activation status of T cells, and the homing rate of DC to and their persistence within lymphoid tissues. However, the major quantitative parameters of the DC–CTL interaction (e.g. the elimination kinetics of DC by CTL, the threshold for T cell activation, and the impact of DC on T cell homing and recirculation) require further analysis.

In this section, we present one possible approach to modeling the interaction of DCs with CTLs. In the presented model, the DC–CTL interaction is described by adapting different theoretical frameworks, such as predator–prey models from population biology and Monod-type kinetics with saturation which are applied in biochemistry. We are considering the underlying processes at the macroscopic level of the whole immune system via a compartmental approach and aim to produce a meaningful mathematical model that is both descriptive and predictive. Using a combination of experimental in vivo work and mathematical modelling, we examine here the systemic aspects of DC–CTL interactions. The interdisciplinary approach presented below is composed of three major segments:

1. initial data collection and model establishment by data assimilation;
2. evaluation of effects of varied parameters in a range that is easily accessible to the model prediction but not experimental measurement;
3. model predictions on DC-based immunization and experimental validation.

4.3.1 The Experimental Model of LCMV gp33-Specific CTL Induction

The experimental murine system based on priming of $CD8^+$ T cells specific for the immunodominant gp33-peptide of the lymphocytic choriomeningitis virus (LCMV) glycoprotein presented by DC proved to be valuable in assessing relevant parameters of CTL induction and maintenance [56, 63, 64]. Reliable input from experimental or clinical research in terms of precise and comprehensive data sets is a core part of an interdisciplinary modelling approach. The data set for model-driven analysis was generated using established protocols [55, 63]. Briefly, major histocompatibility complex (MHC) class I tetramers complexed with the immunodominant CTL epitope (gp33) derived from the glycoprotein of the lymphocytic choriomeningitis virus (LCMV-GP) were used to follow activation of gp33-specific CTLs after immunization with DC. DCs derived from transgenic mice ubiquitously expressing the first 60*aa* of LCMV-GP including gp33 (H8-DC) were injected intravenously into naive C57BL/6 recipient mice. At the specific time points following immunization, the densities of the following cell populations as a function of time t were determined:

- Activated $CD8^+$ $62L^-$ T-cells staining with the gp33-tetramer (tet+) in spleen
- Quiescent $CD8^+CD62L^+tet^+$ cells in spleen $E_m(t)$;
- The availability of adoptively transferred DC for productive interaction with T-cells within secondary lymphoid organs was quantified. To this end 51Cr-labelled H8-

DC were injected i.v. into naive recipient mice, and the accumulated radioactivity was determined in spleen at different time points using established protocols [65].

The data set for homing of adoptively transferred DC from blood to spleen has been published elsewhere [84].

4.3.2 Mathematical Model for DC-Induced Systemic Dynamics of CTL Responses

Mathematical models for the interaction of antigen-presenting cells (APC) and T cells developed so far, consider mainly the stimulatory aspects of the interaction of APC and T cells [67, 68]. However, CTL-mediated killing of the antigen-presenting DC is probably a key process in the downregulation of adaptive immune responses [63, 69, 70]. The positive amplification effect of antigen-presenting DC on the CTL population and the negative feedback from CTL on DC numbers implies that the cell population dynamics of the CTL–DC system in vivo most likely reflects a predator–prey type of interaction (Fig. 4.18). Mathematical modelling facilitates the analysis of the following issues: (i) suitability of the predator–prey-type framework for the dynamics of the DC–CTL system in vivo; (ii) estimation of thresholds for DC-mediated CTL induction and trafficking; (iii) analysis of sensitivity of CTL dynamics to various parameters (e.g. half-life of DC and the initial number of precursor T cells); and (iv) role of TCR avidity in the robustness of CTL priming.

To formulate equations for DC–CTL interaction following i.v. injection, we make the following simplifying biological assumptions. Such a list is also helpful for the evaluation of the modelling results from the viewpoint of the underlying biology. A conceptual model for the predator–prey-type induction/regulation of CD8+ T-cell responses by dendritic cells is shown in Fig. 4.18. Antigen-expressing DC migrate from blood to spleen, where they induce clonal expansion of nave antigen-specific cytotoxic T-lymphocytes (CTL), whereas activated CTL eliminates DC. Arrows indicate the modeled processes. The structure of the model equations is based on the following assumptions:[6]

1. DC do not recirculate from lymphoid organs into the blood after intravenous injection.
2. Adoptively transferred DC are in mature state.
3. DC-mediated induction of antigen-specific CTL is due to their interaction in the spleen.
4. DC do not divide in secondary lymphoid organs.
5. DC decay due to a short lifespan and their killing by activated CTL.

[6](see for details Bocharov et al., (2005): A Mathematical Approach for Optimizing Dendritic Cell-Based Immunotherapy. In: *Adoptive Immunotherapy. Methods and Protocols*, Eds. Ludewig B. and Hoffmann M.W. (Humana Press) **109**: 19–34).

6. The population of antigen-specific CTLs in spleen is split into quiescent (nave or central memory-like) and activated CTL (effector or effector memory-like).
7. CTL recirculate among spleen, blood and peripheral organs (e.g. liver).

To formulate the systemic model, we follow a building block approach and calibrate submodels (1) for initial DC distribution, (2) DC–CTL population dynamics in spleen, and (3) the compartmental dynamics of CTL responses.

4.3.2.1 Initial DC Migration

I.v. injection of DCs leads to one way migration to the peripheral organs, i.e. spleen, liver, lung and others. The rate of chage of the DCs population is described by

$$\frac{d}{dt} D_{Blood}(t) = -\left(\mu_{BS} + \mu_{BL} + \mu_{BLu} + \mu_{BO}\right) \cdot D_{Blood}(t) \tag{4.17}$$

$$\frac{d}{dt} D_{Spleen}(t) = \mu_{BS} \cdot D_{Blood}(t) \frac{Q_{Blood}}{Q_{Spleen}} \tag{4.18}$$

$$\frac{d}{dt} D_{Liver}(t) = \mu_{BL} \cdot D_{Blood}(t) \frac{Q_{Blood}}{Q_{Liver}} \tag{4.19}$$

$$\frac{d}{dt} D_{Lung}(t) = \mu_{BLu} \cdot D_{Blood}(t) \frac{Q_{Blood}}{Q_{Lung}} - \mu_{LuO} \cdot D_{Lung}(t) \tag{4.20}$$

The transfer parameters estimated from experimental data are gived in Table 4.5.

4.3.2.2 DC–CTL Interaction in Spleen

The submodel for DC and CTL interaction in spleen reflects a biological view of the processes as depicted in Fig. 4.18.

The rate of change in the density of DC in the spleen is modelled as

Table 4.5 Transfer rates (hr^{-1}) of DC between Blood-, Spleen-, Liver- and Lung compartments

Organ	Blood	Spleen	Liver	Lung
Blood	−1.124	0	0	0
Spleen	0.12	0	0	0
Liver	0.38	0	0	0
Lung	0.16	0	0	−0.0911

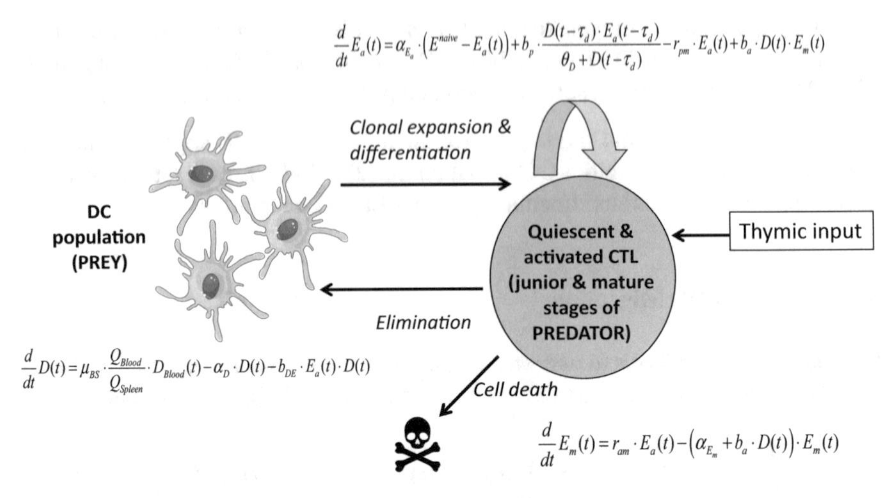

Fig. 4.18 Scheme of the predator–prey type interaction between CTL responses and DCs in spleen used for the model formulation. Arrows indicate the modelled processes which appear as individual terms on the right-hand sides of specified equations. *Cell Pictures taken from Servier Medical Art*

$$\frac{d}{dt}D(t) = \mu_{BS} \cdot \frac{Q_{Blood}}{Q_{Spleen}} \cdot D_{Blood}(t) - \alpha_D \cdot D(t) - b_{DE} \cdot E_a(t) \cdot D(t). \quad (4.21)$$

The first term represents the trafficking of DCs from blood to spleen (Q_{Blood} and Q_{Spleen} being the volumes of the blood and spleen compartments, respectively), and the other two take into account the natural death of the cells and their elimination by activated CTLs.

The dynamics of activated CTLs is modelled by the following equation:

$$\frac{d}{dt}E_a(t) = \alpha_{E_a} \cdot \left(E^{naive} - E_a(t)\right) + b_p \cdot \frac{D(t-\tau_d) \cdot E_a(t-\tau_d)}{\theta_D + D(t-\tau_d)} - r_{am} \cdot E_a(t) + b_a \cdot D(t) \cdot E_m(t) \quad (4.22)$$

The first term considers the homeostasis of naive CTLs in the spleen, the second term represents the DC-induced division of CTLs proceeding at the rate that saturates at a high number of DCs. The time lag between the cognate interaction of CTL with DC represents the duration of pre-programming of CTL for division and differentiation. The last two terms take into account the silencing of activated CTLs into quiescent memory cells (third term) and the activation of the memory cells by DCs.

The equation for the dynamics of quiescent memory CTLs is

$$\frac{d}{dt}E_m(t) = r_{am} \cdot E_a(t) - \left(\alpha_{E_m} + b_a \cdot D(t)\right) \cdot E_m(t) \quad (4.23)$$

which considers the transition of the activated CTLs into the quiescent memory state, the death of memory CTLs at some slow rate, and the activation of memory CTLs depending on the availability of DCs.

The above model for DC–CTL interaction in the spleen was fitted to the experimental data sets to estimate the parameters via a maximum likelihood approach. We assumed that the observational errors of the data follow a log-normal distribution and are independent between cell populations. The relevant information about the model parameters governing the DC–CTL interaction in the spleen is summarized in Table 4.6. The corresponding solution of the model is shown in Fig. 4.19.

The model predicts that the threshold of DC density for half-maximal CTL expansion rate in the spleen is about 200 cells per spleen which explains the rather small effects in the chosen dose range on the magnitude of the CTL response. The amplification factor of the CTL expansion is about 12 cells per day implying that the pre-programming effect probably lasts for three to four divisions. The estimate of per capita CTL-mediated elimination rate of the DC (b_{DE}) suggests that the threshold number of activated CTLs eliminating about 50% of antigen-presenting DCs per day is about 1.4×10^4 per spleen. Furthermore, the model predicts that about 7% of the activated CTLs enter the memory pool.

4.3.2.3 Compartmental Dynamics of CTL Responses

The next step is to extend the spleen-localized model to the systemic dynamics of a DC-induced CTL response according to the scheme shown in Fig. 4.20. The extended model considers the dynamics of DC–CTL interactions in spleen and CTL recirculation between spleen, blood and liver. The trafficking of both activated and quiescent antigen-specific CTLs between the spleen, blood and liver is described in a uniform way as a nonlinear compartmental system. Using the vector notation for the CTL subsets densities in the above compartments $\mathbf{E}_i(t) = \left[E_i^{Blood}(t), E_i^{Spleen}(t), E_i^{Liver}(t)\right]^T$, $i = a, m$

$$\frac{d}{dt}\mathbf{E}_i(t) = \mathbf{M}^{E_i} \cdot \mathbf{E}_i(t)(t) + \mathbf{I}_i(t), \tag{4.24}$$

where $\mathbf{E}_i(t)$ is a state vector of organ distributed CTLs in activated and memory states. Here the compartmental matrix stands for CTL inter-compartmental transfer rates.

$$\mathbf{M}^{E_i} = \begin{pmatrix} -\mu_{BB} & \mu_{SB}(D_{Spleen}(t)) & \mu_{LB} \\ \mu_{BS} & -\mu_{SB}(D_{Spleen}(t)) & 0 \\ \mu_{BL} & 0 & -\mu_{LB} \end{pmatrix}, \text{ with } \mu_{SB}(D_{Spleen}(t)) = \mu_{SB}^* + \frac{\Delta\mu}{1+D_{Spleen}(t)/\theta_{shut}}.$$

Here, the DC-dependent migration rate from the spleen to the blood takes into account the trapping effect. The input/output vector-function $\mathbf{I}_i(t) = \left[0, (\text{division} - \text{death})_{Spleen}, 0\right]^T$ represents the contribution of DC-induced CTL responses in the spleen.

The estimated trafficking rate parameters for CTLs are listed in Table 4.7. The computed curves of CTL dynamics versus the experimental data are shown in Fig. 4.21. A critical feature for the systemic response is that CTL transfer rates from spleen to blood appear to be DC-density dependent. To describe the observed CTL

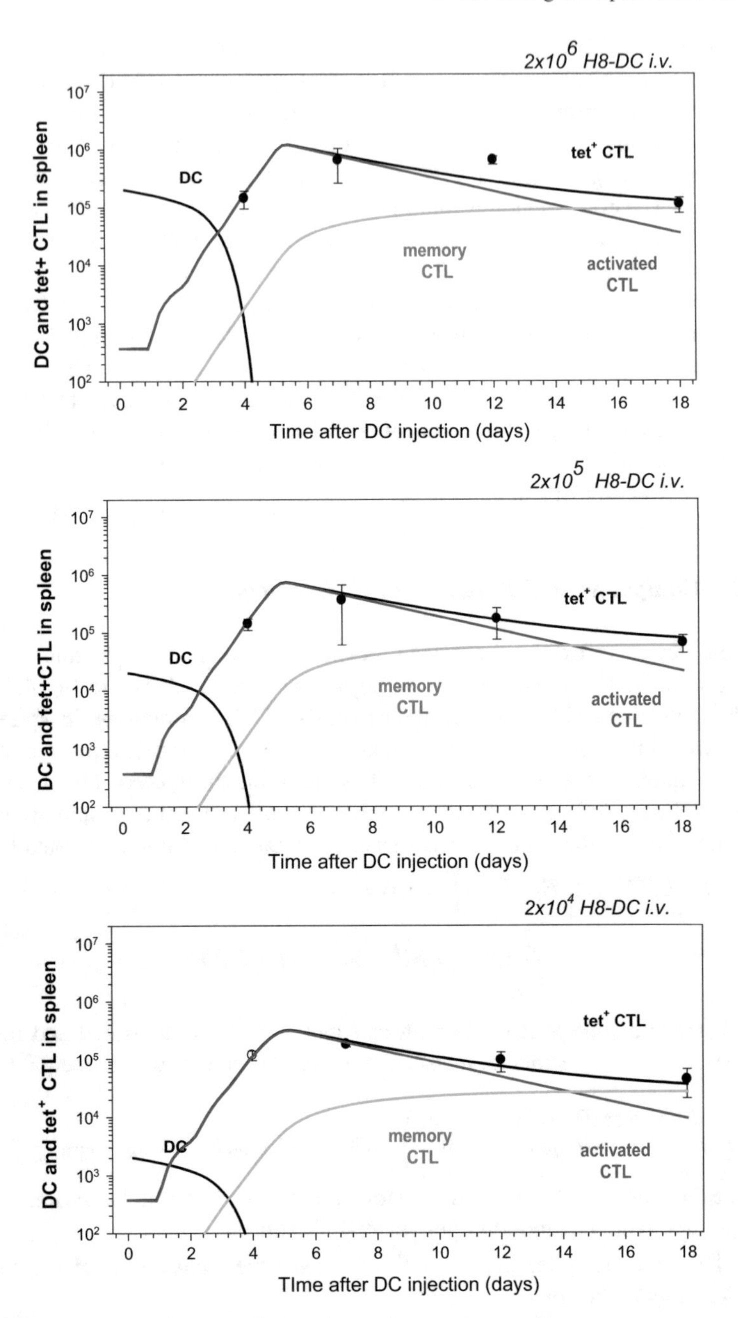

Fig. 4.19 Data versus model description for the population dynamics of DCs and CTLs in spleen induced by i.v. injection of 2×10^4, 2×10^5, 2×10^6 gp33-presenting H8-DCs. The symbols represent averages of 3 mice $\pm$ SD. The lines describe the predicted populations dynamics of total tet+, activated tet+ and quiescent memory tet+ CTL and H8-DC. DC elimination follows a biphasic kinetics, the first slower phase reflects their life-span and the accelerated decay phase results from the killing effect by activated CTLs. (The figure is adapted from Ludewig et al., EJI, 2004)

Table 4.6 The estimated parameters of the mathematical model of H8-DC-induced CTL population dynamics

Parameter	Biological meaning	Units	Best-fit estimate
μ_{BS}^{H8-DC}	Transfer rate of H8-DCs from blood to spleen	1/day	2.832
α_D	Decay rate of gp-33 expressing DCs	1/day	0.23
b_{DE}	Per capita elimination rate of H8-DCs by activated CTLs	ml/(cell day)	0.487×10^{-5}
E^{naive}	The number of naive gp-33-specific CTLs contributing to primary clonal expansion	cell	370
τ_d	Duration of pre-programmed CTL division cycle	day	1
α_{E_a}	Rate constant of activated CTLs death	1/day	0.12
α_{E_m}	Rate constant of resting memory CTLs death	1/day	0.01
b_p	Maximal expansion factor of activated CTLs per day	1/day	12
θ_D	Threshold in DC density in the spleen for half-maximal proliferation rate of CTL	cell/ml	2.12×10^3
r_{am}	Rate constant of reversion of activated CTLs	1/day	0.01
b_a	Activation rate constant of quiescent CTLs by DCs	ml/(cell day)	1.05×10^{-3}
θ_{shut}	Threshold in DC density in the spleen for half-maximal transfer rate of CTL from spleen to blood	cell/ml	13.0

kinetics, one needs to consider the possibility of the DC-dependent retention of T cells. Thus, the model predicts a trapping effect, which reduces the export rate of CTLs to blood by about tenfold above a threshold of about 10 H8-DCs present in

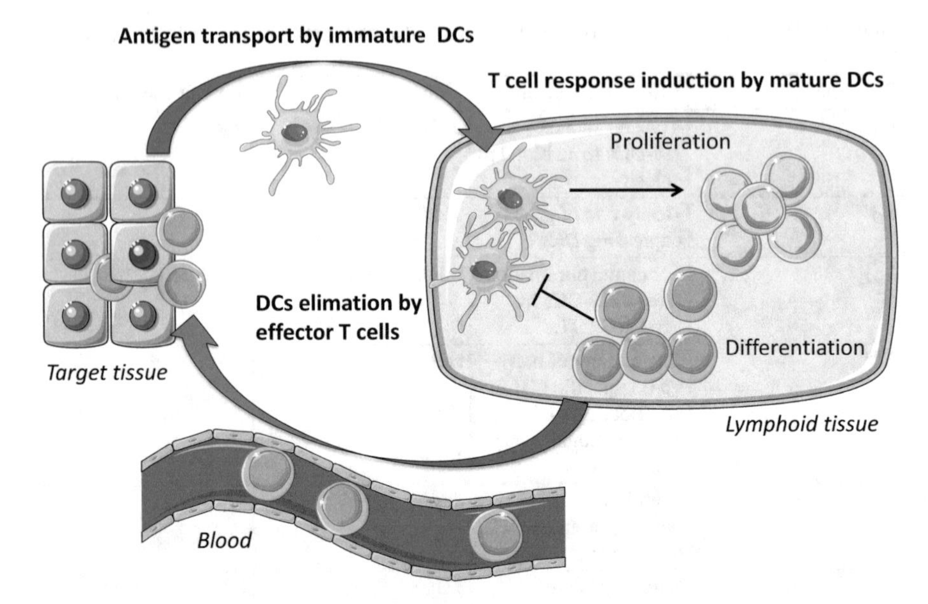

Fig. 4.20 Compartmental scheme of systemic dynamics of adoptively transferred DC and CTL. Antigen-expressing DCs migrate from blood to spleen, where they induce clonal expansion of nave antigen-specific cytotoxic T-lymphocytes, whereas activated CTL eliminate DC. Arrows indicate the modelled processes. *Cell Pictures taken from Servier Medical Art*

Table 4.7 Trafficking rates (day^{-1}) of tet^{+} CTLs between Blood-, Spleen- and Liver compartments

Organ	Blood	Spleen	Liver
Blood	-1	$[0.012, 0.112]$	0.51
Spleen	0.022	$-[0.012, 0.112]$	0
Liver	0.1	0	-0.51

the spleen and equally applies to quiescent and activated CTL. The model predicts that 89% of peptide-specific CTL leave the blood compartment daily to organs other than the spleen and liver.

The sensitivity analysis of the model solutions suggests that T-cell receptor avidity, the half-life of DC, and the rate of CTL-mediated DC elimination are the major control parameters for optimal DC-induced CTL responses. For induction of high avidity CTLs, the number of adoptively transferred DC was of minor importance once a threshold of approximately 200 cells per spleen had been reached. As discussed before, the major objective of DC-based immunization is the maximal expansion and long-term maintenance of high numbers of antigen-specific T cells. Thus, the model can be applied to study the patterns of CTL population dynamics following repeated injection of H8-DC. Two sequential applications of 2×10^4 DCs at days 0, and 40 induce a robust CTL response with only a weak boosting effect. The

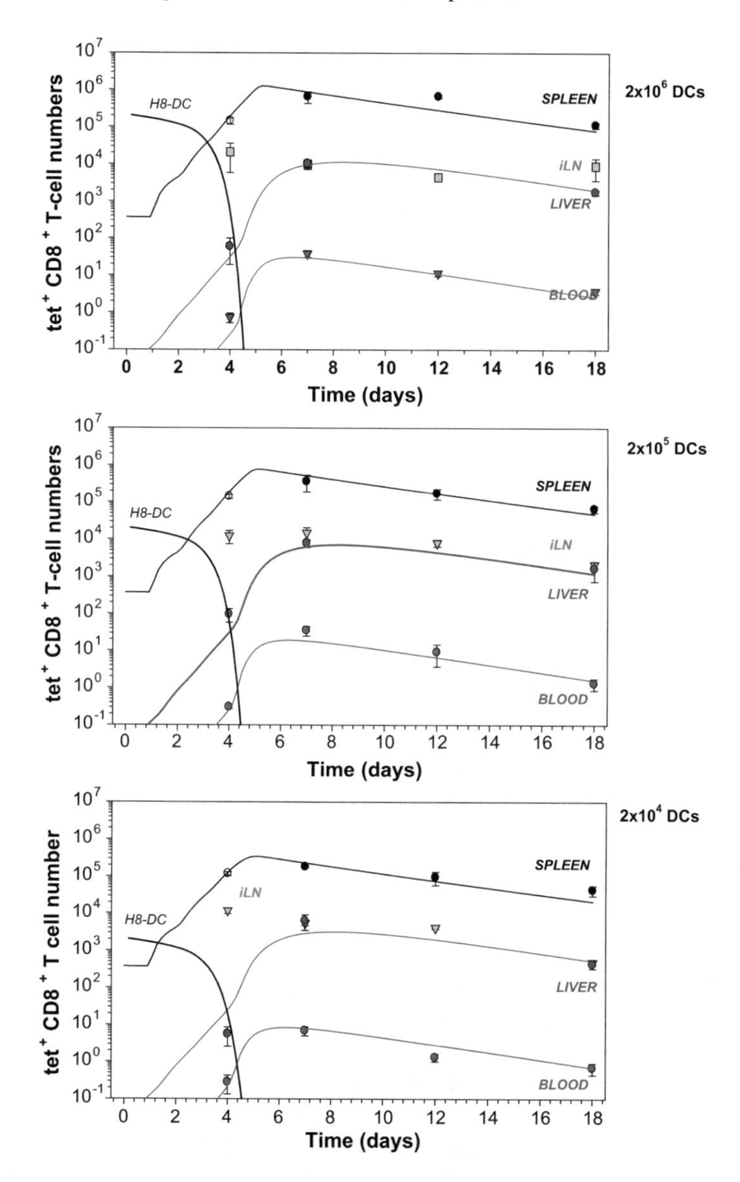

Fig. 4.21 Compartmental dynamics of DC-induced tet$^+$ CTL responses in the spleen, blood and liver. The values for blood indicate the number of tet$^+CD8^+$ T cells/ml. The symbols represent averages of three mice ± SD

model predicts that as long as significant numbers of activated (or memory cells with a faster activation kinetics than that of naive) CTLs persist which ensure rapid elimination of antigen-expressing DCs, any further application of DCs has only a limited enhancement effect. Nevertheless, such repeated DC application is apparently necessary to maintain high levels of activated CTLs.

4.4 MHV Infection: How Robust Is the IFN Type I-Mediated Protection?[6]

Human infections with highly virulent viruses, such as 1918 influenza or SARS-coronavirus, represent major threats to public health. The initial innate immune responses to such viruses have to restrict virus spread before the adaptive immune responses fully develop. Therefore, it is of fundamental practical importance to understand the robustness and fragility of the early protection against such virus infections mediated by the type I interferon response. The inherent complexity of the virus–host system suggests the application of mathematical modelling tools to predict the sensitivity of the kinetics and severity of infection to variations in virus and host parameters.

4.4.1 Immunobiology of MHV Infection

The mouse hepatitis virus (MHV) infection represents a well- understood paradigmatic system for the analysis of type I IFN responses. MHV is a member of the Coronaviridae family that harbour a number of viruses causing severe diseases in animals and humans, such as acute hepatitis, encephalitis, infectious bronchitis, lethal infectious peritonitis and the severe acute respiratory syndrome (SARS) [74, 75]. In systemic MHV infection, spleen and liver represent major target organs [76], and primarily hematopoietic cell-derived type I IFN controls viral replication and virus-induced liver disease [77] as shown schematically in Fig. 4.22.

It has been demonstrated experimentally that pDCs are the major cell population generating IFNα during the initial phase of mouse coronavirus infection [76]. Importantly, mainly macrophages ($M\phi$) and, to a lesser extent conventional DCs, respond most efficiently to the pDC-derived type I IFN and thereby secure containment of MHV within secondary lymphoid organs (SLOs) [78]. Thus, the type I IFN-mediated crosstalk between pDCs and $M\phi$ represents an essential cellular pathway for the protection against MHV-induced liver disease. In systems biology terms, MHV infection triggers a complex array of processes at different biological scales such as protein expression, cellular migration or pathological organ damage. To focus on the front edge of the virus–host interaction, the modelling-based analysis specifically addresses the early dynamics (i.e. the first 48 h) of the type I IFN response to MHV since this is decisive for the outcome of the infection. The reductionists view of the most essential processes underlying the early systemic dynamics

[6]Material of this section uses the results Bocharov et al., PLoS Pathog. 6 (2010), e1001017.

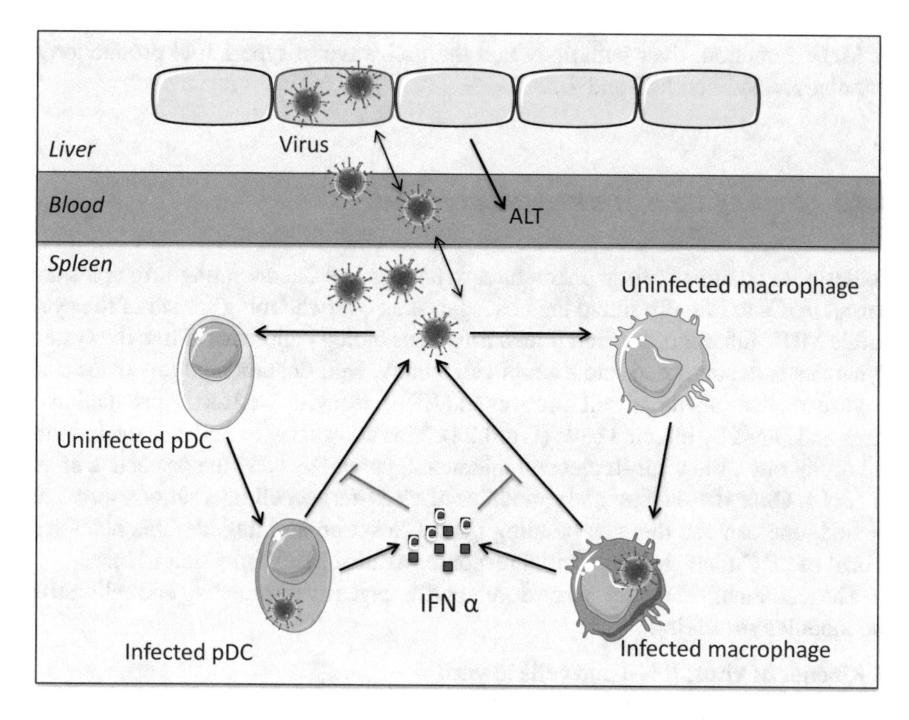

Fig. 4.22 Scheme of type I IFN responses during cytopathic coronavirus infection. Systemic view of the processes determining the early kinetics of MHV infection. (The figure is adapted from Bocharov et al., PLoS pathogens, 2010). *Cell Pictures taken from Servier Medical Art*

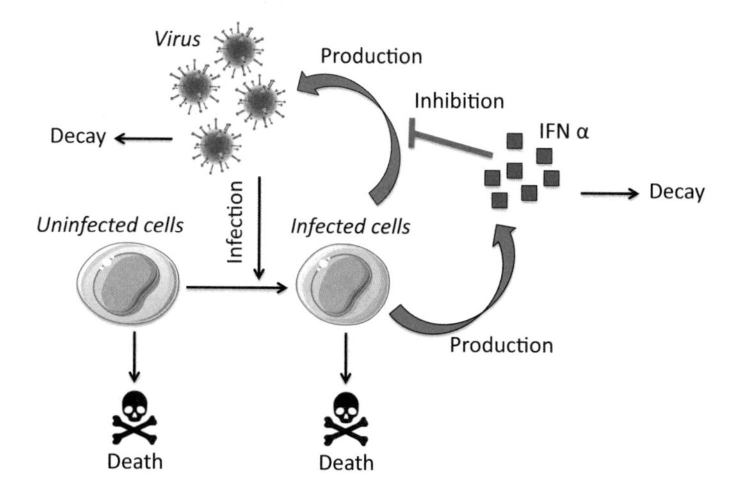

Fig. 4.23 Basic module of mouse hepatitis virus infection and type I interferon response common for pDC and $M\phi$. *Cell Pictures taken from Servier Medical Art*

of MHV infection, liver pathology and the first wave of type I IFN production is summarized in Figs. 4.22 and 4.23.

4.4.2 Setting up a Mathematical Model

To describe quantitatively the structure, dynamics and the operating principles that permit pDCs to initially shield the host against an overwhelming spread of the cytopathic MHV infection, one can follow a systems biology approach. First, the system dynamics is decomposed into a set of elementary, well-documented processes such as virus replication, target cell turnover and IFN-I decay, as well as the production of virus and IFN-I by infected cells (Fig. 4.22). This allows one to quantify the individual decay rates, the virus–target cell interaction parameters and the protective effect of IFN-I. Once these elementary modules of virus–target cell interactions were calibrated, one can use them as building blocks to set up an integrated mathematical model of pDC-mediated type I IFN responses against MHV infection in mice.

The mathematical model can be developed in stages by formulating and calibration the modules specifying

- Kinetics of virus, IFN-I and cells in vitro,
- Basic IFN-I response to infection of target cells,
- Compartmental dynamics of virus growth,
- Systemic model of MHV infection and IFN-I response.

The compartmental model considers the temporal dynamics of

- IFN-I $I(t)$ and unifected/infected pDCs and macrophages $C^{pDC}(t), C^{M\phi}(t), C_V^{pDC}(t), C_V^{M\phi}(t)$ in spleen,
- systemic dynamics of the virus in spleen, blood and liver $V_S(t), V_B(t), V_L(t)$.
- dynamics of liver enzyme AST in blood.

The corresponding equations formulated using a well-established approach (we refer to [3] for further details) read as follows:

$$\frac{dI}{dt}(t) = \rho_I^{pDC} C_V^{pDC}(t - \tau_I^{pDC}) + \rho_I^{M\phi} C_V^{M\phi}(t - \tau_I^{M\phi}) - d_I I(t) \tag{4.25}$$

$$\frac{dC_V^{pDC}}{dt}(t) = \sigma_V^{pDC} V_S(t) C^{pDC}(t) - d_{0CV}^{pDC} C_V^{pDC}(t) \tag{4.26}$$

$$\frac{dC_V^{M\phi}}{dt}(t) = \sigma_V^{M\phi} V_S(t) C^{M\phi}(t) - d_{0CV}^{M\phi} C_V^{M\phi}(t) \tag{4.27}$$

$$\frac{dC^{pDC}}{dt}(t) = -\sigma_V^{pDC} V_S(t) C^{pDC}(t) + d_{0C}^{pDC}\left(C_0^{pDC} - C^{pDC}(t)\right) \tag{4.28}$$

$$\frac{dC^{M\phi}}{dt}(t) = -\sigma_V^{M\phi} V_S(t) C^{M\phi}(t) + d_{0C}^{M\phi}\left(C_0^{M\phi} - C^{M\phi}(t)\right) \tag{4.29}$$

$$\begin{aligned} \frac{dV_S}{dt}(t) = & \frac{\rho_V^{pDC}}{1 + I(t)/\theta_{pDC}} C_V^{pDC}(t - \tau_V^{pDC}) + \frac{\rho_V^{M\phi}}{1 + I(t)/\theta_{M\phi}} C_V^{M\phi}(t - \tau_V^{M\phi}) \\ & - \left(\sigma_V^{pDC} C^{pDC}(t) + \sigma_V^{M\phi} C^{M\phi}(t)\right) V_S(t) - d_V V_S(t) \\ & - \mu_{SB} V_S(t) + \mu_{BS} V_B(t) \frac{Q_B}{Q_S} \end{aligned} \tag{4.30}$$

$$\frac{dV_B}{dt}(t) = \mu_{SB} V_S(t) \frac{Q_S}{Q_B} + \mu_{LB} V_L(t) \frac{Q_L}{Q_B} - (\mu_{BS} + \mu_{BL} + \mu_{BO}) V_B(t) \tag{4.31}$$

$$\frac{dV_L}{dt}(t) = \beta_L V_L(t)\left(1 - V_L(t)/K_L\right) - \mu_{LB} V_L(t) + \mu_{BL} V_B(t) \frac{Q_B}{Q_L} \tag{4.32}$$

$$\frac{dA}{dt}(t) = \rho_A V_L(t) + d_A\left(A^* - A(t)\right). \tag{4.33}$$

The relevant information about the model parameters is summarized in Table 4.8. The best-fit solution is shown in Fig. 4.24.

4.4.3 *Parameter Estimates and Sensitivity Analysis*

The best-fit parameter estimates of the model characterize the concentration of IFN-I which is required to inhibit by twofold the production of virus by the infected cells. It appears that the pDC and Mϕ differ with respect to their sensitivity to the protective effect of interferon, so that the 50% reduction threshold concentrations are about 46 pg/ml and 1 pg/ml, respectively. The per capita type I IFN secretion rate also differs substantially between pDC and Mϕ, being 15586 molec/h and 106 molec/h, respectively. The sensitivity analysis suggests a high protective capacity of single pDCs which protect 10^3–10^4 $M\phi$ from cytopathic viral infection localized to spleen. The model allows one to determine the minimal protective unit of pre-activated pDCs in spleen to be around 200 cells which can rescue the host from severe disease. The modelling results suggest that the spleens capability to function as a sink for the virus produced in peripheral target organs remains operational as long as viral mutations do not permit accelerated growth in peripheral tissues.

Table 4.8 Estimated parameters of the mathematical model MHV infection and type I IFN response

Parameter	Biological meaning	Units	Best-fit estimate
ρ_V^{DC}	Virus production rate by pDC	pfu/cell/h	1.7
$\rho_V^{M\phi}$	Virus production rate by $M\phi$	pfu/cell/h	36.7
ρ_I^{DC}	Type I IFN production rate by pDC	pg/cell/h	4.4×10^{-4}
$\rho_I^{M\phi}$	Type I IFN production rate by $M\phi$	pg/cell/h	1.0×10^{-6}
θ_{pDC}	The threshold for 50% reduction of virus production rate by type I IFN	pg/ml	45.8
$\theta_{M\phi}$	The threshold for 50% reduction of virus production rate by type I IFN	pg/ml	0.97
σ_V^{DC}	Infection rate of pDC	cell/pfu/h	1.3×10^{-6}
$\sigma_V^{M\phi}$	Infection rate of $M\phi$	cell/pfu/h	0.9×10^{-7}
τ_V^{pDC}	Virus production delay by pDC	h	5.96
$\tau_V^{M\phi}$	Virus production delay by pDC	h	5.99
τ_I^{pDC}	Type I IFN production delay by pDC	h	5.77
$\tau_I^{M\phi}$	Type I IFN production delay by $M\phi$	h	5.8
d_{0CV}^{pDC}, k_{CV}^{pDC}	Gompertz death rate parameters for infected pDC	1/h	0.2, 0.087
$d_{0CV}^{M\phi}$, $k_{CV}^{M\phi}$	Gompertz death rate parameters for infected $M\phi$	1/h	0.049, 0.057
μ_{BS}	Virus transfer rate from blood to spleen	1/h	3.46
μ_{BL}	Virus transfer rate from blood to liver	1/h	0.018
μ_{SB}	Virus transfer rate from spleen to blood	1/h	0.91
μ_{LB}	Virus transfer rate from liver to blood	1/h	0.61

(continued)

Table 4.8 (continued)

Parameter	Biological meaning	Units	Best-fit estimate
μ_{BO}	Virus elimination rate from blood	1/h	1.22
β_L	Virus growth rate in liver	pfu/ml/h	0.78
K_L	Carrying capacity of the liver	pfu/ml	10^7
ρ_A	Rate constant of ALT release into blood	IU/l	0.68×10^{-3}
d_A	Decay rate of ALT release in blood	1/h	0.16
A_*	Physiological level of ALT in blood	IU/l	25

Importantly, the mathematical model of MHV infection can be used to evaluate the limits of protection against severe disease for increasing virus replication rates.

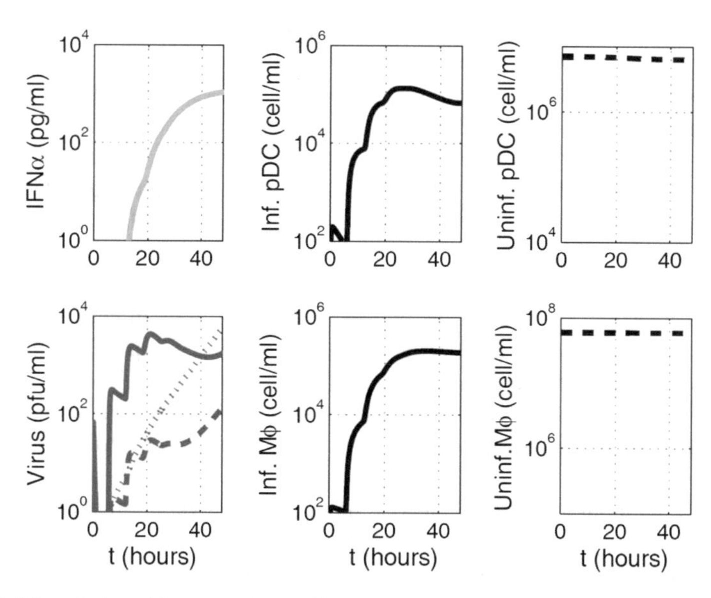

Fig. 4.24 The solution of the compartmental model describing the kinetics of **a** interferon response and the population dynamics of **b** virus in spleen (solid line), liver (dotted line) and blood (dashed line), **c** and **e** uninfected/infected pDCs, **d** and **f** uninfected/infected macrophages. The parameters of the interferon response were estimated from in vitro and in vivo (spleen) data. The compartmental approach assumes an instantaneous mixing of IFN-I implying that the concentration across the SLO (spleen) is uniform. Reprinted from Mathematical Modelling of Natural Phenomena, Vol. 6, Bocharov et al., Reaction-Diffusion Modelling of Interferon Distribution in Secondary Lymphoid Organs, Pages 13–26, Copyright © 2011, with permission from EDP Sciences

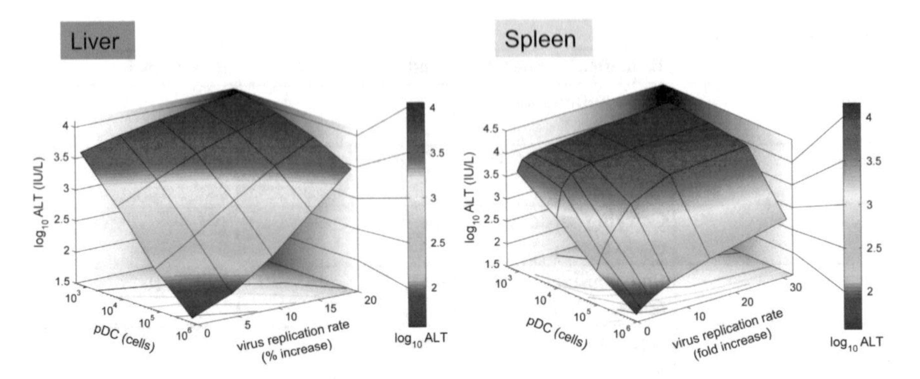

Fig. 4.25 Effect of virus growth rates on pDC-mediated protection against disease. **a** Sensitivity of the disease severity to variations in pDC numbers (cells per spleen) and the global increase of viral replication rate in the liver (% increase). Disease severity is determined as peak ALT levels in serum within 48 h post-infection following i.v. infection with 50 pfu. **b** Determination of the systems robustness against disease with respect to variations in pDC numbers (cells per spleen) and increasing viral replication rates restricted to $M\phi$ in the spleen (Note: fold increase). (The figure is reprinted from Bocharov et al. PLoS pathogens, 2010)

Since various MHV strains display significant differences in their ability to replicate in different organs, two complementary scenarios were considered: the increase in virus growth rate in the peripheral organs (liver) versus secondary lymphoid organs (spleen). Figure 4.25 A shows that pDCs in spleen provide very limited protection against severe disease for faster replicating strains of the virus in hepatocytes. Indeed, only a 15% increase in the growth rate of MHV in the liver leads to infection with ALT levels rising to 10^3 IU/L within 2 days. The decrease of pDC numbers in spleen makes the situation more fragile to even smaller increases in the virus growth rate. On the contrary, pDCs provide a robust protection against severe disease when the virulence-enhancing mutation leads to faster replication only in target cells located in spleen, i.e. splenic pDCs protect against severe disease for up to 30-fold increase in the viral replication rate in splenic $M\phi$ (See Fig. 4.25b). Taken together, these analyses indicate that the spleen represents a robust sink system able to cope with substantially enhanced virus production as long as this gain of viral fitness remains restricted to this SLO.

Overall, the modelling results suggest that the pDC population in spleen ensures a robust protection against virus variants which substantially downmodulate type I IFN secretion. However, the ability of pDCs to protect against severe disease caused by virus variants exhibiting an enhanced liver tropism and higher replication rates appears to be rather limited. Taken together, this system immunology analysis suggests that antiviral therapy against cytopathic viruses should primarily limit viral replication within peripheral target organs.

4.5 Identifying a Feedback Regulating Proliferation and Differentiation of CD4$^+$ T Cells[8]

In response to antigens, specific T-cell clones rapidly increase in size and then steeply decline, approaching relatively stable frequencies higher than those of the naive cell population. It was discovered by W.E. Paul's team (see data presented in [79]) that there is a log-linear relation between the CD4$^+$ T-cell precursor number (PN) and the factor of expansion (FE), with a slope of ~0.5 over a range of 3–30,000 antigen-specific precursors per mouse. The experimental results suggested an inhibition mechanism of precursor expansion either by competition for specific antigen-presenting cells or by the action of other antigen-specific cells in the same microenvironment. Mathematical modelling can be used to identify the specific functions underlying the feedback regulation of the observed clonal dynamics.

As it was discussed in Chap. 2, the role of immunological theories in specifying mathematical models is essential. To formulate a mathematical model which describes and explains the observed findings, i.e. the data on CD4$^+$ T-cell expansion for various precursor numbers and the accompanying data sets from BrdU- and CFSE labeling kinetics, we considered the feedback-regulated balance of growth and differentiation concept by Grossman and Paul [80–82]. It was assumed that the most differentiated effectors (or memory cells) limit the growth of less differentiated effectors, locally, by increasing the rate of differentiation of the latter cells in a dose-dependent manner [4]. The biological scheme and the sequence of functionally distinct stages in cell development underlying the equations of the mathematical model is shown schematically in Fig. 4.26. Cell proliferation and differentiation rates were assumed to be regulated in a feedback fashion, i.e. they depend on the number of differentiated cells.

The population dynamics of the above four subsets of CD4$^+$ T cells was modelled using a system of ODEs. The core mathematical model was used for data assimilation either directly (data on the kinetics of clonal expansion and contraction) or in two extended forms in which the cell subsets were further subdivided into unlabelled and labelled compartments, to describe the BrdU-labeling data and CFSE dilution data, respectively. A number of data fitting and analysis methods, including the maximum likelihood approach, Akaike information criteria, statistical model comparison methods and sensitivity analyses were used to identify a parsimonious model of the kinetics of antigen-driven CD4$^+$ T-cell expansion (we refer to [4] for details). The set of core model equations with feedback regulation is represented in the following form:

$$\frac{\mathrm{d}}{\mathrm{d}t}X_1(t) = p_1 X_1(t) - \left(\alpha_1 + \alpha_{12}\frac{1}{1 + (Z_1(t) + Z_2(t))/\theta_{x1}} Z_2\right) X_1, \qquad (4.34)$$

[8] Material of this subsection uses the results of our studies from Proceedings of the National Academy of Sciences of the United States of America (PNAS USA), Vol. 108, Bocharov et al., Feedback regulation of proliferation vs. differentiation rates explains the dependence of CD4 T-cell expansion on precursor number, Pages 3318–3323, Copyright © 2011 with permission from PNAS USA.

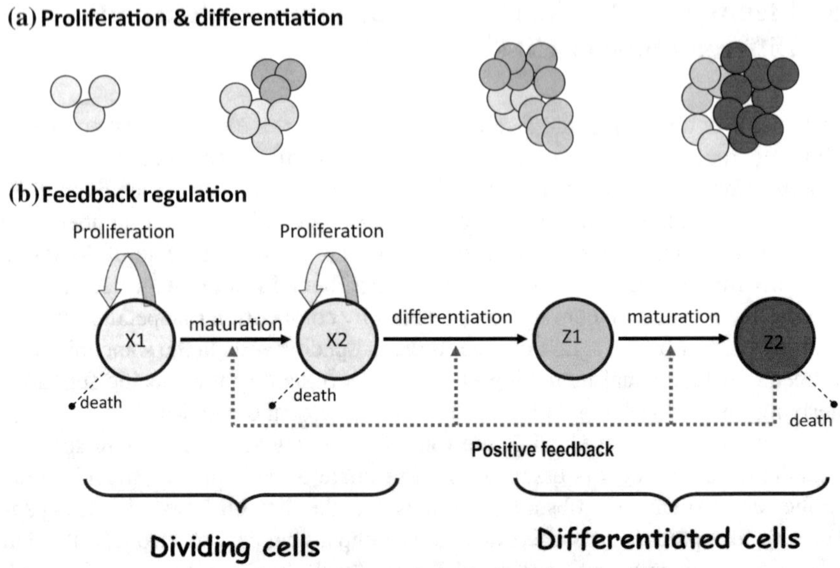

(a) Proliferation & differentiation

(b) Feedback regulation

Proliferation
Proliferation
X1
maturation
X2
differentiation
Z1
maturation
Z2
death
death
death
Positive feedback
Dividing cells
Differentiated cells

Fig. 4.26 Biological scheme of the concept of feedback-regulated balance of growth and differentiation of cells by Z. Grossman by W. Paul. **a** Heterogeneity of the proliferation and differentiating clones. **b** A simple view of the sequence of functionally distinct stages in cell development. Cell proliferation and differentiation rates are assumed to be regulated in a feedback fashion, i.e. they depend on the number of differentiated cells. Two subsets X1 and X2 represent the proliferating cell population. X1 is less mature than X2. The differentiated cell populations Z1 and Z2 can not divide. The population of more differentiated cells Z2 controls the balance of proliferation and differentiation of X1 and X2 subsets. Reprinted from Proceedings of the National Academy of Sciences of the United States of America (PNAS USA), Vol. 108, Quiel et al., Antigen-stimulated CD4 T-cell expansion is inversely and log-linearly related to precursor number, Pages 3312–3317, Copyright © 2011 with permission from PNAS USA

$$\frac{\mathrm{d}}{\mathrm{d}t}X_2(t) = p_2X_2(t) + \left(\alpha_1 + \alpha_{12}\frac{1}{1+(Z_1(t)+Z_2(t))/\theta_{x1}}Z_2\right)X_1 - (\alpha_2 + \alpha_{22}Z_2)\,X_2, \quad (4.35)$$

$$\frac{\mathrm{d}}{\mathrm{d}t}Z_1(t) = (\alpha_2 + \alpha_{22}Z_2)\,X_2 - \beta_1 Z_1, \quad (4.36)$$

$$\frac{\mathrm{d}}{\mathrm{d}t}Z_2(t) = \beta_1 Z_1 - \delta Z_2. \quad (4.37)$$

The consistency of the model with data on clonal expansion of $CD4^+$ T cells starting from 300 precursors in the LNs at the time of immunization and 3×10^4 cells along with the evolving structure of the clones is illustrated in Fig. 4.27. The model gives a precise quantitative relation between the factor of expansion (FE) and the precursor number (PN) as follows: $FE = PN^{0.48} \times 3.981$.

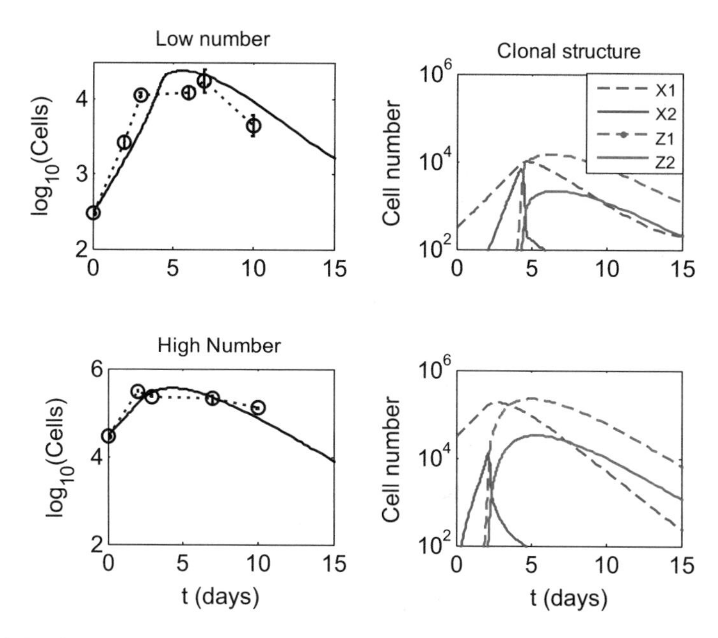

Fig. 4.27 Model-based data assimilation and parameter estimation. The kinetics of clonal expansion and contraction for different initial numbers of transferred antigen-specific precursor CD4$^+$ T cells. The time evolution of the total number of cells (Left) and evolution of the clonal structure (Right) are shown. Upper row, 300 antigen-specific CD4$^+$ T cells in the LNs at the time of immunization; Lower row, 3×10^4 antigen-specific precursor CD4$^+$ T cells at immunization. Reprinted from Proceedings of the National Academy of Sciences of the United States of America (PNAS USA), Vol. 108, Bocharov et al., Feedback regulation of proliferation vs. differentiation rates explains the dependence of CD4 T-cell expansion on precursor number, Pages 3318–3323, Copyright © 2011 with permission from PNAS USA

One can conclude that the feedback-regulated balance of growth and differentiation hypothesis, although requiring definite experimental characterization of the hypothetical cell phenotypes and molecules involved in the identified regulation, can explain the kinetics of CD4$^+$ T-cell responses to antigenic stimulation. We note that a mathematical model based on a different hypothesis (e.g. 'grazing of peptide-MHC complexes') was proposed to explain the same phenomenon although in a semi-quantitative manner [83]. However, no evidence of its consistency with all available data sets that were described and analysed in [4, 79] was presented.

In conclusion, while a multitude of mathematical models can be generated to describe any given immunological phenomenon, it is crucial to always link it to available experimental data. If model and data are in good agreement, then the model may help to generate new hypothesis of underlying mechanisms and provide further testable predictions. In addition, as outlined in Sect. 4.5, a model may also strongly support a novel hypothesis that was brought up ad hoc from immunological considerations.

References

1. Bocharov, G.A. Modelling the dynamics of LCMV infection in mice: conventional and exhaustive CTL responses. *J. Theor. Biol.*, **192** (1998) 283–308.
2. Luzyanina T, Engelborghs K, Ehl S, Klenerman P, Bocharov G. Low level viral persistence after infection with LCMV: a quantitative insight through numerical bifurcation analysis. Math Biosci. 2001 ;173(1):1–23.
3. Bocharov G, Zst R, Cervantes-Barragan L, Luzyanina T, Chiglintsev E, Chereshnev VA, Thiel V, Ludewig B. A systems immunology approach to plasmacytoid dendritic cell function in cytopathic virus infections. PLoS Pathog. 2010 ;6(7):e1001017.
4. Bocharov G, Quiel J, Luzyanina T, Alon H, Chiglintsev E, Chereshnev V, Meier-Schellersheim M, Paul WE, Grossman Z. Feedback regulation of proliferation vs. differentiation rates explains the dependence of CD4 T-cell expansion on precursor number. Proc Natl Acad Sci U S A. 2011 ;108(8):3318–3323
5. Doherty PC, Zinkernagel RM: H-2 compatibility is required for T-cell-mediated lysis of target cells infected with lymphocytic choriomeningitis virus. J Exp Med 1975, 141:502–507.
6. Zinkernagel RM, Doherty PC: Restriction of in vitro T cell-mediated cytotoxicity in lymphocytic choriomeningitis within a syngeneic or semiallogeneic system. Nature 1974, 248:701–702.
7. Kagi D, Ledermann B, Burki K, Seiler P, Odermatt B, Olsen KJ, Podack ER, Zinkernagel RM, Hengartner H: Cytotoxicity mediated by T cells and natural killer cells is greatly impaired in perforin-deficient mice. Nature 1994, 369:31–37.
8. Masson D, Tschopp J: Isolation of a lytic, pore-forming protein (perforin) from cytolytic T-lymphocytes. J Biol Chem 1985, 260:9069–9072.
9. Ehl S, Klenerman P, Zinkernagel RM, Bocharov G: The impact of variation in the number of CD8(+) T-cell precursors on the outcome of virus infection. Cell Immunol 1998, 189:67–73.
10. Waggoner SN, Cornberg M, Selin LK, Welsh RM: Natural killer cells act as rheostats modulating antiviral T cells. Nature 2012, 481:394–398.
11. Karrer U, Althage A, Odermatt B, Roberts CW, Korsmeyer SJ, Miyawaki S, Hengartner H, Zinkernagel RM: On the key role of secondary lymphoid organs in antiviral immune responses studied in alymphoplastic (aly/aly) and spleenless (Hox11(-)/-) mutant mice. J Exp Med 1997, 185:2157–2170.
12. Cole GA, Nathanson N, Prendergast RA: Requirement for theta-bearing cells in lymphocytic choriomeningitis virus-induced central nervous system disease. Nature 1972, 238:335–337.
13. Kim JV, Kang SS, Dustin ML, McGavern DB: Myelomonocytic cell recruitment causes fatal CNS vascular injury during acute viral meningitis. Nature 2009, 457:191–195.
14. Riviere Y, Gresser I, Guillon JC, Tovey MG: Inhibition by anti-interferon serum of lymphocytic choriomeningitis virus disease in suckling mice. Proc Natl Acad Sci U S A 1977, 74:2135–2139.
15. Chen HD, Fraire AE, Joris I, Brehm MA, Welsh RM, Selin LK: Memory CD8+ T cells in heterologous antiviral immunity and immunopathology in the lung. Nat Immunol 2001, 2:1067–1076.
16. Ohashi PS, Oehen S, Buerki K, Pircher H, Ohashi CT, Odermatt B, Malissen B, Zinkernagel RM, Hengartner H: Ablation of "tolerance" and induction of diabetes by virus infection in viral antigen transgenic mice. Cell 1991, 65:305–317.
17. Oldstone MB, Nerenberg M, Southern P, Price J, Lewicki H: Virus infection triggers insulin-dependent diabetes mellitus in a transgenic model: role of anti-self (virus) immune response. Cell 1991, 65:319–331.
18. Barber DL, Wherry EJ, Masopust D, Zhu B, Allison JP, Sharpe AH, Freeman GJ, Ahmed R: Restoring function in exhausted CD8 T cells during chronic viral infection. Nature 2006, 439:682–687.
19. Moskophidis D, Lechner F, Pircher H, Zinkernagel RM: Virus persistence in acutely infected immunocompetent mice by exhaustion of antiviral cytotoxic effector T cells. Nature 1993, 362:758–761.

20. Wherry EJ, Ha SJ, Kaech SM, Haining WN, Sarkar S, Kalia V, Subramaniam S, Blattman JN, Barber DL, Ahmed R: Molecular signature of CD8+ T cell exhaustion during chronic viral infection. Immunity 2007, 27:670–684.
21. Zajac AJ, Blattman JN, Murali-Krishna K, Sourdive DJ, Suresh M, Altman JD, Ahmed R: Viral immune evasion due to persistence of activated T cells without effector function. J Exp Med 1998, 188:2205–2213.
22. Leavy O: Tumour immunology: A triple blow for cancer. Nat Rev Immunol 2015, 15:265.
23. Trautmann L, Janbazian L, Chomont N, Said EA, Gimmig S, Bessette B, Boulassel MR, Delwart E, Sepulveda H, Balderas RS, et al: Upregulation of PD-1 expression on HIV-specific CD8+ T cells leads to reversible immune dysfunction. Nat Med 2006, 12:1198–1202.
24. Velu V, Titanji K, Zhu B, Husain S, Pladevega A, Lai L, Vanderford TH, Chennareddi L, Silvestri G, Freeman GJ, et al: Enhancing SIV-specific immunity in vivo by PD-1 blockade. Nature 2009, 458:206–210.
25. Moskophidis D, Lechner F, Pircher H, Zinkernagel RM. Virus persistence in acutely infected immunocompetent mice by exhaustion of antiviral cytotoxic effector T cells. Nature. 1993;362(6422):758–761
26. Wherry EJ, Kurachi M: Molecular and cellular insights into T cell exhaustion. Nat Rev Immunol 2015, 15:486–499.
27. A. Ciurea, P. Klenerman, L. Hunziker, E. Horvath, B. Odermatt, A. F. Ochsenbein, H. Hengartner, and R. M. Zinkernagel. Persistence of lymphocytic choriomeningitis virus at very low levels in immune mice. *Proc. Natl. Acad. Sci. USA*, 96:11964–11969, 1999.
28. A. J. Zajac, J. N. Blattman, K. Murali-Krishna, D. J. D. Sourdive, M. Suresh, J. D. Altman, and R. Ahmed. Viral immune evasion due to persistence of activated T cells without effector function. *J. Exp. Med.*, 188:2205–2213, 1998.
29. R. Ahmed, L. A. Morrison, and D. M. Knipe. Viral persistence. In N. Nathanson et al., editor, *Viral Pathogenesis*, pages 181–205. Lippincott-Raven Publishers, Philadelphia, 1997.
30. M. B. A. Oldstone. Viral persistence. *Cell*, 56:517–520, 1989.
31. D. Tortorella, B. E. Gewurz, M. H. Furman, D. J. Schust, and H. L. Ploegh. Viral subversion of the immune system. *Ann. Rev. Immunol.*, 18:861–926, 2000.
32. R. Ahmed, B. D. Jamieson, and D. D. Porter. Immune therapy of a persistent and disseminated viral infection. *J. Virol.*, 61:3920–3929, 1987.
33. O. Planz, S. Ehl, E. Furrer, E. Horvath, M.-A. Bründler, H. Hengartner, and R. M. Zinkernagel. A critical role of neutralizing-antibody-producing B cells, $CD4^+$ T cells and interferons in persistent and acute infections of mice with lymphocytic choriomeningitis virus: Implications for adoptive immunotherapy of virus carriers. *Proc. Natl. Acad. Sci. USA*, 94:6874–6879, 1997.
34. K. Engelborghs. *DDE-BIFTOOL: a Matlab package for bifurcation analysis of delay differential equations*. Department of Computer Science, Katholieke Universiteit Leuven, Belgium, March 2000. Report TW 305, (http://www.cs.kuleuven.ac.be/~koen/delay/ddebiftool.shtml).
35. K. Engelborghs. *Numerical bifurcation analysis of delay differential equations*. Ph.D. thesis, Department of Computer Science, Katholieke Universiteit Leuven, Belgium, 2000.
36. K. Engelborghs and D. Roose. On stability of LMS methods and characteristic roots of delay differential equations. Submitted, 2000.
37. K. Engelborghs and E. J. Doedel. Stability of piecewise polynomial collocation methods for computing periodic solutions of delay differential equations. Submitted, 2000.
38. K. Engelborghs, T. Luzyanina, K. in't Hout, and D. Roose. Collocation methods for the computation of periodic solutions of delay differential equations. *SIAM J. Sci. Comput.*, 22:1593–1609, 2000.
39. F. Lehmann-Grube. *Lymphocytic Choriomeningitis Virus*. New York:Springer-Verlag, 1971.
40. Mims, C.A. (1995). Mims Pathogenesis of Infectious Disease. London: Academic Press.
41. Zinkernagel, R.M. and Hengartner H. (1997). Antiviral immunity. Immunol. Today 18, 258–260.
42. Kägi, D., Odermatt, B., Seiler, P., Zinkernagel, R. M., Mak, T.W. Hengartner, H. (1997). Reduced incidence and delayed onset of diabetis in perforindeficient nonobese diabetic mice. J. Exp. Med. 186, 989–997.

43. Ludewig, B., Odermatt, B., Landmann, S. Hengartner, H. and Zinkernagel, R.M. (1998). Dendritic cells induce autoimmune diabetis and maintain disease via de novo formation of local lymphoid tissue. J. Exp. Med. 188, 1–9.
44. Binder, D., van den Broek, M. F., Kägi, D., Bluethmann, Fehr, J., Hengartner, H. and Zinkernagel, R.M. (1998). Aplastic anemia rescued by exhaustion of cytokine-secreting CD8+ T cells in persistent infection with lymphocytic choriomeningitis virus. J. Exp. Med. 187, 1903–1920
45. Oldstone, M.B.A., Blount, P., Southern, P.J. and Lampert, P.W. (1986). Cytoimmunotherapy for persistent virus infection reveals a unique clearance pattern from the central nervous system. Nature, 321, 239–243.
46. Zinkernagel, R.M. (1993). Immunity to viruses. In: Fundamental Immunology, 3rd. Edn. (Paul, W., ed.), Chap. 34, pp. 1211–1250. New York: Raven Press.
47. Zinkernagel, R.M., Haenseler, E., Leist, T., Cerny, A., Hengartner, H. and Althage, A. (1986). T cell-mediated hepatitis in mice infected with lymphocytic choriomeningitis virus. J. Exp. Med. 164, 1075–1092.
48. Zinkernagel, R.M., Planz, O., Ehl, S., Battegay, M., Odermatt, B., Klenerman, P. and Hengartner, H. (1999). General and specific immunosuppression caused by antiviral T-cell responses. Immunol. Reviews 168,305–315.
49. Odermatt, B., Eppler, M., Leist, T.P., Hengartner, H. and Zinkernagel, R.M. (1991) Virus-triggered acquired immunodeficiency by cytotoxic T-cell-dependent destruction of antigen-presenting cells and lymph node follicle structure. Proc. Natl. Acad. Sci. USA, 88, 8252–8256.
50. Jacquez, J.A. and Simon, C.P. (1993). Qualitative theory of compartmental systems. SIAM Review, 35, 43–79.
51. Bocharov G, Klenerman P, Ehl S. Modelling the dynamics of LCMV infection in mice: II. Compartmental structure and immunopathology. J Theor Biol. 2003 Apr 7;221(3):349–378. Erratum in: J Theor Biol. 2004 Jan 7;226(1):123
52. Pardoll, D. M., Spinning molecular immunology into successful immunotherapy. Nat. Rev. Immunol. 2002. 2: 227–238.
53. Steinman, R. M. and Pope, M., Exploiting dendritic cells to improve vaccine efficacy. J. Clin. Invest 2002. 109: 1519–1526.
54. Schuler, G., Schuler-Thurner, B. and Steinman, R. M., The use of dendritic cells in cancer immunotherapy. Curr. Opin. Immunol. 2003. 15: 138–147.
55. Ludewig, B., Ehl, S., Karrer, U., Odermatt, B., Hengartner, H. and Zinkernagel, R. M., Dendritic cells efficiently induce protective antiviral immunity. J. Virol. 1998. 72: 3812–3818.
56. Ludewig, B., Ochsenbein, A. F., Odermatt, B., Paulin, D., Hengartner, H. and Zinkernagel, R. M., Immunotherapy with dendritic cells directed against tumor antigens shared with normal host cells results in severe autoimmune disease. J. Exp. Med. 2000. 191: 795–804.
57. Ludewig, B., Barchiesi, F., Pericin, M., Zinkernagel, R. M., Hengartner, H. and Schwendener, R. A., In vivo antigen loading and activation of dendritic cells via a liposomal peptide vaccine mediates protective antiviral and anti-tumour immunity. Vaccine 2000. 19: 23–32.
58. Nair, S. K., Boczkowski, D., Morse, M., Cumming, R. I., Lyerly, H. K. and Gilboa, E., Induction of primary carcinoembryonic antigen (CEA)-specific cytotoxic T lymphocytes in vitro using human dendritic cells transfected with RNA. Nat. Biotechnol. 1998. 16: 364–369.
59. Rea, D., Havenga, M. J., van Den Assem, M., Sutmuller, R. P., Lemckert, A., Hoeben, R. C., Bout, A., Melief, C. J. and Offringa, R., Highly efficient transduction of human monocytederived dendritic cells with subgroup B fiber-modified adenovirus vectors enhances transgene-encoded antigen presentation to cytotoxic T cells. J. Immunol. 2001. 166: 5236–5244.
60. Hsu, F. J., Benike, C., Fagnoni, F., Liles, T. M., Czerwinski, D., Taidi, B., Engleman, E. G. and Levy, R., Vaccination of patients with B cell lymphoma using autologous antigen-pulsed dendritic cells. Nat. Med. 1996. 2: 52–58.
61. Nestle, F. O., Alijagic, S., Gilliet, M., Sun, Y., Grabbe, S., Dummer, R., Burg, G. and Schadendorf, D., Vaccination of melanoma patients with peptide- or tumor lysate-pulsed dendritic cells. Nat. Med. 1998. 4: 328–332.
62. Fong, L. and Engleman, E. G., Dendritic cells in cancer immunotherapy. Annu. Rev. Immunol. 2000. 18: 245–273.

63. Ludewig, B., Bonilla, W. V., Dumrese, T., Odermatt, B., Zinkernagel, R. M. and Hengartner, H., Perforin-independent regulation of dendritic cell homeostasis by CD8(+) T cells in vivo: implications for adaptive immunotherapy. Eur. J. Immunol. 2001. 31: 1772–1779.
64. Ludewig, B., Oehen, S., Barchiesi, F., Schwendener, R. A., Hengartner, H. and Zinkernagel, R. M., Protective antiviral cytotoxic T cell memory is most efficiently maintained by restimulation via dendritic cells. J. Immunol. 1999. 163: 1839–1844.
65. Dunn, G. P., Bruce, A. T., Ikeda, H., Old, L. J. and Schreiber, R. D., Cancer immunoediting: from immunosurveillance to tumor escape. Nat. Immunol. 2002. 3: 991–998.
66. Brossart, P., Zobywalski, A., Grunebach, F., Behnke, L., Stuhler, G., Reichardt, V. L., Kanz, L. and Brugger, W., Tumor necrosis factor alpha and CD40 ligand antagonize the inhibitory effects of interleukin 10 on T cell stimulatory capacity of dendritic cells. Cancer Res. 2000. 60: 4485–4492.
67. De Boer, R. J. and Perelson, A. S., Towards a general function describing T cell proliferation. J. Theor. Biol. 1995. 175: 567–576.
68. Borghans, J. A., Taams, L. S., Wauben, M. H. and De Boer, R. J., Competition for antigenic sites during T cell proliferation: a mathematical interpretation of in vitro data. Proc. Natl. Acad. Sci. USA 1999. 96: 10782–10787.
69. Ronchese, F. and Hermans, I. F., Killing of dendritic cells: a life cut short or a purposeful death? J. Exp. Med. 2001. 194: F23–F26.
70. Ludewig, B., Krebs, P., Junt, T. and Bocharov, G., Dendritic cell homeostasis in the regulation of self-reactivity. Curr. Pharm. Des. 2003. 9: 221–231.
71. Perelson, A. S., Modelling viral and immune system dynamics. Nat. Rev. Immunol. 2002. 2: 28–36.
72. Chakraborty, A. K., Dustin, M. L. and Shaw, A. S., In silico models for cellular and molecular immunology: successes, promises and challenges. Nat. Immunol. 2003. 4: 933–936.
73. Komarova, N. L., Barnes, E., Klenerman, P. and Wodarz, D., Boosting immunity by antiviral drug therapy: a simple relationship among timing, efficacy, and success. Proc. Natl. Acad. Sci. USA 2003. 100: 1855–1860.
74. Barchet W, Cella M, Colonna M (2005) Plasmacytoid dendritic cellsvirus experts of innate immunity. Semin Immunol 17: 253–261.
75. Perlman S, Netland J (2009) Coronaviruses post-SARS: update on replication and pathogenesis. Nat Rev Microbiol 7: 439–450.
76. Cervantes-Barragan L, Zust R, Weber F, Spiegel M, Lang KS, et al. (2007) Control of coronavirus infection through plasmacytoid dendritic-cell-derived type I interferon. Blood 109: 1131–1137.
77. Lang PA, Cervantes-Barragan L, Verschoor A, Navarini AA, Recher M, et al. (2009) Hematopoietic cell-derived interferon controls viral replication and virusinduced disease. Blood 113: 1045–1052.
78. Cervantes-Barragan L, Kalinke U, Zust R, Konig M, Reizis B, et al. (2009) Type I IFN-mediated protection of macrophages and dendritic cells secures control of murine coronavirus infection. J Immunol 182: 1099–1106.
79. Quiel J, Caucheteux S, Laurence A, Singh NJ, Bocharov G, Ben-Sasson SZ, Grossman Z, Paul WE. Antigen-stimulated CD4 T-cell expansion is inversely and log-linearly related to precursor number. Proc Natl Acad Sci U S A. 2011 ;108(8):3312–3317
80. Grossman Z, Paul WE: Dynamic tuning of lymphocytes: physiological basis, mechanisms, and function. Annu Rev Immunol 2015, 33:677–713.
81. Grossman, Z., Min, B., Meier-Schellersheim, M. and Paul, W. E., Concomitant regulation of T cell activation and homeostasis. Nat. Rev. Immunol. 2004. 4: 7–15.
82. Grossman Z. Recognition of self and regulation of specificity at the level of cell populations. Immunol Rev (1984) 79:119–138.

83. De Boer RJ, Perelson AS. Antigen-stimulated CD4 T cell expansion can be limited by their grazing of peptide-MHC complexes. J Immunol. 2013 Jun 1;190(11):5454–5458. https://doi.org/10.4049/jimmunol.1203569.
84. Ludewig B, Krebs P, Junt T, Metters H, Ford NJ, Anderson RM, Bocharov G. Determining control parameters for dendritic cell-cytotoxic T lymphocyte interaction. Eur J Immunol. 2004;34(9):2407–2418.

Chapter 5
Modelling of Human Infections

In this chapter we illustrate the application of mathematical models and computational analyses tools of various complexity to the description and explanation of some observed phenotypes of viral infections in humans, such as HIV and HBV infections. Specifically, we try to gain a deeper understanding of the sensitivity of infection dynamics to growth rate and the efficacy of antigen presentation by APCs, the phenomenon of spontaneous recovery from HBV infection and the kinetic determinants of a low-level (i.e. below the detection threshold) HBV persistence. The material of this chapter is based on our previous work published in [3, 12–14, 33].

5.1 Outcome of Virus Infections as a 'Numbers Game'

Application of mathematical models for analysis and prediction of the mechanisms underlying unfavourable outcomes of viral infections in humans presents a formidable challenge. This is due to the genetic diversity of individuals as well as limited access to dynamic processes in vivo. Virus infections of humans are characterized by a spectrum of courses and outcomes which can be categorized as subclinical, acute with recovery, chronic and lethal infection phenotypes as presented in Fig. 5.1.

Research in viral immunology centres around the questions 'Why' and 'How to cure' unfavourable infectious disease. From the clinician's perspective, the differences in the disease course and outcomes depend on characteristics such as (see Fig. 5.2)

- Health condition of the infected individual (e.g. Tx-patient, newborn etc., age),
- Immunopathology,
- Cytopathicity of virus,
- Persistence,
- Tropism,

G. Bocharov et al., *Mathematical Immunology of Virus Infections*,
https://doi.org/10.1007/978-3-319-72317-4_5

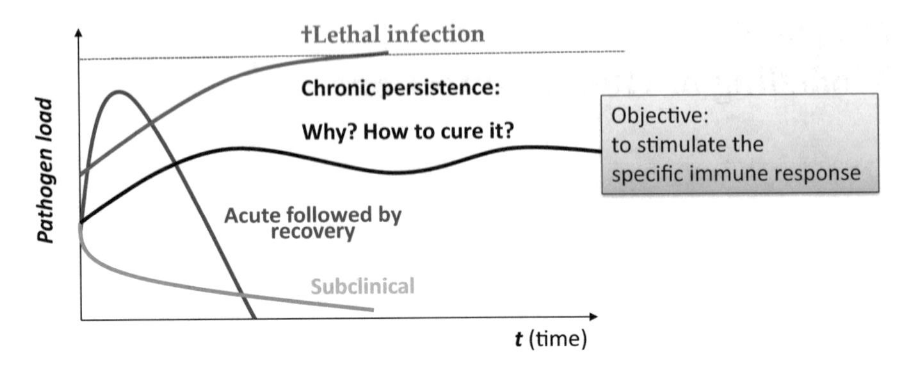

Fig. 5.1 Basic dynamic patterns of infectious diseases: subclinical, acute with recovery, chronic and lethal infection

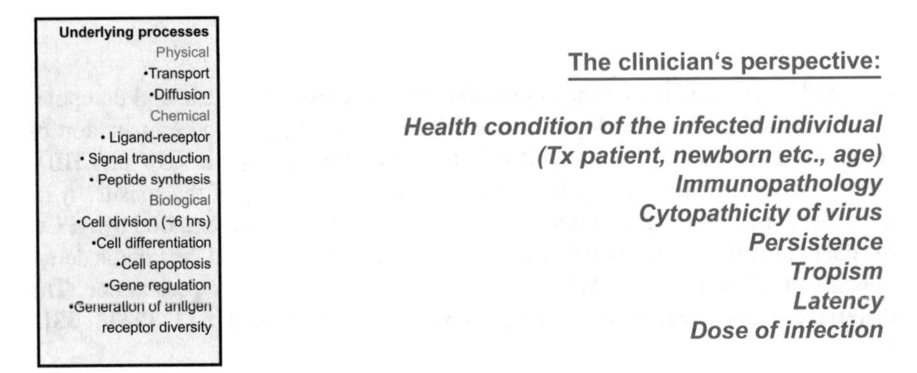

Fig. 5.2 Coordinatization problem in infectious disease modelling: view of the disease from a clinician's and mechanistic perspectives

- Latency,
- Dose of infection,

with most of them being difficult to quantify [1]. From a mechanistic point of view, the phenotype of virus–host organism interaction is determined by a set of physical, chemical and biological processes (Fig. 5.2) which are in principle amenable to a mathematical description. A general framework for understanding the regulation (sensitivity) of the immune response by the antigen growth-accumulation kinetics has been proposed by Grossman and Paul [2] based upon the notion of perturbation. The mathematical models allow one to explore a dynamic interplay between virus and host parameters in the outcome of infection (Fig. 5.3). Once the mathematical

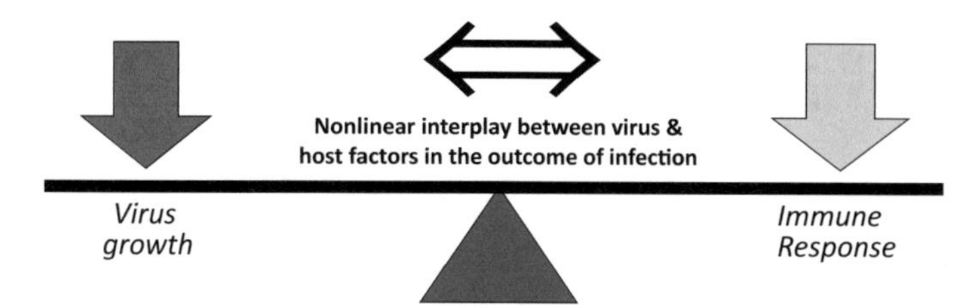

Fig. 5.3 Schematics view of the viruses and the host organism as competitors for resources of survival. There exists a nonlinear interplay between virus and host factors in the outcome of infection

model is formulated, it can be used to study theoretically the regulation of disease dynamics from a quantitative and causal perspective, i.e. in terms of 'numbers game' [1]. The effect of pathogen replication rate on the outcome of virus infection is a straightforward example of a controversial issue that can be resolved using mathematical modelling. Indeed, earlier experimental studies with LCMV infection in mice suggested that a faster virus replication is an advantage for a virus in overcoming the immune system control and establishing persistent infection by exhaustion [4]. However, the mathematical model based steady-state stability analyses suggested that slow virus replication also favours long-term persistence [6] consistent with the sensitivity theory of Grossman and Paul [2]. The contradiction was partly resolved by additional experiments as described in Chap. 4, which showed that the dependence of CTL expansion depends nonlinearly on the growth rate of the pathogen, suggesting that slower replicating (non-cytopathic) viruses have an advantage in establishing persistence.

5.2 Reference Curves: HIV and Memory T-Cell Decay Under HAART

Antigen-specific memory T cells[1] are of fundamental importance for the control of microbes, particularly persistent infections such as HIV, and hepatitis B and C viruses. The longevity of virus-specific memory T cells in humans and, in particular, their dependence on persisting antigen are poorly known and understood. To estimate turnover of such memory T cells, we analysed the data on the population kinetics of HIV-1 Gag-specific and cytomegalovirus (CMV)-specific $CD4^+$ and $CD8^+$ T cells in a cohort of HIV-1-infected individuals after highly active antiretroviral therapy (HAART) [3]. We used a simplest mathematical model in the form of an exponential (decay/growth) kinetics

$$\frac{d}{dt}N(t) = b \cdot N(t), \tag{5.2.1}$$

[1]Material of this section uses the results of our studies from AIDS Res Hum Retroviruses, Vol. 23, Sester et al., Maintenance of HIV-Specific Central and Effector Memory CD4 and CD8 T Cells Requires Antigen Persistence, Pages 549–553, Copyright © 2007 Mary Ann Liebert, Inc.

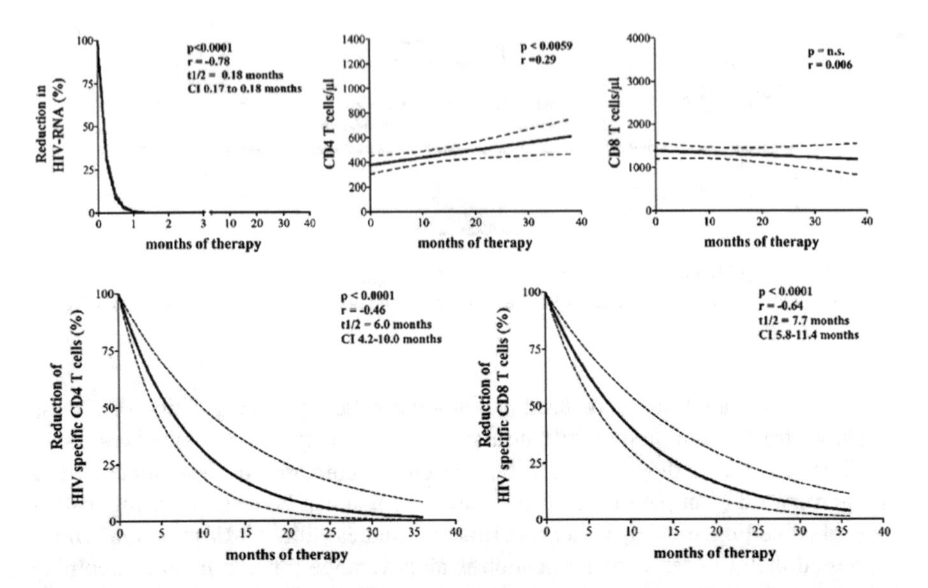

Fig. 5.4 Effect of HAART on antiviral responses in HIV infection. Best-fit regression curves are shown. Upper row: HIV viral load normalized in each individual as a percentage of the respective value at the beginning of HAART; absolute number of $CD4^+$ T cells (centre) $CD8^+$ T cells (right). Lower row: The values for absolute numbers of HIV-specific $CD4^+$ T cells (left) and $CD8^+$ T cells (right) during HAART in each individual were normalized as a percentage of the respective values at the beginning of HAART. Solid line—the exponential model, dotted lines—the 95% CIs of the uncertainty in dynamics. (The figure is reprinted from Sester et al., AIDS Res Hum Retroviruses, (2007) 23: 549–553

where $N(t)$ stands for the viral load, or $CD4^+$ and $CD8^+$ T cell numbers. Figure 5.4 shows the effect of HAART on HIV load and HIV-specific $CD4^+$ T cells and $CD8^+$ T cells summarizing the data in the form of exponential regression. The best-fit value and 95% confidence intervals for half-live parameter $\tau_{1/2} = log2/b$ are estimated. The model predicts the following mean half-lives for the key characteristics of HIV infection:

- HIV-RNA in blood $\tau_{1/2}^{VL} = 0.176$ months, $CI_{95\%} = [0.169, 0.184]$;
- HIV-specific CD4 T cells in blood $\tau_{1/2}^{CD4} = 6.0$ months, $CI_{95\%} = [4.2, 10]$;
- HIV-specific CD8 T cells in blood $\tau_{1/2}^{CD8} = 7.7$ months, $CI_{95\%} = [5.8, 11.4]$;

With respect to the HIV-specific immunity under HAART, the HIV load is reduced so low that the HIV-specific $CD4^+$ T-cell memory effectively collapses. This loss of help plus the reduction of the antigen load may in concert lead to the loss of the $CD8^+$ T-cell memory pool with slightly slower kinetics. Although the decrease of HIV-specific CTL responses under HAART was known, the above analysis shows that both global HIV-specific $CD4^+$ and $CD8^+$ T memory cell responses decay in a manner directly coupled to plasma viral load.

In contrast to HIV-specific responses, the cytomegalovirus (CMV)-specific T-cell responses in CMV-positive individuals are shown to be robust to HAART. The

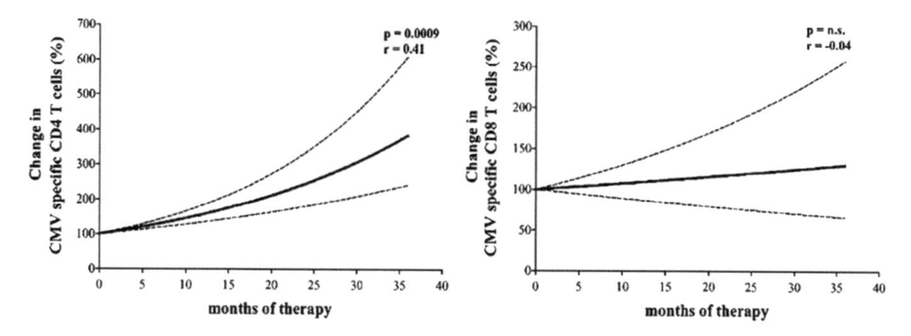

Fig. 5.5 Effect of HAART on absolute CMV-specific T-cell subsets. The values for absolute numbers of CMV-specific CD4$^+$ T cells (left) and CD8$^+$ T cells (right) during HAART in each individual were normalized as a percentage of the respective values at the beginning of HAART. Solid line—the exponential kinetics model, dotted lines—the 95% CIs. (The figure is reprinted from Sester et al., AIDS Res Hum Retroviruses, (2007) 23: 549–553

corresponding reference curves are shown in Fig. 5.5. In general, the maintenance of T cell memory represents a dynamic equilibrium between cell proliferation and cell death. Inhibition of HIV replication by HAART would then revert this balance to a normal equilibrium and lead to an increase in CMV-specific CD4$^+$ and CD8$^+$ T cells. Alternatively, instead of being an effect of mere cell death, the production rate of CMV-specific cells may be lower before the treatment because of activation threshold tuning resulting from chronic low-level stimulation by cytokines and/or self-antigens. This may reduce responsiveness of resting memory cells to signals, including those involved in homeostasis. Consequently, normalization of the immune activation level after HAART may have facilitated a slow increase of the CMV-specific population to the normal steady state.

The above reference curves, although being quite simple in terms of the underlying model, can be used to monitor the effect of HAART on the restoration on immune function and robustness of specific T cell immunity in HIV-infected individuals under therapy. Note that earlier, similar methodology was developed by G.I. Marchuk for management of recovery from HBV infection [6].

5.3 Chronic HBV Infection

Infection with human hepatitis B virus (HBV),[2] affecting about 400 million persons worldwide [26], represents a dynamic process with a spectrum of clinical outcomes ranging from acute infection followed by virus clearance to chronic persistence of

---

[2]Material of Sects. 5.3.1 and 5.3.2 uses the results of our studies from Journal of Virology, Vol. 78, Bocharov et al., Underwhelming the immune response: effect of slow virus growth on CD8+-T-lymphocyte responses, Pages 2247–2254, Copyright © 2004 by the American Society for Microbiology. Material of Sect. 5.3.3 uses the results of our studies from Journal of Computational and Applied Mathematics, Vol. 184, Luzyanina et al., Numerical bifurcation analysis of immunological models with time delays, Pages 165–176, Copyright © 2005, with permission from Elsevier.

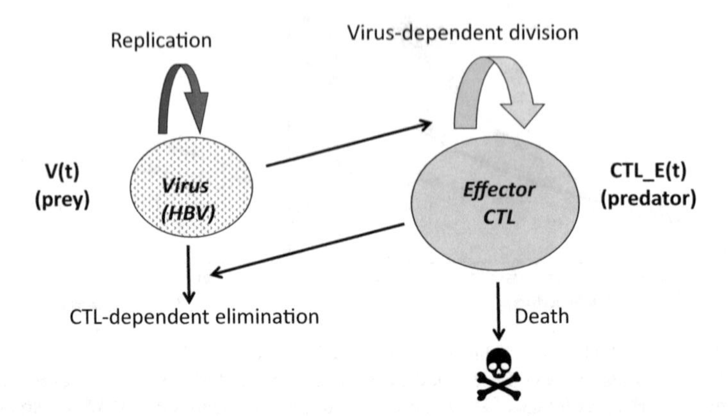

Fig. 5.6 Schematic view of the mechanistic framework for the dynamic analysis of the HBV-CTL interaction representing the amplification mode

the virus. A persistent HBV infection can lead to high morbidity and mortality due to the development of end-stage liver diseases [16]. The basis for the inadequate immune response that is characteristic of the onset of chronic HBV infection is not well understood [15].

Although protection against infection is a multifactorial phenomenon depending on both the innate and adaptive immune mechanisms, CTLs play a vital role in the control of human HBV infection via both cytolytic and non-cytolytic mechanisms. The dynamics of virus infections can be biologically characterized in terms of the interacting populations of viruses and effector CTLs (see Fig. 5.6). In the amplification mode of the immune response, the virus acts as a positive regulator of the CTL population, whereas CTLs function to eliminate the virus population, so their mutual interaction dynamics can be viewed and formally described as a predator (CTL)-prey (virus)-type system. Our analysis of the limited set of HBV data are based on a simple predator–prey-type model, which differs from the classical Lotka and Volterra model in that the rate of growth of the predator population (CTLs) saturates at a defined predator density. This allows us to more accurately mimic the kinetic patterns of virus and CTLs observed in a patient during acute HBV infection, as shown in Fig. 5.7.

5.3.1 Deterministic Model of HBV Infection

The dynamics and outcome of HBV infection is driven by a complex interaction between the virus and the host immune response. The HBV-specific $CD8^+$ cytotoxic T-cell (CTL) response plays a fundamental role in viral clearance. The ability to clear the virus after infection is associated with the presence of a strong virus-specific CTL response [24]. Earlier [14] a delay-differential equation based model was proposed to describe the acute phase of HBV infection. The available patient's data characterized the population dynamics of HBV (HBV DNA copies/ml) and HBV-specific CTL (cell/ml) in blood, see Fig. 5.7. As it was discussed in Chap. 3, such a low-resolution data do not support complex models with a number of parameters above some threshold. Therefore, we consider a reduced complexity DDE-based model which describes the interaction between the virus and the CTL response as a predator (CTL)—prey (virus) type system. Using $V(t)$ and $E(t)$ for population densities of virus and CTL respectively, we describe the rate of changes of the densities by the following system of delay differential equations:

$$\begin{aligned} \dot{V}(t) &= \beta V(t)(1 - V(t)/K) - \gamma V(t)E(t), \\ \dot{E}(t) &= bV(t-\tau)E(t-\tau)/(\theta + V(t)) - \alpha E(t) + C. \end{aligned} \tag{5.3.2}$$

The equations of the deterministic model specify the rate of changes of the population densities of the hepatitis B virus $V(t)$ and virus-specific cytotoxic T lymphocytes $E(t)$ in the acute phase of the infection. This simple model is based on (i) a Verhulst–Pearl logistic form for virus growth; (ii) second-order virus elimination kinetics by CTLs; (iii) the Holling type II response curve for CTLs expansion with a time lag representing cell division time and antigen-independent production/death of CTLs in the immune system (homeostasis).

The parameters of the model $(\beta, \gamma, K, b, \tau, \theta, \alpha, C)$ were estimated following the maximum likelihood approach from the data on acute HBV infection. The relevant information about the model parameters is given in Table 5.1. The best-fit solution is shown in Fig. 5.7a.

5.3.2 *Sneaking Through Phenomenon*

Since the experimental discovery of T-cell exhaustion phenomenon in LCMV infection [4], the speed of virus replication has typically been seen as an advantage for a virus in overcoming the ability of the immune system to control its population growth. From the other side, according to the 'balance of growth and differentiation' concept, the immune system tends to respond to strong perturbations caused by rapid increases in antigen appearance and by inflammation, which are characteristic of acute infections, but adapts to and/or tolerates slow changes. Mathematical studies of the asymptotic stability of the steady states of the basic infectious disease model [6] suggested that slow virus replication favours the long-term persistence. The frequencies of virus-specific CTLs that characterize chronic infections in humans by non-cytopathic viruses, such as hepatitis B virus (HBV) (and hepatitis C virus) are generally very low in the face of high virus loads compared to the frequencies in those patients who successfully resolve the infection. However, it is not clear to what extent this reflects differences in the initial status of the antiviral response or subsequent

Table 5.1 The biological meaning of the model parameters, their best-fit estimates for the acute phase of HBV infection and plausible ranges, i.e. 95% confidence intervals

Parameter	Biological meaning	Units	Best-fit estimates for acute infection	Plausible ranges 95% CIs
β	Replication rate constant of viruses	1/day	0.27	[0.27, 0.32]
γ	Rate constant of virus clearance due to CTLs	ml/(copies day)	6.2×10^{-4}	$[10^{-3}, 3.4 \times 10^{-3}]$
K	Virus carrying capacity	copies/ml	1.5×10^{10}	$\geq 0.4 \times 10^{10}$
b	Rate constant of CTL stimulation	1/day	0.15	[0.09, 0.15]
θ	Viral load saturation in CTL expansion rate	copies/ml	1.2×10^7	$[0.34 \times 10^2, 10^6]$
τ	Duration of CTL division cycle	day	0.6	fixed ad hoc [0.4 – 1]
α	Rate constant of CTL death	1/day	0.035	[0.021, 0.063]
C	Rate of CTL export from thymus	cell/(ml day) export from thymus	0.23	fixed ad hoc [0, 0.23]

exhaustion. Under some circumstances, however, the ability of such viruses to persist correlates with acute CTL responses which appear to be weaker from the outset. To examine the impact of the magnitude of the CTL response to virus growth rate, a mathematical model of CTL response to HBV infection can be fruitfully applied.

The major point of the modelling process was to examine the sensitivity of the immune response to a decrease in the HBV growth rate compared to the acute infection. To investigate the virus growth effect on CTL expansion, we conducted mathematical analyses based on the model simulating CTL responses of patients to infection with HBV mutants with different doubling times. The replication rate estimates (τ_d ranging from 2.2 to 5.8 days) for acute HBV infection are derived from a study of a cohort of seven patients performed by Whalley et al. [9]. The results summarized in Fig. 5.7b, c show that a decrease in the virus population doubling time proportionally increases the peak virus number, the magnitude of the CTL response and the overall efficacy of virus elimination. An HBV infection with a slowly replicating strain, with the doubling time, set at ~14.7 days, induces only a weak response, and the virus tends to persist.

Patients may have different major histocompatibility complex backgrounds, resulting in various patterns of responsiveness to a given virus strain. Variations in the clonal burst size and dynamics of virus-specific CTLs might reflect differ-

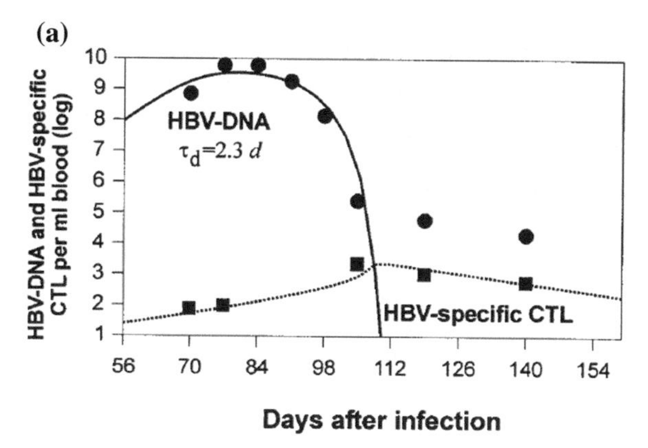

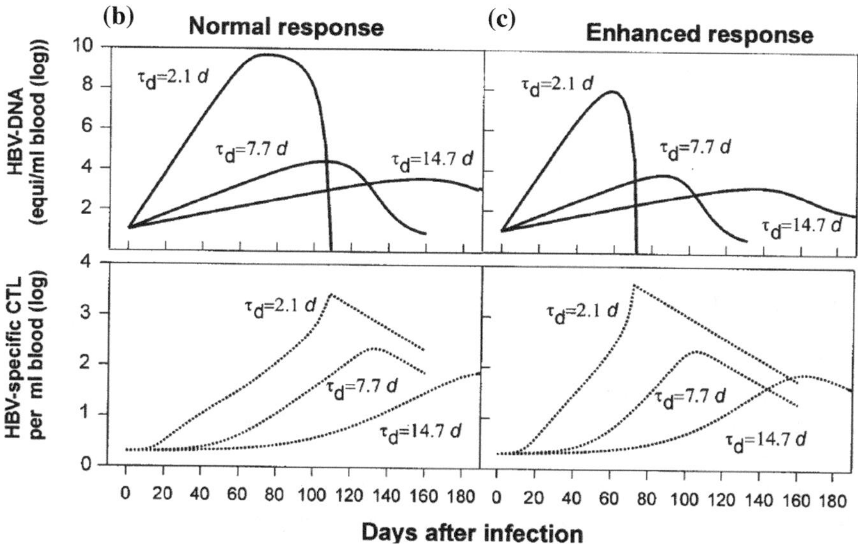

Fig. 5.7 Model predictions of HBV-CTL dynamics for different virus population doubling times. **a** Clinical data for patient 1 [8] and the corresponding mathematical model simulation. d, days. **b** and **c** Viremia and CTL numbers in blood of a normal responder (**b**) and a high responder (**c**). A reduction in the HBV growth rate during the incubation period leads to a weaker CTL expansion and underwhelming of the host immune response. (The figure is reprinted from Bocharov et al. Journal of Virology (American Society for Microbiology), 2004, 78: 2247–2254)

ences in the initial CTL precursor frequency and/or in the activation thresholds of lymphocytes. The latter are also determined by the genetic makeup of the host. To simulate the difference in the HBV dynamics between a high responder and a normal responder, we increased the CTL population division rate by 50%. It has been recorded [5] that minor quantitative differences in the lymphocyte responsiveness parameter may result in large changes in the magnitude of the response due to nonlinear amplification, akin to the growth of cells. The overall dynamics of infection shifts towards a lower viremia, a stronger CTL response and earlier virus elimination (Fig. 5.7c). However, the slowly replicating strain ($\tau_d \sim 14.7$ days) still has moderate growth but elicits a weak and transient CTL response that decreases after day 200 (not shown).

The above modelling analyses suggest that transition from acute to chronic HBV infection might be explained (at least in part) by a decrease in the replication rate of the evolving virus population. Overall, more slowly replicating viruses may evoke weaker cellular immune responses and therefore enhance their likelihood of persistence. Thus, such viruses may sneak through immune surveillance by underwhelming the immune response of the host, rather than overwhelming it, in order to establish a persistent infection.

5.3.3 Low-Level HBV Persistence

Clinical data suggest that HBV[3] can persist at various levels rather than being completely eliminated from the host. The low-level HBV persistence (i.e. below the detection limit of conventional assays of about $10^2 - 10^3$ HBV DNA copies/ml of serum) is relevant for sustaining the immunological memory (called 'infection immunity' [10]) but can have negative consequences for the host, e.g. the possibility of viral reactivation during immunosuppression and the likelihood that these individuals or their organs may be infectious to others [7]. The mechanism of the viral persistence is not understood [7] as the quantitative parameters of the low-level HBV persistence are outside the range of the analytical techniques existing in clinical practice. Therefore, one can use the mathematical model for HBV-CTL dynamics to explore the kinetic basis of low-level HBV persistence.

Time integration of the initial value problem (5.3.2) with the estimated parameters from acute infection data shows that the corresponding solution presented in Fig. 5.8 is a periodic solution with $V(t)$ oscillating between (almost) zero and $3.5 \cdot 10^9$ copies/ml. We are interested to know the necessary conditions, in terms of the turnover rates for virus and CTLs, allowing the coexistence of a small-scale virus population and memory CTL population either as an equilibrium state or as an oscillatory pattern. In the context of dynamical system analysis, the low-level virus persistence corresponds to a stable steady-state solution or to a stable periodic solution of model (5.3.2) with V, respectively $V(t)$, below 10^3 copies/ml. We study the model predictions related to the above issue through a numerical bifurcation analysis of model (5.3.2) with the package DDE-BIFTOOL (see Sect. 4.2.3.3 for details).

It is necessary to note that model parameters characterizing the virus and CTL kinetics might change in transition from the acute to the memory phase of HBV infection, reflecting, for example, the effect of humoral immunity at later stages

[3] Material of this section uses the results of our studies from Journal of Computational and Applied Mathematics, Vol. 184, Luzyanina et al., Numerical bifurcation analysis of immunological models with time delays, Pages 165–176, Copyright © 2005, with permission from Elsevier.

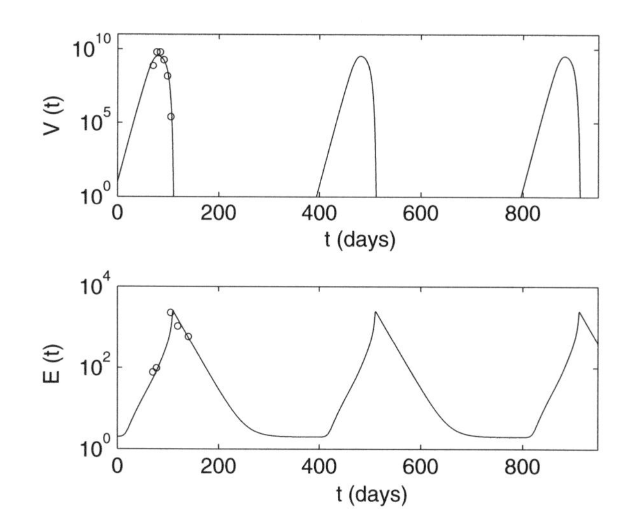

Fig. 5.8 Solutions of model (5.3.2) with the parameters estimated for the acute phase of HBV infection (Table 5.1). The initial data used: $V(t) = 0$, $t \in [-\tau, 0)$, $V(0) = 10$ copies/ml, $E(t) = 2$ cell/ml, $t \in [-\tau, 0]$. $V(t) < 1$ is not depicted. Clinical data are denoted by $\circ$. (Reprinted from Journal of Computational and Applied Mathematics, Vol. 184, Luzyanina et al., Numerical bifurcation analysis of immunological models with time delays, Pages 165–176, Copyright © 2005, with permission from Elsevier)

of the acute infection as well as maturation of initially naive CTLs to the memory phenotype. Therefore, we focus our analysis on the virus replication rate β and the CTL death rate α. We start with the stability analysis of steady-state solutions of (5.3.2) and proceed by analysing oscillatory patterns in virus and memory CTL persistence.

5.3.3.1 Steady States Representing Infection Immunity

To start the analysis, we set the virus growth rate $\beta = 0.1$ and all other parameters as in Table 5.1. We examine the dependence of the corresponding steady state on the parameter α, $\alpha \in [0.001, 0.12]$. The corresponding branch of steady-state solutions (cf. Fig. 5.9) was computed using continuation and its stability was assessed by computing the rightmost characteristic roots λ with $\Re(\lambda) > -13$. For $\alpha = 0.12$, the steady state is unstable, $\Re(\lambda_{1,2}) \approx 2.5 \cdot 10^{-3}$, and it is stabilized through a Hopf bifurcation at $\alpha \approx 0.031$. Further, at $\alpha = 1.75 \cdot 10^{-3}$, the computed solution intersects the solution $V = 0$, $E = C/\alpha$. At this point, a real characteristic root equals zero indicating a transcritical bifurcation.

Note that a pair of dominant complex conjugate characteristic roots with a very small (positive or negative) real part is a distinctive property of steady-state solutions of system (5.3.2) observed for the model parameters within their plausible biological ranges (see [12] for details). Mathematically, this implies a very slow and oscillatory transition either to a new stable state of the system or to its previous state after a perturbation of its current state. The latter is consistent with the nature of HBV infection characterized by a slow kinetics in both the acute and memory phase of the infection.

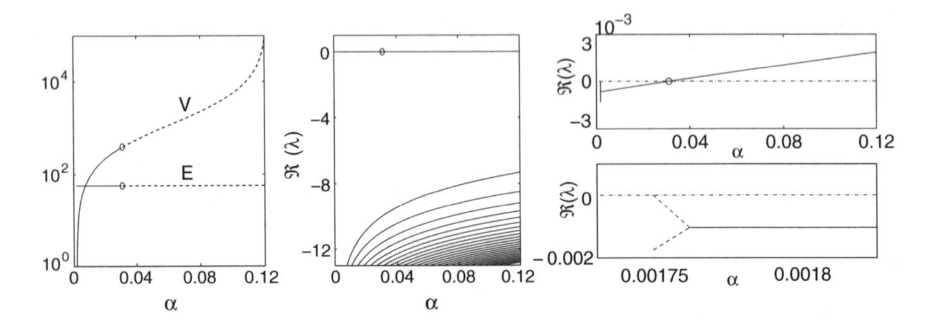

Fig. 5.9 Left: Stable (−) and unstable (−−) solutions V and E along a branch of steady-state solutions of (5.3.2) versus parameter α. Middle: Real part of the rightmost roots of the characteristic equation along the same branch. Right: Two enlargements of the middle figure. Hopf bifurcation (○) at $\alpha \approx 0.031$. Complex and real roots are denoted by solid, respectively dashed lines. (Reprinted from Journal of Computational and Applied Mathematics, Vol. 184, Luzyanina et al., Numerical bifurcation analysis of immunological models with time delays, Pages 165–176, Copyright © 2005, with permission from Elsevier)

Under the condition $V \ll K$, which is fulfilled in the case of low level viral persistence, an approximate steady-state density of the populations can be obtained from the nonlinear algebraic system for steady-state solutions of (5.3.2),

$$\begin{aligned} &\beta V(1 - V/K) - \gamma EV = 0, \\ &bVE/(\theta_{sat} + V) - \alpha E + C = 0, \end{aligned} \tag{5.3.3}$$

as

$$V \approx \frac{\theta(\alpha\beta - C\gamma)}{\beta(b - \alpha) + C\gamma}, \quad E \approx \frac{\beta}{\gamma}. \tag{5.3.4}$$

These approximations are in agreement with the numerical results presented in Fig. 5.9(left) corresponding to low values of V.

Starting from the found Hopf point, we computed a branch of Hopf bifurcation points in the (β, α)-plane, see Fig. 5.10(left). Using a sequence of similar continuations, we computed two more branches of Hopf points corresponding to $\tau = 0.4$ and $\tau = 1$. This figure also shows the curve of transcritical bifurcation points. Since no other bifurcations were found, the stability region of the steady state is bounded by the corresponding Hopf curve and the transcritical bifurcation curve. The regions in the (β, α)-plane where virus persists below the detection limit are depicted in Fig. 5.10(left). The analysis predicts that (i) the value of V is almost independent of β unless β gets close to 0 and (ii) the virus can persist at a very low level if the death rate of CTL is small enough. Steady-state values of V and E along the branch of Hopf points ($\tau = 0.6$) as a function of parameter β are shown in Fig. 5.10(right). Notice that for each value of β, there is an associate value of α on the branch of Hopf points.

These results specify a quantitative connection between the parameters β, α and τ necessary to ensure a stable coexistence of low-level HBV population and memory

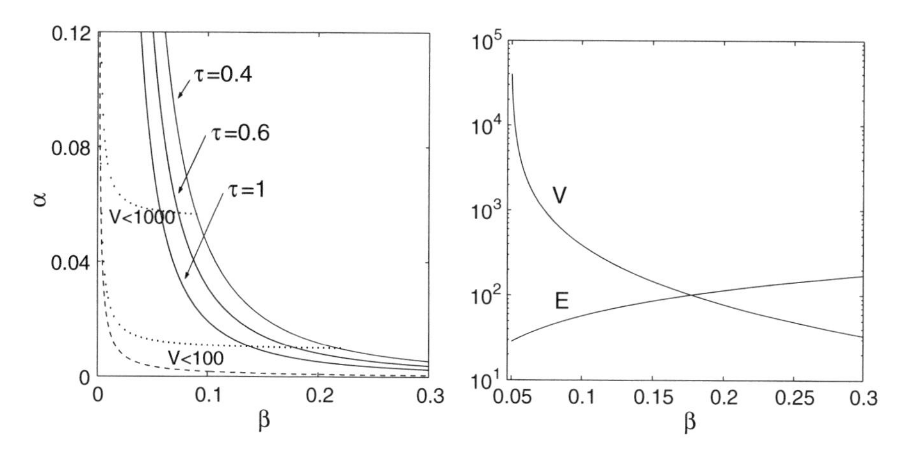

Fig. 5.10 *Left* In the (β, α)-plane, branches of Hopf points (–) for $\tau = 0.4, 0.6, 1$, and the curve of transcritical bifurcation points (– –). Dotted curves indicate the upper bound of regions with $V < 10^2$ and $V < 10^3$. *Right* Steady-state solutions V and E along the Hopf curve shown in the left figure for $\tau = 0.6$. (Reprinted from Journal of Computational and Applied Mathematics, Vol. 184, Luzyanina et al., Numerical bifurcation analysis of immunological models with time delays, Pages 165–176, Copyright © 2005, with permission from Elsevier)

CTLs as an equilibrium state. In particular, the model predicts that such coexistence is possible for relatively small values of the rates of virus replication and CTL death. The impact of other virus and CTL parameters (γ, b, C) on low-level viral persistence can be investigated in a similar way.

5.3.3.2 Oscillatory Patterns of Persistence

From Hopf bifurcation points at which the (steady state) value of V is low, branches of periodic solutions emanate which represent another mode of a low-level viral persistence. Figure 5.11 depicts a branch of periodic solutions emanating from the Hopf point $(\beta = 0.3, \ \alpha \approx 0.0038, \ \tau = 0.6)$ as a function of the parameter β. The variation of solutions along this branch is characterized by their maximal and minimal values over the period for each computed point on the branch, i.e. $V_{\max}(\beta) = \max_{t \in [0,T]} V(t, \beta)$, $V_{\min}(\beta) = \min_{t \in [0,T]} V(t, \beta)$, etc. As β increases from its Hopf point value, the amplitude of oscillations of $V(t)$ grows very rapidly and a subtle change in β (from 0.3 to 0.317) leads to 'pulse' oscillations in virus population size, exceeding the detection level, see Fig. 5.12. Note that periodic solutions are depicted on the time interval [0, 1], i.e. after time is scaled by the factor T^{-1} with T being the period of the solution.

In a similar way (starting from different Hopf points), branches of periodic solutions can be computed as function of any parameter of the model. The results show a high sensitivity of the amplitude of oscillations of the virus population to changes in the parameters, especially in β and α. Hence, the model predicts that oscilla-

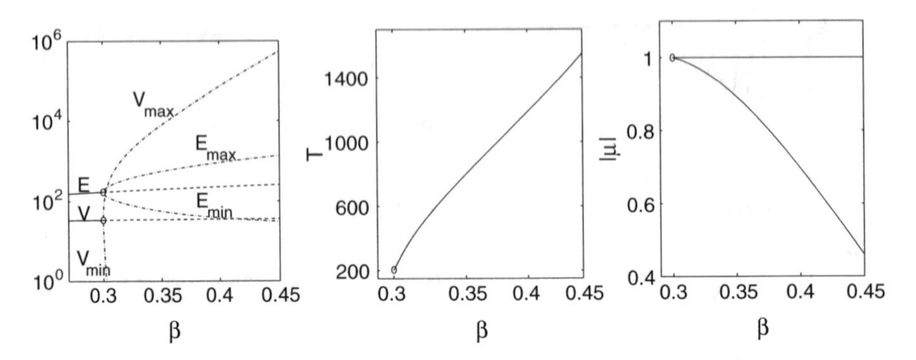

Fig. 5.11 Evolution of maximal and minimal values of V and E (left), period T (middle) and moduli of the computed dominant Floquet multipliers μ (right) along a branch of periodic solutions of (5.3.2) emanating from a Hopf point (∘) versus parameter β. Branches of stable (−) and unstable (−−) steady-state solutions and branch of stable periodic solutions (−·). $\alpha \approx 0.0038$. (Reprinted from Journal of Computational and Applied Mathematics, Vol. 184, Luzyanina et al., Numerical bifurcation analysis of immunological models with time delays, Pages 165–176, Copyright © 2005, with permission from Elsevier)

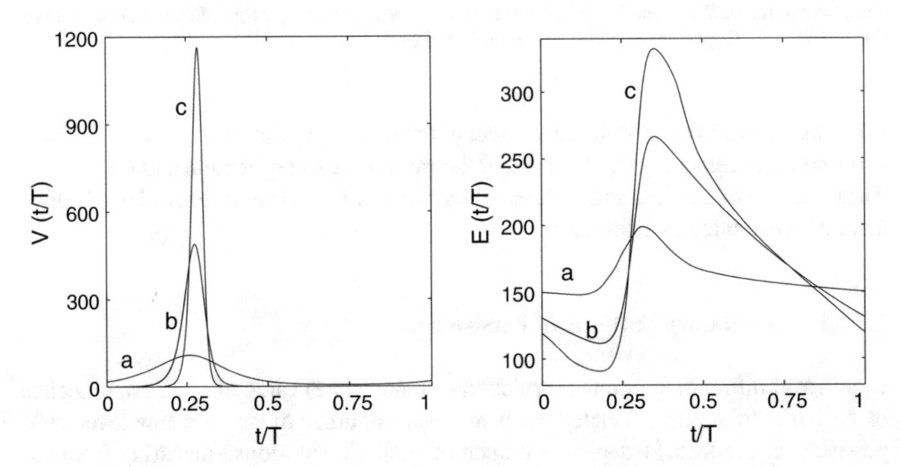

Fig. 5.12 Sensitivity of periodic solutions to changes in virus growth rate β. Solutions V and E corresponding to three points on the branch of periodic solutions shown in Fig. 5.11: $\beta = 0.301$ (**a**), $\beta = 0.308$ (**b**), $\beta = 0.317$ (**c**). Values of the period T (days): $T \approx 221.7$ (**a**), $T \approx 344.1$ (**b**), $T \approx 469.0$ (**c**). (Reprinted from Journal of Computational and Applied Mathematics, Vol. 184, Luzyanina et al., Numerical bifurcation analysis of immunological models with time delays, Pages 165–176, Copyright © 2005, with permission from Elsevier)

tory patterns in low-level viral persistence (with virus population varying below 10^3 copies/ml) are possible within very narrow intervals of the rates of virus replication and CTLs death.

5.4 Spontaneous Recovery from Chronic HBV Infection

One of the intriguing issues in the pathogenesis of chronic HBV infection[4] is related to the phenomenon of spontaneous recovery observed in some patients [11, 15, 16]. In the previous section, we examined the emergence of HBV persistence through numerical bifurcation analysis of the deterministic mathematical model. In the following section, we elaborate a computational methodology for analysing the sensitivity of the virus infection dynamics to random perturbations in the virus replication and immune responses parameters. The practical details of the implementation of the stochastic ODE models in the analysis of spontaneous recovery are presented. These include the effect of sampling of the parameter space, the number of simulation runs needed for a robust estimation of the mean and the variance of the spontaneous recovery pattern, the impact of the noise intensity and the noise type on the response of the models.

5.4.1 *Stochastic Framework for Modelling HBV Infection*

Although persistent HBV infection is associated with ineffective CTL responses, the onset of chronic infections is considered to be multifactorial with the potential contributing factors including mutational variations of HBV, alterations in the innate- and B-cell responses [15]. The variations can affect either the replication of the virus or its elimination kinetics, e.g. via parameters β and γ, respectively. One way to explore their impact on the dynamics of HBV infection could be the extension of the deterministic description of the virus-CTL interaction to include the stochastic forcing either in an additive or multiplicative way. The resulting stochastic models offer a more realistic representation for studying the long-term kinetics of the HBV-immune system interaction. Specifically, we consider two stochastic extensions of the deterministic ODEs version of model (5.3.2) (in which the delay in proliferation of CTLs is taken into account implicitly by the proliferation rate parameter b) assuming that the virus dynamics equation is subject to either (i) a nonspecific rapidly fluctuating forcing term, or (ii) additionally to (i), the virus replication/elimination processes are fluctuating in the presence of ongoing CTL response.

5.4.1.1 Additive Noise Extension

Introducing an additive noise with an intensity $\sigma > 0$ to the first equation of (5.3.2), we obtain the following system of stochastic differential equations of Itô type (see for definitions [18]):

[4]Material of Sect. 5.4 uses the results of our studies from Mathematics and Computers in Simulation, Vol. 96, Luzyanina and Bocharov, Stochastic modeling of the impact of random forcing on persistent hepatitis B virus infection, Pages 54–65, Copyright © 2014, with permission from Elsevier.

$$\begin{aligned} \mathrm{d}V(t) &= \Big(\beta V(t)(1 - V(t)/K) - \gamma V(t)E(t)\Big)\mathrm{d}t + \sigma \mathrm{d}W(t), \\ \mathrm{d}E(t) &= \Big(bV(t)E(t)/(\theta + V(t)) - \alpha E(t) + C\Big)\mathrm{d}t, \end{aligned} \tag{5.4.5}$$

where $W(t),\ t > 0$, is a standard Wiener process, i.e. a Gaussian process with independent increments such that $W(0) = 0$ w.p. (with probability) 1, $E(W(t)) = 0,\ Var(W(t) - W(s)) = t - s$ for all $0 \leq s \leq t$. The solution $\{V(t),\ E(t),\ t \in [0, t_{final}]\}$ of (5.4.5) is a two-dimensional Itô process. The equivalent stochastic integral equations read

$$\begin{aligned} V(t) &= V(0) + \textstyle\int_0^t (\beta V(s)(1 - V(s)/K) - \gamma V(s)E(s))\mathrm{d}s + \int_0^t \sigma \mathrm{d}W(s), \\ E(t) &= E(0) + \textstyle\int_0^t (bV(s)E(s)/(\theta + V(s)) - \alpha E(s) + C)\mathrm{d}s, \end{aligned} \tag{5.4.6}$$

where the first integral in both equations is a regular Riemann–Stieltjes integral and the second integral in the first equation is a stochastic integral in Itô form.

5.4.1.2 Multiplicative Noise Extension

To investigate influence of fluctuating (versus constant) rates β and γ on the model solutions, we randomize these parameters as follows. Let $p \in [p_l,\ p_u]$ be a parameter being randomized and $p_l,\ p_u$ are its low and upper bounds. We assume that p varies randomly according to the equation $p(t) = \tilde{p} + \sigma_p \xi(t)$, where $\tilde{p}$ is the value of p around which we randomize, $\xi(t)$ is a white noise process (i.e. $\xi(t)$ is a standard Gaussian random variable for each t) and $\sigma_p > 0$ is the intensity of the noise. We choose σ_p as $\sigma_p = \min((\tilde{p} - p_l),\ (p_u - \tilde{p}))/3$ to ensure that p remains in the interval $[p_l,\ p_u]$ with probability 0.997. This implies that about 99.7% of values drawn from a normal distribution are within three standard deviations σ_p away from the mean (3-sigma rule).

Randomizing the parameters β and γ in (5.4.5) will result in the following system of stochastic differential equations:

$$\begin{aligned} \mathrm{d}V(t) &= \Big(\beta V(t)(1 - V(t)/K) - \gamma V(t)E(t)\Big)\mathrm{d}t + \\ &\quad \sigma_\beta V(t)(1 - V(t)/K)\mathrm{d}W_1(t) - \sigma_\gamma V(t)E(t)\mathrm{d}W_2(t) + \sigma \mathrm{d}W_3(t), \\ \mathrm{d}E(t) &= \Big(bV(t)E(t)/(\theta + V(t)) - \alpha E(t) + C\Big)\mathrm{d}t, \end{aligned} \tag{5.4.7}$$

where $W_1(t),\ W_2(t)$ and $W_3(t)$ are independent standard Wiener processes, constituting a three-dimensional Wiener process $W(t) = (W_1(t),\ W_2(t),\ W_3(t))$. The equivalent stochastic integral equations are

$$\begin{aligned} V(t) &= V(0) + \textstyle\int_0^t (\beta V(s)(1 - V(s)/K) - \gamma V(s)E(s))\mathrm{d}s + \\ &\quad \textstyle\int_0^t \sigma_\beta V(s)(1 - V(s)/K)\mathrm{d}W_1(s) - \\ &\quad \textstyle\int_0^t \sigma_\gamma V(s)E(s)\mathrm{d}W_2(s) + \int_0^t \sigma \mathrm{d}W_3(s), \\ E(t) &= E(0) + \textstyle\int_0^t (bV(s)E(s)/(\theta + V(s)) - \alpha E(s) + C)\mathrm{d}s. \end{aligned} \tag{5.4.8}$$

To avoid dealing with very large and small numbers in computations, we rewrite stochastic models (5.4.5) and (5.4.7) using the logarithmic transform $\tilde{V} = \ln V,\ \tilde{E} = \ln E$ of the state variables. Then both stochastic models read

$$\begin{aligned} d\tilde{V}(t) &= \Big(\beta(1 - \exp(\tilde{V}(t))/K) - \gamma \exp(\tilde{E}(t))\Big)dt + \\ &\quad \sigma \exp(-\tilde{V}(t))dW(t), \\ d\tilde{E}(t) &= \Big(b \exp(\tilde{V}(t))/(\theta + \exp(\tilde{V}(t))) - \alpha + C \exp(-\tilde{E}(t)\Big)dt \end{aligned} \tag{5.4.9}$$

and

$$\begin{aligned} d\tilde{V}(t) &= \Big(\beta(1 - \exp(\tilde{V}(t))/K) - \gamma \exp(\tilde{E}(t))\Big)dt + \\ &\quad \sigma_\beta(1 - \exp(\tilde{V}(t))/K)dW_1(t) - \\ &\quad \sigma_\gamma \exp(\tilde{E}(t))dW_2(t) + \sigma \exp(-\tilde{V}(t))dW_3(t), \\ d\tilde{E}(t) &= \Big(b \exp(\tilde{V}(t))/(\theta + \exp(\tilde{V}(t)) - \alpha + C \exp(-\tilde{E}(t))\Big)dt. \end{aligned} \tag{5.4.10}$$

Note that after this transformation, the random forcing in model (5.4.9) becomes a multiplicative noise.

5.4.2 *Quantitative Spectrum of Chronic HBV Infection*

The persistent HBV infection can take various clinical forms, which are classified (e.g. asymptomatic carrier, chronic infection) according to the observable characteristics of the virus–host interaction, such as viral load, the strength and the patterns of immune responses and liver damage. Studies in HBV-infected chimpanzees suggested that low virus doses favour the establishment of persistence, while higher doses favour viral clearance [15]. Chronic hepatitis B infection is a highly heterogeneous disease with the viral load ranging from 10^3 to 10^9 HBV DNA copies/ml [24]. In treated patients, the withdrawal of antiviral therapy resulted in HBV rebound from $10^2 - 10^3$ copies/ml to $10^9 - 10^{10}$ copies/ml [23]. It was found that completely different levels of HBV replication can coexist with slightly different numbers in the circulation and comparable numbers of intrahepatic HBV-specific CTL [19]. This feature was interpreted, using the theoretical framework provided in [21], to reflect the differences in CTL responsiveness which denote the rate of expansion and antiviral efficacy of the virus-specific CTLs. Efficient CTL responsiveness would result in a lowering of the viral load and smaller CTL abundance with the lower antigenic stimulation. Dysfunctional CTLs are considered to be a hallmark of chronic HBV infection [23]. Notice that the parameter, representing the functionality of CTL in the considered mathematical model, is the rate constant γ of HBV elimination from the host. The clinical outcome of persistent HBV infection may result in spontaneous recovery [11, 16, 23] with the underlying mechanisms remaining to be identified.

The biological objective of our study is to investigate the sensitivity of persistent HBV infection to the fluctuating variations in virus production/elimination processes. Based on the above consideration, we examine two variants of chronic HBV infection:

1. LVLTA: the low viral load and low cytotoxic T lymphocyte abundance (LVLTA) infection and
2. HVLTA: the high viral load and high cytotoxic T lymphocyte abundance (HVLTA) infection.

According to the results of longitudinal analysis of CTL responses in chronic HBV infection [24], the threshold level of viral load, separating the two types, is set to 10^7 HBV DNA copies/ml. The threshold for CTL abundance in blood is taken to be 10^3 HBV-specific cells/ml, which is the peak number of CTLs in acute resolving HBV infection [14]. Therefore, the first type of chronic HBV infection (LVLTA) is characterized by smaller values of virus and CTL populations:

$$V_{average} := \frac{1}{T}\int_0^T V(t)\mathrm{d}t < 10^7, \quad E_{average} := \frac{1}{T}\int_0^T E(t)\mathrm{d}t < 10^3, \qquad (5.4.11)$$

where $[0, T]$ is a time interval of the chronic infection and $\{V(t),\ E(t),\ t \in [0, T]\}$ is a solution of the deterministic model (5.3.2). In our study, we use $T = 730$ days (2 years). For the second type of chronic HBV infection (HVLTA), we assume that

$$V_{average} > 10^7, \quad E_{average} > 10^3. \qquad (5.4.12)$$

We also assume that once the viral load variable of the model, $V(t)$, drops below the clinical detection threshold of 10^2 copies/ml, i.e. $V(t) < 10^2$ at some time t, the corresponding solution can be interpreted as a spontaneous recovery case. We examine the robustness of the two variants of the persistent HBV infection subject to random fluctuating forcing, using the above two stochastic versions of the mathematical model of HBV infection.

5.4.3 Numerical Methods

For every specified set of the model parameter values, we solve the deterministic model (5.3.2) and compute $V_{average}$ and $E_{average}$ to decide whether this parameter set results in LVLTA- or HVLTA chronic HBV infection. For the corresponding simulations, the MATLAB code `ode45` [20] is used.

5.4.3.1 Simulation of the Stochastic Models

In order to compute the trajectories of solutions to the SDE model (5.4.9), we used the solvers of the SDE Toolbox [22] intended for simulation of sample paths of solutions to SDEs with diagonal noise. Two methods to solve SDEs are implemented in this Toolbox: the Euler–Maruyama method (strong order $1/2$) and the Milstein method (strong order 1). We used the Milstein scheme and modified the code to reduce computational effort. In order to solve model (5.4.10) with three-dimensional non-diagonal noise, we implemented the approach proposed in [17] to approximate double Itô integrals appearing in the Milstein scheme for non-diagonal noise.

Error analysis

If we know a solution $X(t)$ to an SDE explicitly, we can calculate the error of its approximation $\tilde{X}(t)$ using the absolute error criterion $\varepsilon = E(|\tilde{X}(T) - X(T)|)$ [18]. This is the expectation of the absolute value of the difference between the approximation and the Itô process at time T, which gives a measure of closeness at the end of the time interval $[0, T]$.

Since the exact solutions to models (5.4.9) and (5.4.10) are not known explicitly, we estimate ε statistically using computer experiments, see, e.g. [17, 18]. For this, we solve SDE models (5.4.9) and (5.4.10) over M Wiener paths, using step sizes $h = 2^{-k}$, $k = 2, 3, \ldots, k_{\max}$. Note that for each h, the same M sample paths of the Wiener process must be used as for $h = 2^{-k_{\max}}$. We denote the value of the j-th trajectory at time T, simulated with the step size h^{-k}, by $X_{j,k}$, $j = 1, \ldots, M$, $k = 2, 3, \ldots, k_{\max}$. The solutions $X_{j,k_{\max}}$ are regarded as 'exact'. Then, for each $k = 2, 3, \ldots, k_{\max} - 1$, we estimate the absolute error by the statistic

$$\tilde{\varepsilon}_k = \frac{1}{M} \sum_{j=1}^{M} |X_{j,k} - X_{j,k_{\max}}|. \tag{5.4.13}$$

The computed estimates $\tilde{\varepsilon}_k$ against $h = 2^{-k}$ for SDE models (5.4.9) and (5.4.10) are shown on a log–log scale in Fig. 5.13. This figure provides an idea about dependence of the accuracy of the computed solutions on the value of the step size h. For all results presented in the next sections, we used $h = 10^{-2}$ and $h = 2 \times 10^{-2}$ to compute the trajectories of solutions to models (5.4.9) and (5.4.10), respectively. The larger step size in case of model (5.4.10) is due to much higher (compared to model

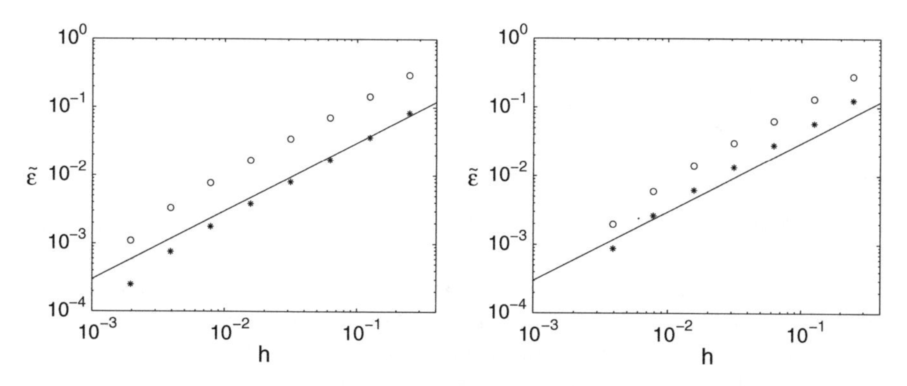

Fig. 5.13 Computed estimates $\tilde{\varepsilon}$ of the absolute error: (∘) for $\tilde{V}$ and (∗) for $\tilde{E}$. The time interval used is [0, 730] days, $M = 100$. The values of the model parameters: $\beta = 0.066$, $\gamma = 4.6 \times 10^{-4}$, $b = 0.14$, $\theta = 6.7 \times 10^6$, $\alpha = 0.039$, $C = 0.21$, $K = 1.5 \times 10^{10}$. The initial values: $\tilde{V}(0) = 15.5$, $\tilde{E}(0) = 5.89$. The solid line indicates the slope 1. Left: Model (5.4.9) with $\sigma = 10^4$; $k_{\max} = 10$. Right: Model (5.4.10) with $\sigma = 10^4$, $\sigma_\beta = 0.022$, $\sigma_\gamma = 1.52 \times 10^{-4}$; $k_{\max} = 9$. (Reprinted from Mathematics and Computers in Simulation, Vol. 96, Luzyanina and Bocharov, Stochastic modeling of the impact of random forcing on persistent hepatitis B virus infection, Pages 54–65, Copyright © 2014, with permission from Elsevier)

(5.4.9)) computational efforts caused by the need to approximate double Itô integrals in the Milstein scheme. These step sizes correspond to the absolute error estimate about 10^{-2} for the computed solutions. The corresponding relative error is less than 0.3%. Note that these figures also indicate the order 1 of the Milstein scheme.

5.4.3.2 Sampling of the Model Parameters and Initial Data

We need to find sets of the values of the model parameters and initial data for which solutions of the deterministic model correspond to the first or the second variant of chronic HBV infection. One way is to split biologically admissible ranges of the model parameters and initial data into a number of small subintervals and compute model solutions for all possible combinations of the parameters from each subinterval. This way is computationally expensive. In order to construct samples of parameters and initial data in an efficient way, we use the statistical method of Latin Hypercube Sampling (LHS). For this, we first determine biologically admissible ranges of the model parameters and initial values and subintervals of these ranges necessary to apply the LHS.

Biologically consistent ranges of the model parameters and initial values

As biologically admissible ranges of the model parameters $\beta,\ \gamma,\ b,\ \theta,\ C$, characterizing chronic HBV infection, we use the intervals $[0,\ p^*]$, where p^* is the best-fit value of the parameter p, estimated for the acute phase of HBV, see Table 5.1. Doing so, we take into account that chronic HBV infection is characterized by smaller values of these kinetic parameters compared to acute resolving HBV infection. For the parameter α, we use the interval $[0,\ 10 \times \alpha^*]$, to take into account that the rate constant of CTL death can be higher during chronic HBV infection. The parameter K, characterizing virus carrying capacity, is assumed to be independent of the type of HBV infection. Therefore, it is fixed at its best-fit value estimated for acute HBV, cf. Table 5.1. For the initial values of the SDE models, we use the following intervals: $V(0) \in [10^6,\ 10^7]$, $E(0) \in [10,\ 10^3]$.

Subintervals of the model parameters and initial values

We include the initial data $V(0)$ and $E(0)$ in the list of the model parameters being sampled and call the vector $\mathbf{p} = (\beta,\ \gamma,\ b,\ \theta,\ C,\ \alpha,\ V(0),\ E(0))$ the vector of parameters. We assume that the model parameters and initial values have triangular, respectively, uniform probability density functions. Let $\mathbf{p}_l$ and $\mathbf{p}_u$ denote vectors of the low and upper bounds of the elements of $\mathbf{p}$, respectively. According to the Latin Hypercube Sampling algorithm, we divide the admissible ranges of all elements of $\mathbf{p}$ into N non-overlapping equiprobable intervals $[\mathbf{p}_{\min,j},\ \mathbf{p}_{\max,j}],\ j = 1,\ldots,N$. The limits $\mathbf{p}_{\min,j}$ and $\mathbf{p}_{\max,j}$ of each interval are ascertained as follows. We set $\mathbf{p}_{\min,1} = \mathbf{p}_l,\ \mathbf{p}_{\max,N} = \mathbf{p}_u$. Other interval limits are calculated by using the corresponding cumulative distribution function F, i.e. $\mathbf{p}_{\max,j} = F^{-1}(F(\mathbf{p}_{\min,j}) + 1/N)$, and setting the minimum value for the next interval ($\mathbf{p}_{\min,j+1}$) to be equal to the maximum value for the previous interval ($\mathbf{p}_{\max,j}$) and repeating the whole process.

Latin Hypercube Sampling

Next, we generate an LHS table as an $(N \times n_p)$ matrix, where $n_p = 8$ is the number of the sampled parameters (6 model parameters and 2 initial data). For this, we use the MATLAB code `lhsdesign` [20] which generates a latin hypercube sample containing N values on each of n_p parameters. For each parameter, the N values are randomly distributed with one from each interval $(0, 1/N),\ (1/N, 2/N), \ldots, (1 - 1/N, 1)$. After the LHS matrix is created, its elements are replaced by the values of the parameters using the corresponding cumulative distribution function.

5.4.3.3 Choice of the Number of Subintervals of the Model Parameters in the LHS Method

To the best of our knowledge, no practical recommendations for the sample size N of LHS samplings have been proposed in the literature. Therefore, we performed computational experiments with various $N = 25,\ 50,\ 100,\ 200,\ 300,\ 400$ to determine the value of N that is sufficient for our analysis, and to decide on robustness of results.

Let the range of each model parameter be divided into N subintervals. Using the Latin hypercube algorithm, we obtain N sets of the model parameter values. We compute N solutions of the deterministic model, using these sets of parameters, and count the number (n) of solutions satisfying condition (5.4.11), i.e. the number of solutions resulting in the chronic LVLTA infection. For each of these n sets of the model parameters, we compute M trajectories of the stochastic model (5.4.9) with $\sigma = 10^4$, i.e. we compute Mn trajectories. Next, we count the number of trajectories (r) satisfying $\tilde{V}_i < \ln(100)$ for a certain time $t = ih$. We refer these trajectories as spontaneous recovery, see an example in Fig. 5.14. We are interested in the percentage

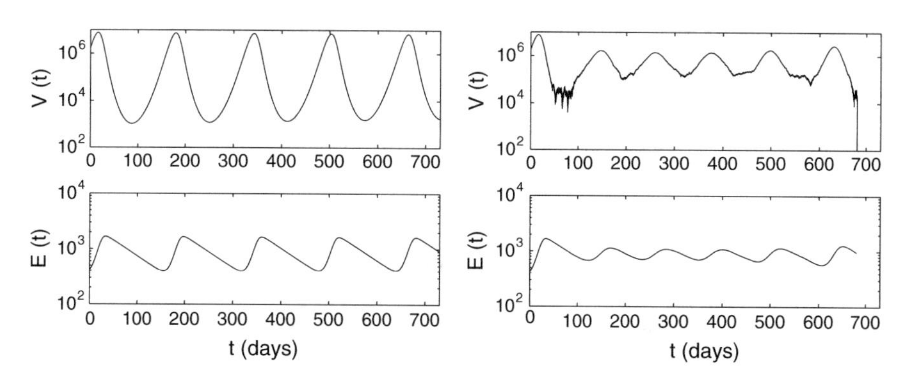

Fig. 5.14 Left: Solutions $V(t)$ and $E(t)$ of the deterministic model (5.3.2) corresponding to the case of chronic LVLTA infection. Right: Trajectories $V(t)$ and $E(t)$ of the stochastic model (5.4.5) with $\sigma = 10^4$ and the same parameter values as the ones used to compute the deterministic solution shown in the left figure. We refer this dynamic pattern as a spontaneous recovery since $V(t) < 10^2$ for $t \geq 680.7$ days. (Reprinted from Mathematics and Computers in Simulation, Vol. 96, Luzyanina and Bocharov, Stochastic modeling of the impact of random forcing on persistent hepatitis B virus infection, Pages 54–65, Copyright © 2014, with permission from Elsevier)

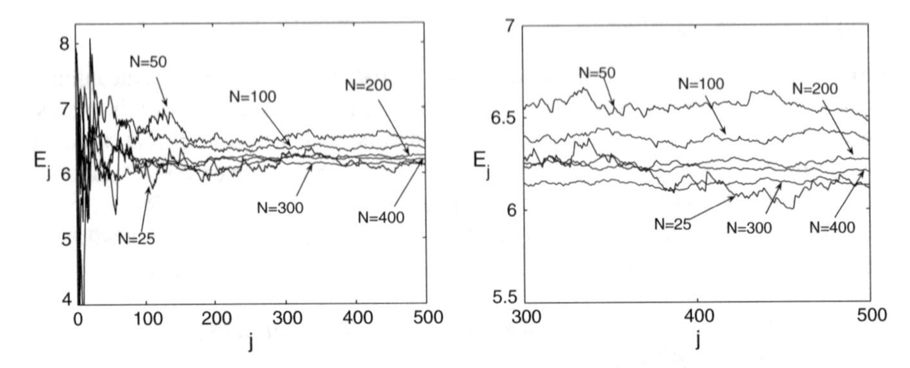

Fig. 5.15 Choice of the number of subintervals N of the model parameters in the LHS method. The mean E_j of the percentage of recoveries over $j = 1, 2, \ldots, 500$ observations. The right figure is a zoom of the left figure. (Reprinted from Mathematics and Computers in Simulation, Vol. 96, Luzyanina and Bocharov, Stochastic modeling of the impact of random forcing on persistent hepatitis B virus infection, Pages 54–65, Copyright © 2014, with permission from Elsevier)

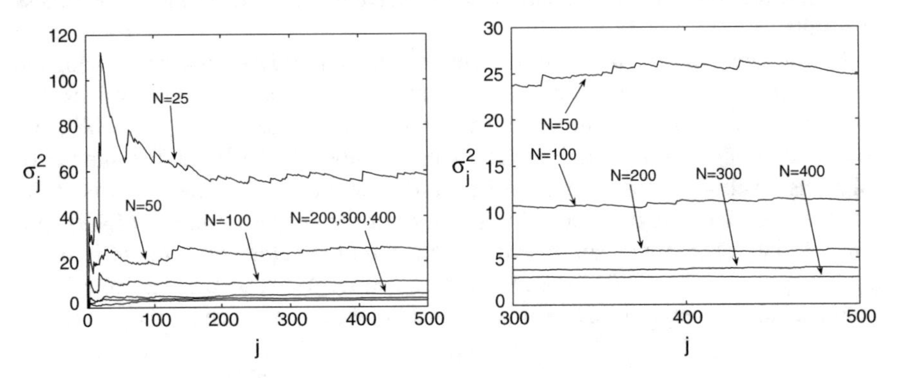

Fig. 5.16 Choice of the number of subintervals N of the model parameters in the LHS method. The variance σ_j^2 of the percentage of recoveries over $j = 1, 2, \ldots, 500$ observations. The right figure is a zoom of the left figure. (Reprinted from Mathematics and Computers in Simulation, Vol. 96, Luzyanina and Bocharov, Stochastic modeling of the impact of random forcing on persistent hepatitis B virus infection, Pages 54–65, Copyright © 2014, with permission from Elsevier)

of the recoveries, $p = 100r/(Mn)$, its mean and variance over j observations, $j = 1, \ldots, J$. The latter are computed as follows. For a fixed N, we repeat the above steps J times and, for each observation j, $j = 1, \ldots, J$, we compute n_j, r_j and p_j as described above. Then, we compute the mean of the percentage of recoveries over j observations $E_j = (\sum_{i=1}^{j} p_i)/j$, $j = 1, 2, \ldots, J$, and the corresponding variance of the percentage of recoveries $\sigma_j^2 = (\sum_{i=1}^{j}(p_i - E_j)^2)/j$, $j = 2, \ldots, J$.

Comparative plots of the computed mean E_j and variance σ_j^2 (model 5.4.5) for $N = 25, 50, 100, 200, 300, 400$ and $j = 1, \ldots, J = 500$, are depicted in Figs. 5.15 and 5.16. We observe that the higher N is, the faster is the convergence of E_j and σ_j^2. The histograms of the percentage of recoveries, presented in Fig. 5.17, indicate

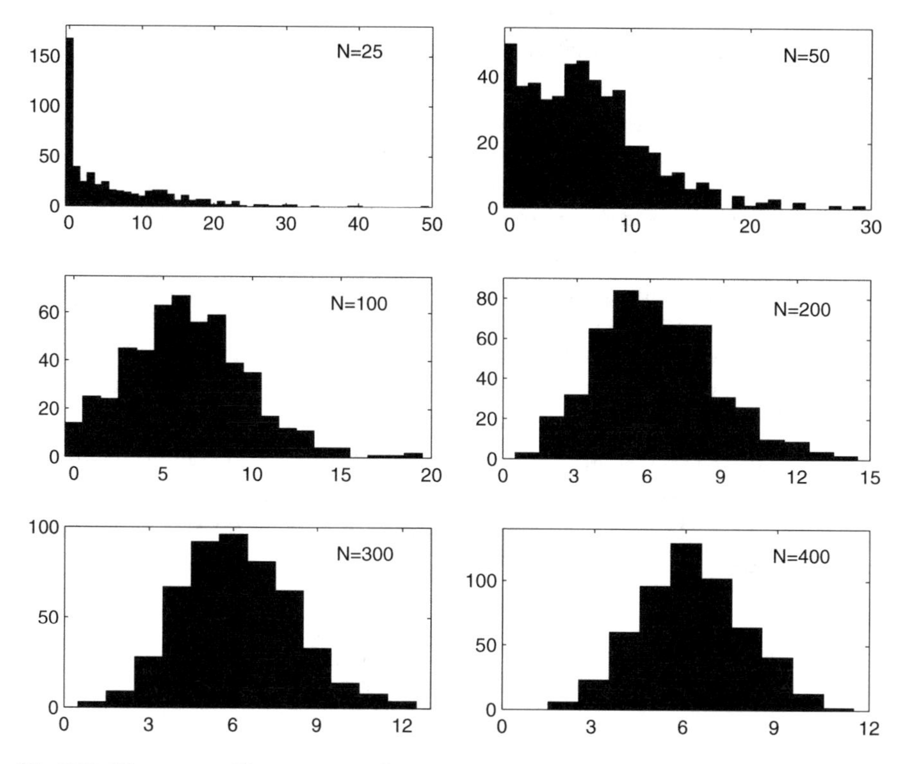

Fig. 5.17 Histograms of the percentage of recoveries (p_j) over 500 observations for different values of the number (N) of subintervals of the model parameters in the LHS method used. (Reprinted from Mathematics and Computers in Simulation, Vol. 96, Luzyanina and Bocharov, Stochastic modeling of the impact of random forcing on persistent hepatitis B virus infection, Pages 54–65, Copyright © 2014, with permission from Elsevier)

that the mean E_{500} and the variance σ_{500}^2 are well localized for $N \geq 100$. We also observe that the distribution of the percentage of recoveries for N large enough ($N = 200, 300, 400$ in Fig. 5.17) becomes close to a normal distribution with a mean about $E_{500} \approx 6$. However, this is not fulfilled for smaller values of the noise intensity factor σ as we show in the next section.

Based on the results shown in Figs. 5.15, 5.16 and 5.17, we conclude that $N = 25$ and $N = 50$ are certainly not sufficient to estimate consistently the sets of the model parameters corresponding to the chronic phase of HBV infection. On the other hand, for $N = 200, 300, 400$ and large j, we observe that values of E_j and σ_j^2 are located in a rather narrow band, cf. Figs. 5.15(right) and 5.16(right). We choose $N = 200$ for our further investigations to reduce the computational cost. We also choose $J = 500$ observations to be made, and we use E_{500} and σ_{500}^2 to characterize the percentage of recoveries in experiments presented in the next section. Clearly, E_{500} and σ_{500}^2 approximate the mean and the variance of the percentage of recoveries with a certain accuracy. As follows from Fig. 5.15, E_{500} is estimated with the relative accuracy of $1 - 3\%$, which is sufficient for the purpose of our study.

5.4.4 Determinants of Spontaneous Recovery

Using $N = 200$, we repeat the LHS algorithm $J = 500$ times. For each observation $j = 1, 2, \ldots, J$ we analyse the corresponding N sets of the model parameters and initial data to select such sets that the solution of the deterministic model (5.3.2) with these parameters satisfies either condition (5.4.11) (LVLTA) or (5.4.12) (HVLTA). We denote these sets by $S_{lv}^{(j)}$ and $S_{hv}^{(j)}$ and the corresponding number of sets by $N_{lv}^{(j)}$ and $N_{hv}^{(j)}$, respectively, i.e. $S_{lv}^{(j)} \in \mathbb{R}^{8\times N_{lv}^{(j)}}$, $S_{hv}^{(j)} \in \mathbb{R}^{8\times N_{hv}^{(j)}}$, $j = 1, 2, \ldots, J$. Histograms of N_{lv} and N_{hv} are presented in Fig. 5.18. We observe that while the average number of 'chronic persistent' outcomes of the deterministic model is about 25 (out of 200), the average number of 'chronic active' outcomes is only about 5.5.

For each set from $S_{lv}^{(j)}$ and $S_{hv}^{(j)}$, we compute M solutions (trajectories) of a stochastic model, i.e. we compute $MN_{lv}^{(j)}$ and $MN_{hv}^{(j)}$ trajectories. We count the number of trajectories satisfying $V_i < 10^2$ for a certain time $t = ih$, for sets $S_{lv}^{(j)}$ and $S_{hv}^{(j)}$ separately. Let $N_{lvr}^{(j)}$ and $N_{hvr}^{(j)}$ denote the number of such trajectories, respectively. Then the values

$$R_{lvr} = \frac{1}{500}\sum_{j=1}^{500} 100\frac{N_{lvr}^{(j)}}{MN_{lv}^{(j)}}, \quad \sigma_{lvr}^2 = \frac{1}{500}\sum_{i=1}^{500}(p_i - R_{lvr})^2 \tag{5.4.14}$$

and

$$R_{hvr} = \frac{1}{500}\sum_{j=1}^{500} 100\frac{N_{hvr}^{(j)}}{MN_{hv}^{(j)}}, \quad \sigma_{hvr}^2 = \frac{1}{500}\sum_{i=1}^{500}(p_i - R_{hvr})^2 \tag{5.4.15}$$

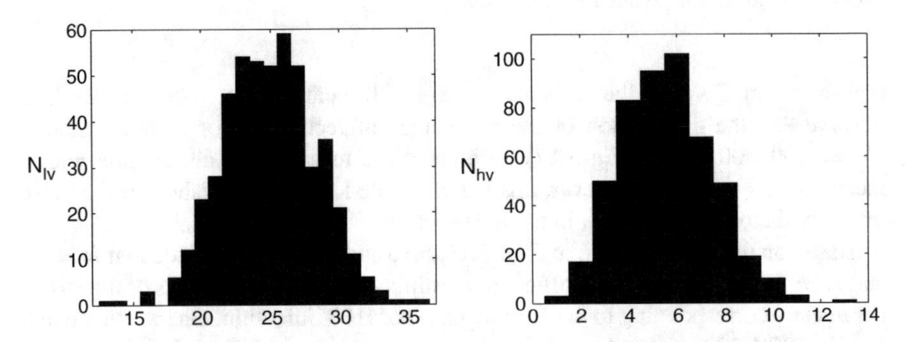

Fig. 5.18 Histograms of the number of sets of the model parameters corresponding to the LVLTA chronic infection ($N_{lv}^{(j)}$, left) and the HVLTA chronic infection ($N_{hv}^{(j)}$, right). Each $N_{lv}^{(j)}$ and $N_{hv}^{(j)}$, $j = 1, 2, \ldots, 500$, is a number out of $N = 200$ sets of the parameters obtained by the LHS method. (Reprinted from Mathematics and Computers in Simulation, Vol. 96, Luzyanina and Bocharov, Stochastic modeling of the impact of random forcing on persistent hepatitis B virus infection, Pages 54–65, Copyright © 2014, with permission from Elsevier)

Table 5.2 Influence of the noise intensity σ on the mean and the variance of the percentage of recoveries. (Reprinted from Mathematics and Computers in Simulation, Vol. 96, Luzyanina and Bocharov, Stochastic modeling of the impact of random forcing on persistent hepatitis B virus infection, Pages 54–65, Copyright © 2014, with permission from Elsevier)

Model	Intensity σ of additive noise	R_{lvr}	σ^2_{lvr}	R_{hvr}	σ^2_{hvr}
Model (5.4.9)	10	0.4	0.7		
	10^2	1.7	3.0	1.8	30.6
	10^3	3.6	5.2	3.6	31.4
	10^4	6.3	5.8	4.9	28.6
Model (5.4.10)	10	1.3	1.9		
	10^2	2.6	4.6	3.7	31.3
	10^3	4.5	5.8	5.4	40.6
	10^4	7.6	5.9	6.6	40.1

approximate the mean value and the variance of the percentage of recoveries in cases of LVLTA and HVLTA chronic infections, respectively.

Obviously, the noise intensity factors σ, σ_β and σ_γ affect values of R_{lvr} and R_{hvr}. The values of σ_β and σ_γ are uniquely determined by the 3-sigma rule as described in Sect. 5.4.1. To have an idea about the suitable range of the intensity σ of additive noise, we estimated ranges of the virus population $V(t)$ and the deterministic term in the first equation of SDE model (5.4.5) for both LVLTA and HVLTA chronic HBV infections. In case of the chronic infection with LVLTA, these ranges are $(10^2,\ 10^8]$ and $[-10^6,\ 10^6]$, respectively, while in case of HVLTA chronic infection, the corresponding intervals are $(10^2,\ 10^{10}]$ and $[-10^9,\ 10^9]$. So we used $\sigma = 10^d$ with $d = 1, 2, 3, 4$ in case of LVLTA- and $d = 2, 3, 4$ in case of HVLTA chronic HBV infections. The influence of σ on the values of R_{lvr}, R_{hvr}, σ^2_{lvr} and σ^2_{hvr} is summarized in Table 5.2.

Note that all computations were done for the time interval [0, 730] days (2 years). In (5.4.14) and (5.4.15), we used $M = 10^3$ and $M = 2\times10^2$ for the models (5.4.9) and (5.4.10), respectively,to reduce the CPU time needed for much more computationally expensive model (5.4.10). The difference in the approximations to the mean value computed with $M = 10^3$ and $M = 200$ is within the approximation error.

We observe in Table 5.2 that the percentage of recoveries for both low viral load and high viral load variants of chronic HBV infection decreases with the decreasing noise intensity σ. Also, as σ decreases, the number of zero values of N_{lvr} and N_{hvr} increases, see histograms in Figs. 5.19, 5.20, 5.21 and 5.22. The percentage of zero values of N_{lvr} in the case of additive noise (model (5.4.9)) vs. the case of multiplicative noise (model (5.4.10)) is 1% vs. 0.2% ($\sigma = 10^3$), 18% vs. 2% ($\sigma = 10^2$) and 70% vs. 6% ($\sigma = 10$). The percentage of zero values of N_{hvr} in the case of additive noise (model (5.4.9)) vs. the case of multiplicative noise (model (5.4.10)) is 16% vs. 10% ($\sigma = 10^4$), 50% vs. 21% ($\sigma = 10^3$) and 76% vs. 32% ($\sigma = 10^2$).

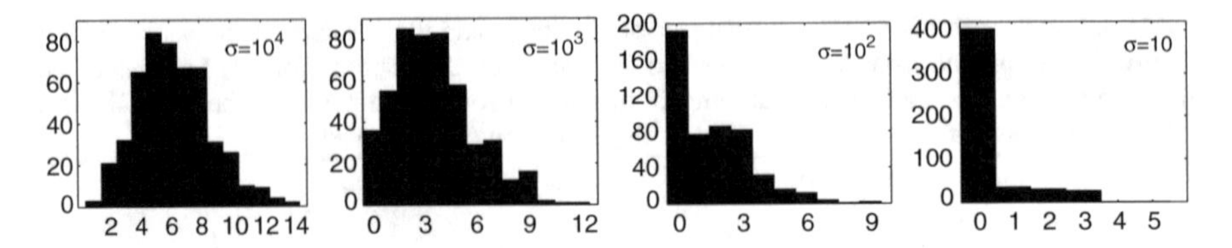

Fig. 5.19 Histograms of the percentage of recoveries over 500 observations. Model (5.4.9), case LVLTA chronic infection. (Reprinted from Mathematics and Computers in Simulation, Vol. 96, Luzyanina and Bocharov, Stochastic modeling of the impact of random forcing on persistent hepatitis B virus infection, Pages 54–65, Copyright © 2014, with permission from Elsevier)

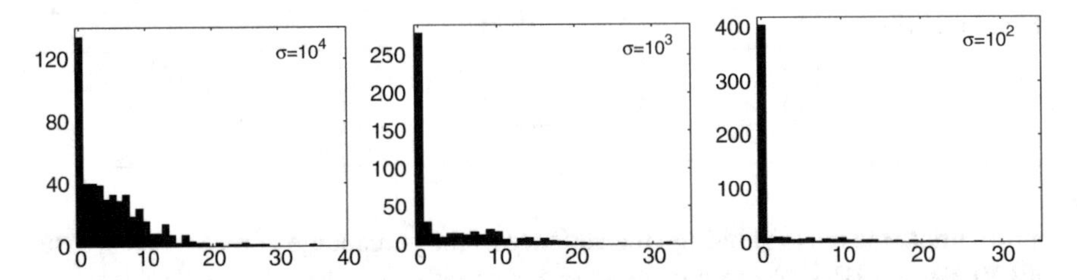

Fig. 5.20 Histograms of the percentage of recoveries over 500 observations. Model (5.4.9), case HVLTA chronic infection. (Reprinted from Mathematics and Computers in Simulation, Vol. 96, Luzyanina and Bocharov, Stochastic modeling of the impact of random forcing on persistent hepatitis B virus infection, Pages 54–65, Copyright © 2014, with permission from Elsevier)

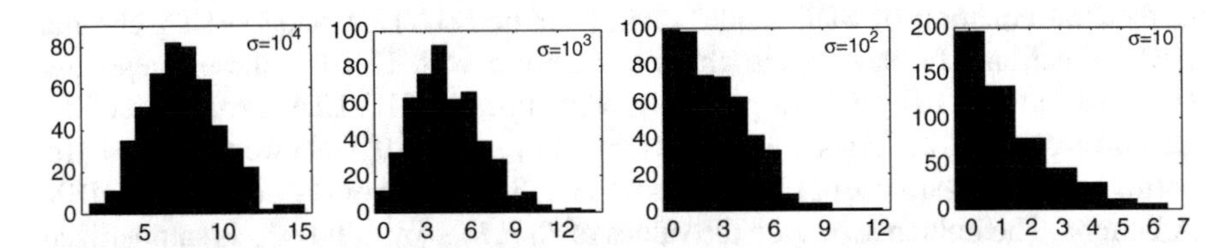

Fig. 5.21 Histograms of the percentage of recoveries over 500 observations. Model (5.4.10), the case of LVLTA chronic infection. (Reprinted from Mathematics and Computers in Simulation, Vol. 96, Luzyanina and Bocharov, Stochastic modeling of the impact of random forcing on persistent hepatitis B virus infection, Pages 54–65, Copyright © 2014, with permission from Elsevier)

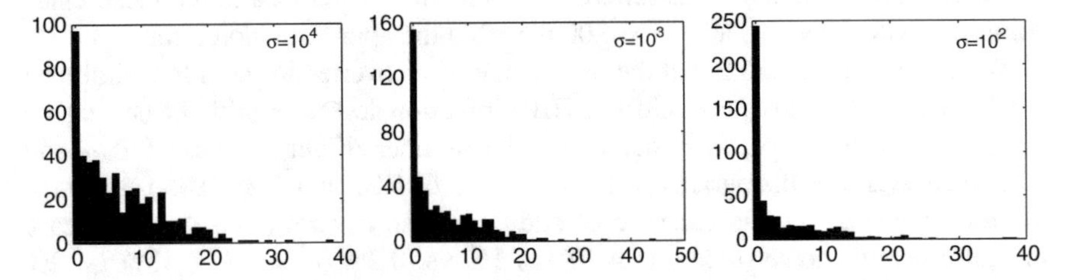

Fig. 5.22 Histograms of the percentage of recoveries over 500 observations. Model (5.4.10), the case of HVLTA chronic infection. (Reprinted from Mathematics and Computers in Simulation, Vol. 96, Luzyanina and Bocharov, Stochastic modeling of the impact of random forcing on persistent hepatitis B virus infection, Pages 54–65, Copyright © 2014, with permission from Elsevier)

Since the chronic HBV infection with HVLTA is characterized by higher values of virus and CTL numbers, one would expect that R_{hvr} must be smaller than R_{lvr} for any noise intensity σ. However, only for $\sigma = 10^4$ we observe that $R_{hvr} < R_{lvr}$. For smaller values of σ, we observe that R_{hvr} and R_{lvr} have close values in case of model (5.4.9) and $R_{hvr} > R_{lvr}$ in case of model (5.4.10). This can be explained as follows. When computing the percentage of recoveries, the impact of the most sensitive to noise trajectories is smoothed in case of LVLTA chronic infection due to a relatively high value of N_{lv} compared to N_{hv}, see histograms in Fig. 5.18, while the impact of such solutions in case of HVLTA chronic infection is much more pronounced, as confirmed by large values of the variance σ^2_{hvr}, cf. Table 5.2.

Overall, the stochastic model derived analyses predict that

- the relative frequency of spontaneous recovery from persistent HBV infection is rather low, less than 10%, for the random forcing intensity varying within three orders of magnitude;
- the increase of the fluctuation intensity by tenfold approximately doubles the recovery frequency;
- the persistent infection with low viral load/CTL abundance is driven to recovery more frequently than the high viral load/CTL number type once the intensity of the random perturbations exceeds a certain level;
- a combination of a nonspecific rapidly fluctuating forcing term with fluctuating virus replication and elimination processes leads to more frequent recoveries as compared to the additive noise scenario.

We considered the stochastic version of the deterministic ODE model in which the random forcing appears in the equation of virus dynamics. The effect of the intrinsic variability inherent in the CTL response (e.g. due to bystander activation) can be studied an a similar way.

5.5 Pathogenesis of Chronic HBV Infection via Adjoint Equations Sensitivity Analysis

The major problem in understanding the nature of immunity[5] is related to the difficulty of dealing conceptually with the complexity of the immune system in vivo [1, 26, 28]. Mathematical immunology is dealing with increasingly complex models of immune phenomena formulated with ordinary or delay differential equations.

One of the earliest frameworks for a mathematical analysis of the basic patterns of infectious disease dynamics within an individual was the so-called 'basic model of infectious disease' by G.I.Marchuk in 1974 [29]. It is based on the following assumptions:

---

[5] Material of Sect. 5.5 uses the results of our studies from Journal of Computational and Applied Mathematics, Vol. 184, Marchuk et al., Adjoint equations and analysis of complex systems: Application to virus infection modelling, Pages 177–204, Copyright © 2005, with permission from Elsevier.

G.I. Bell: A mathematical description for replicating antigen (1973)

- *Lotka-Volterra*-type of equations
- ***Predator-Prey view***

G.I.Marchuk Mathematical model of infectious disease (1974)

- *Original system of delay differential equations*
- *Target organ damage*
- *Competition between the virus population and the host for the survival resources*

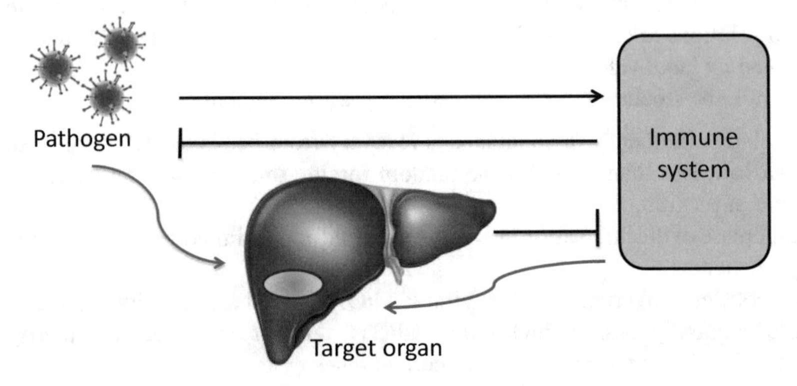

Fig. 5.23 Conceptual views of infectious disease: predator–prey versus antigen and tissue damage dependent regulation immune responses

1. an infectious disease is considered as the immune response of an organism to multiplying pathogenic antigens (virus, bacteria);
2. the damage of the sensitive tissue cells (virus-infected cells) provides a basis for clinical illness;
3. an outcome of the disease depends on the competition between the pathogen and the host for resources of survival;
4. all model equations are derived on the basis of mass balance relations for the components of the virus–host interaction on a short time interval dt;
5. interactions between several components of the process on dt are considered to be additive and proportional to their concentrations multiplied by the length of dt.

The view of the virus and the antiviral immune response as competitors for resources of survival and the consideration of the target organ damage represent the features which made this model unique. Indeed, in other early models (e.g. by G.I Bell [39]), the complexity of virus–host interaction was reduced to the predator (immune system)–prey (pathogen)-type system (see Fig. 5.23).

The steady-state stability analysis of the basic Marchuk model suggested that viruses which cause chronic infections should display a slow growth rate. This prediction resembles the 'sneaking through' phenomenon established in tumour immunology by Lloyd Old in the early 1960s. The most outstanding prediction with the model was that some chronic infections can be treated via exacerbation—a treatment strategy that has been clinically successful [29]. It is worth noticing that although the negative feedback of the virus infection on the specific immune response was initially associated in the basic model with cytopathic viruses which damage directly the target organ (e.g. influenza), later developments in immunology (e.g. exhaustion in LCMV system) had established that weakly or non-cytopathic viruses may also suppress immune responses via anergy and apoptosis (see the corresponding model in Chap. 4.).

5.5.1 Mathematical Model of Antiviral Immune Response

The basic mathematical model of an infectious disease was extended jointly by Marchuk (mathematician) and Petrov (immunologist), who formulated the delay differential equations model for antiviral immune responses [29, 31]. The time-dependent state variables are as follows:

(1) free virus population— $V_f(t)$;
(2) antigen-presenting cell population— $M_V(t)$;
(3) T helper 1 cell population— $H_E(t)$;
(4) T helper 2 cell population— $H_B(t)$;
(5) cytotoxic T lymphocyte (CTL) population— $E(t)$;
(6) B-cell population— $B(t)$;
(7) plasma cell population— $P(t)$;
(8) antibody population— $F(t)$;
(9) virus-infected sensitive tissue cell population— $C_V(t)$;
(10) destroyed sensitive tissue cells— $m(t)$.

The model equations take account of the most relevant processes of the virus–host interaction summarized by the following conceptual view in Fig. 5.24. The reaction of the immune system to any virus infection is mediated by both T-cell (cellular) and B-cell (humoral) immune responses. It is accepted that the recovery from primary virus infection is provided by cytotoxic T lymphocytes' mediated killing of infected cells and destruction of sensitive tissue (by cytotoxicity and/or recruitment of inflammatory cells) to prevent further virus replication. The humoral immunity is composed of B cells, plasma cells and antibodies. Antibodies neutralize free virus particles in particular body fluids (blood, lymph or mucosa). The immune response is triggered in lymph nodes by virus antigens being presented in the context of MHC class I or class II molecules on the surface of macrophages or other antigen-presenting cells. Helper T cells play a critical role in the generation of both the CTL and humoral immunity. The T- and B-cell responses are considered as an antigen-driven selection of virus antigen-specific clones leading to expansion by a series of division and differentiation events. The population dynamics of the state variables is modelled by the following system of DDEs specified in a modular manner as follows:

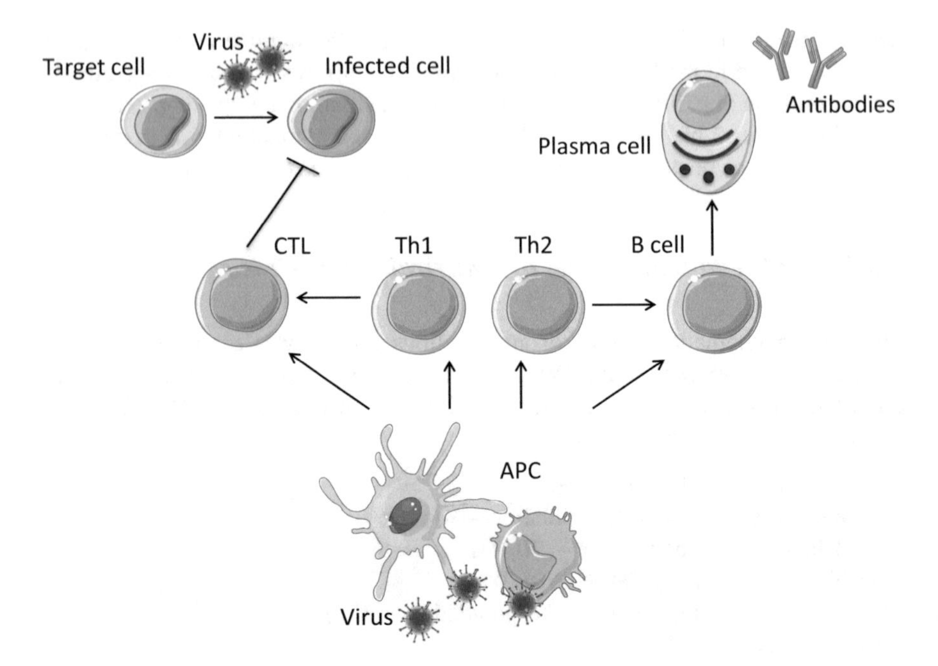

Fig. 5.24 Scheme of the populations and processes underlying the mathematical model of antiviral immune response

Virus-, infected target cells-, destroyed target cell compartments

$$\frac{d}{dt}V_f(t) = \nu C_V(t) + nb_{CE}C_V(t)E(t) - \gamma_{VF}V_f(t)F(t) - \gamma_{VM}V_f(t) \\ - \gamma_{VC}V_f(t)(C^0 - C_V(t) - m(t)), \quad (5.5.16)$$

$$\frac{d}{dt}C_V(t) = \sigma V_f(C^0 - C_V(t) - m(t)) - b_{CE}C_V(t)E(t) - b_m C_V(t), \quad (5.5.17)$$

$$\frac{d}{dt}m(t) = b_{CE}C_V(t)E(t) + b_m C_V(t) - \alpha_m m(t), \quad \xi(m) = 1 - m/C^0. \quad (5.5.18)$$

Antigen-presenting cells-, Th1-helper cells-, cytotoxic T lymphocytes compartments:

$$\frac{d}{dt}M_V(t) = \gamma_{MV}M^0V_f(t) - \alpha_M M_V(t), \quad (5.5.19)$$

$$\frac{d}{dt}H_E(t) = b_H^E[\xi(m)\rho_H^E M_V(t-\tau_H^E)H_E(t-\tau_H^E) - M_V(t)H_E(t)] \\ - b_p^{H_E}M_V(t)H_E(t)E(t) + \alpha_H^E(H_E^0 - H_E(t)), \quad (5.5.20)$$

$$\frac{d}{dt}E(t) = b_p^E[\xi(m)\rho_E M_V(t-\tau_E)H_E(t-\tau_E)E(t-\tau_E) - M_V(t)H_E(t)E(t)]$$
$$- b_{EC}C_V(t)E(t) + \alpha_E(E^0 - E(t)). \quad (5.5.21)$$

Th2-helper cells-, B-lymphocytes-, Plasma cells-, Antibodies compartments:

$$\frac{d}{dt}H_B(t) = b_H^B[\xi(m)\rho_H^B M_V(t-\tau_H^B)H_B(t-\tau_H^B) - M_V(t)H_B(t)]$$
$$- b_p^{H_B} M_V(t)H_B(t)B(t) + \alpha_H^B(H_{B-}^0 H_B(t)), \quad (5.5.22)$$

$$\frac{d}{dt}B(t) = b_p^B[\xi(m)\rho_B M_V(t-\tau_B)H_B(t-\tau_B)B(t-\tau_B) - M_V(t)H_B(t)B(t)]$$
$$+ \alpha_B(B^0 - B(t)), \quad (5.5.23)$$

$$\frac{d}{dt}P(t) = b_p^P\xi(m)\rho_P M_V(t-\tau_P)H_B(t-\tau_P)B(t-\tau_P)$$
$$+ \alpha_P(P^0 - P(t)), \quad (5.5.24)$$

$$\frac{d}{dt}F(t) = \rho_F P(t) - \gamma_{FV}F(t)V_f(t) - \alpha_F F(t). \quad (5.5.25)$$

Time lags are introduced to allow for the duration of proliferation and differentiation of lymphocytes. The negative effect of target organ damage on the immune reaction is described phenomenologically by a decreasing function $\xi(m)$. The homeostasis of the target tissue cells and virus-specific lymphocytes in an uninfected host is taken into account by considering constant input terms.

Consideration of a limited set of processes in the model (although assumed to be most essential), as compared to the full reality is a part of a reductionist approach to the mathematical analysis of virus–host interactions. However, even for the above reduced model, one can hardly expect data to be collectable which simultaneously characterize the evolution of the state variables represented by $\mathbf{y} \equiv [V_f, C_V, m, M_V, H_E, E, H_B, B, P, F]^T$. The key aspect of our approach to the adjustment between the model and numerous partial data on virus infections like hepatitis B or influenza A was the derivation of a consistent data set called 'the generalized picture of infection' [29, 31]. The concept provides a theoretical definition of the typical kinetics of virologic, immunological and pathologic processes occurring after virus invasion in normal individuals (e.g. those without immunodeficiences). The solution of the model corresponding to the 'normal infection course' (see [32]) can be used as a reference trajectory in the analysis of unfavourable disease outcomes (e.g. chronic or lethal infections). This provides a sensible way to deal with the complexity of the infection in an individual patient under study on the one hand and a paucity of available data on the other. In this case, partial information about the solution in the form of functionals of solutions, like

- the cumulative viral load, $J(\mathbf{y}) \equiv J(V_f) = \int\limits_0^T V_f(t)dt$,
- the severity of clinical disease $J(\mathbf{y}) \equiv J(m) = \int\limits_0^T (m_{chronic}(t) - m_{normal}(t))^2 dt$,

can still provide enough information to differentiate the causal relationships between the deviations in the virus–host parameters from the reference ones and the outcome of an infection.

5.5.2 Sensitivity of Functionals to Deviations of Parameters from 'norms'

Adjoint equation methods provide a powerful method to examine the sensitivity of objective functionals, depending on the model solution, to deviations of the model parameters from 'norms'. Consider the functional $J(\mathbf{y}) = \int\limits_{t_0}^T \langle \mathbf{y}, \mathbf{y} \rangle dt$, where $\langle \cdot, \cdot \rangle$ is a scalar product in R^N and $\mathbf{y}(t) = \mathbf{y}(t, \mathbf{p}) \in R_+^N$ is the solution to the initial value problem (IVP)

$$\frac{d}{dt}\mathbf{y}(t) = \mathbf{f}\left(t, \mathbf{y}(t), \mathbf{y}(t-\tau), \mathbf{p}\right), \quad t \in [t_0, T] \tag{5.5.26}$$

$$\mathbf{y}(t) = \boldsymbol{\phi}(t, \mathbf{p}),\ t \in [t_0 - \tau, t_0], \tag{5.5.27}$$

with parameter vector $\mathbf{p} = [p_1, p_2, \cdots, p_L]^T \in R_+^L$. The right-hand side function $\mathbf{f}$ is a smooth function of its arguments and is Lipshitz-continuous with respect to $\mathbf{y}$ and $\mathbf{z} \equiv \mathbf{y}(t - \boldsymbol{\tau})$, where τ is a constant time lag. $\boldsymbol{\phi}(t, \mathbf{p})$ is an initial function.

Let $\widehat{\mathbf{y}} \equiv \mathbf{y}(t, \widehat{\mathbf{p}})$ be an unperturbed trajectory and let the model parameters be changed by $\delta\mathbf{p} = \widehat{\mathbf{p}} - \mathbf{p}$. The corresponding solution can be represented as

$$\mathbf{y}(t, \mathbf{p}) \approx \mathbf{y}(t, \widehat{\mathbf{p}}) + \sum_{i=1}^{L} \mathbf{s}_i(t, \widehat{\mathbf{p}})\delta p_i,$$

with the sensitivity vector function

$$\mathbf{s}_i(t, \widehat{\mathbf{p}}) \equiv \frac{\partial \mathbf{y}(t, \widehat{\mathbf{p}})}{\partial p_i},$$

being a solution to

$$\mathscr{A}(\mathbf{y}(t,\widehat{\mathbf{p}}),\widehat{\mathbf{p}})\mathbf{s}_i(t,\widehat{\mathbf{p}}) = \frac{\partial \mathbf{f}}{\partial p_i}, \quad t \in [t_0, T], \quad \mathbf{s}_i(t,\widehat{\mathbf{p}}) = \frac{\partial \boldsymbol{\phi}}{\partial p_i}, \quad t \in [t_0 - \tau, t_0].$$

Here, the differential matrix operator

$$\mathscr{A} \equiv \frac{d}{dt} - \left[\frac{\partial \mathbf{f}}{\partial \mathbf{y}}\right] - \left[\frac{\partial \mathbf{f}}{\partial \mathbf{y}_\tau}\right] D_\tau$$

specifies a linearized system of the model equations. $[\cdot]$ denotes a matrix of partial derivatives and D_τ is the backward time shift operator. Then, the first-order variation of the functional is

$$\delta J(\widehat{\mathbf{y}}) = 2\sum_{i=1}^{L} \int_{t_0}^{T} \langle \widehat{\mathbf{y}}, \mathbf{s}_i(t,\widehat{\mathbf{p}})\rangle dt \delta p_i.$$

Using the Lagrange identity, one can introduce the operator $\mathscr{A}^*$ adjoint to $\mathscr{A}$

$$(\mathscr{A}\mathbf{s}, \mathbf{w}) = \left(\mathbf{s}, \mathscr{A}^*\mathbf{w}\right)$$

with the domain $\mathscr{D}(\mathscr{A}^*)$. Consider the set $\mathscr{D}_1$ of continuous functions $\mathbf{w}(t)$ on $\Omega = [t_0, T+\tau]$ such that (i) they are $C^1(\Omega)$–piecewise and $\mathbf{w}'(t) \in L_2(\Omega)$, and (ii) $\mathbf{w}(t) = 0,\ t \in [T, T+\tau]$.
Then

$$(\mathscr{A}\mathbf{s}, \mathbf{w}) \equiv \int_{t_0}^{T} \langle \mathscr{A}\mathbf{s}, \mathbf{w}\rangle dt = \int_{t_0}^{T} \langle \mathbf{s}, \mathscr{A}^*\mathbf{w}\rangle dt,$$

where $\mathscr{A}^*(\widehat{\mathbf{y}},\widehat{\mathbf{p}}) \equiv -\frac{d}{dt} - \left[\frac{\partial \mathbf{f}}{\partial \mathbf{y}}\right]^T - \left[\frac{\partial \mathbf{f}}{\partial \mathbf{y}_\tau}\right]^T_{t+\tau} D_{-\tau}.$

The first-order variation of the functional $J(\mathbf{y})$ can be represented in the form

$$\delta J = 2\sum_{i=1}^{L} \int_{t_0}^{T} \langle \mathbf{w}, \frac{\partial \mathbf{f}}{\partial p_i}\delta p_i\rangle dt,$$

using the solution $\mathbf{w}(t,\mathbf{p})$ to the adjoint problem

$$\mathscr{A}^*(\widehat{\mathbf{y}},\widehat{\mathbf{p}})\mathbf{w} \equiv -\frac{d}{dt}\mathbf{w}(t) - \left[\frac{\partial \mathbf{f}}{\partial \mathbf{y}}\right]^T \mathbf{w}(t) - \left[\frac{\partial \mathbf{f}}{\partial \mathbf{y}_\tau}\right]^T_{t+\tau} \mathbf{w}(t+\tau) = \mathbf{y}(t,\widehat{\mathbf{p}}),$$
$$t_0 \le t \le T, \quad \mathbf{w}(t) = 0, \ t \in [T, T+\tau]. \tag{5.5.28}$$

The gradient of the functional can be calculated by integrating numerically the ODEs

$$\frac{d}{dt}\left(\frac{\partial J}{\partial p_i}\right) = 2\langle \mathbf{w}(t), \frac{\partial \mathbf{f}}{\partial p_i}(t)\rangle, \quad t \in [t_0, T], \quad \frac{\partial J}{\partial p_i}|_{t_0} = 0. \qquad (5.5.29)$$

5.5.3 *Numerical Treatment*

We obtained the numerical solution of the model using the $DIFSUB - DDE$ code developed for DDEs with constant time lags [25]. The Nordsieck vector form for the solution to the difference problem

$$\bar{y}_n = \left[y_n, h_n y_n', \ldots, \frac{h_n^p y_n^{(p)}}{p!} \right]^T \in R^{p+1}$$

provides a 'natural' built-in p–th order interpolating polynomial in a neighbourhood of the mesh point t_n

$$\pi_{q,n}(\bar{y}_n, t - \tau) = C\left(\frac{t - \tau - t_n}{h_n}\right)\bar{y}_n,$$

where $C(\alpha) = [1, \alpha, \ldots, \alpha^p]$, $(t - \tau) \in (t_{n-1}, t_n]$. The computation of the gradient of the functional splits into three distinct stages

(1). Solve numerically the *direct* problem for which the set of jump discontinuity points includes $\{t_0 + j\tau\}_{j=0}^{r+1}$:

$$\frac{d}{dt}\mathbf{y}(t) = \mathbf{f}\left(t, \mathbf{y}(t), \mathbf{y}(t - \tau), \mathbf{p}\right), \quad t \in [t_0, T] \qquad (5.5.30)$$

$$\mathbf{y}(t) = \boldsymbol{\phi}(t, \mathbf{p}), \; t \in [t_0 - \tau, t_0]. \qquad (5.5.31)$$

Keep the Nordsick history array $\mathbf{y}_n$ at the whole set of mesh points t_n on $[t_0, T]$: $t_0 < t_1 < \cdots < t_N = T$; $h_n = t_n - t_{n-1}$, $n = 1(1)N$.

(2). Solve the *adjoint* problem from T to 0 for a system of equations from linearized around the unperturbed solution specified by parameter vector $\widehat{\mathbf{p}}$

$$\mathscr{A}^*(\widehat{\mathbf{y}}, \widehat{\mathbf{p}})\mathbf{w} \equiv -\frac{d}{dt}\mathbf{w}(t) - \left[\frac{\partial \mathbf{f}}{\partial \mathbf{y}}\right]^T \mathbf{w}(t) - \left[\frac{\partial \mathbf{f}}{\partial \mathbf{y}_\tau}\right]^T_{t+\tau} \mathbf{w}(t+\tau) = \mathbf{y}(t, \widehat{\mathbf{p}}),$$
$$t_0 \le t \le T, \quad \mathbf{w}(t) = 0, \; t \in [T, T+\tau], \tag{5.5.32}$$

with negative stepsizes. The Nordsieck's vector allows to approximate the solution to the main problem at any given time. Keep the adjoint solution $\bar{\mathbf{w}}(t)$ at the mesh points t_{n^*}. The set of jump discontinuity points is $\{t_0 + j\tau\}_{j=0}^{r+1} \cup \{T - j\tau\}_{j=0}^{r+1}$.

(3). Compute the ***gradient*** of the functional by integrating the set of L ODEs:

$$\frac{d}{dt}\left(\frac{\partial J}{\partial p_i}\right) = 2\langle \mathbf{w}(t), \frac{\partial \mathbf{f}}{\partial p_i}(t)\rangle, \quad t \in [t_0, T], \quad \frac{\partial J}{\partial p_i}|_{t_0} = 0, \; i = 1(1)L. \tag{5.5.33}$$

5.5.4 *Adjoint Equations for the Antiviral Immune Response Model*

Consider the following notation for the state vector function of an adjoint problem:

$$\mathbf{w}(t) \equiv \left(V_f^*(t), M_V^*(t), H_E^*(t), H_B^*(t), E^*(t), B^*(t), P^*(t), F^*(t), C_V^*(t), m^*(t)\right)^T.$$

Then, the adjoint equations look as follows:

$$\begin{aligned}\frac{d}{dt}V_f^*(t) &= (\gamma_{VF}F(t) + \gamma_{VM} + \gamma_{VC}(C^0 - C_V(t) - m(t)))V_f^*(t)\\ &\quad - \sigma(C^0 - C_V(t) - m(t))C_V^*(t) - \gamma_{MV}M_V^*(t)\\ &\quad + \gamma_{FV}F(t)F^*(t),\end{aligned} \tag{5.5.34}$$

$$\begin{aligned}\frac{d}{dt}M_V^*(t) &= \alpha_M M_V^*(t) - \rho_H^E b_H^E H_E(t)\xi(m(t+\tau_H^E))H_E^*(t+\tau_H^E)\\ &\quad - \rho_H^B b_H^B H_B(t)\xi(m(t+\tau_H^B))H_B^*(t+\tau_H^B)\\ &\quad - \rho_E b_p^E H_E(t)E(t)\xi(m(t+\tau_E))E^*(t+\tau_E)\\ &\quad - \rho_B b_p^B H_B(t)B(t)\xi(m(t+\tau_B))B^*(t+\tau_B)\\ &\quad - b_p^P H_B(t)B(t)\xi(m(t+\tau_P))P^*(t+\tau_P) + (b_H^E H_E(t) + b_p^{H_E} H_E(t)E(t))H_E^*(t)\\ &\quad + (b_H^B H_B(t) + b_p^{H_B} H_B(t)B(t))H_B^*(t) + b_p^E H_E(t)E(t)E^*(t)\\ &\quad + b_p^B H_B(t)B(t)B^*(t),\end{aligned} \tag{5.5.35}$$

$$\begin{aligned}\frac{d}{dt}H_E^*(t) &= -\rho_H^E b_H^E M_V(t)\xi(m(t+\tau_H^E))H_E^*(t+\tau_H^E)\\&+ (b_H^E M_V(t) + b_p^{H_E} M_V(t)E(t) + \alpha_H^E)H_E^*(t)\\&- \rho_E b_p^E M_V(t)E(t)\xi(m(t+\tau_E))E^*(t+\tau_E) + b_p^E M_V(t)E(t)E^*(t),\end{aligned} \quad (5.5.36)$$

$$\begin{aligned}\frac{d}{dt}H_B^*(t) &= -\rho_H^B b_H^B M_V(t)\xi(m(t+\tau_H^B))H_B^*(t+\tau_H^B)\\&+ (b_H^B M_V(t) + b_p^{H_B} M_V(t)B(t) + \alpha_H^B)H_B^*(t)\\&- \rho_B b_p^B M_V(t)B(t)\xi(m(t+\tau_B))B^*(t+\tau_B) + b_p^B M_V(t)B(t)B^*(t)\\&- b_p^P M_V(t)B(t)\xi(m(t+\tau_P))P^*(t+\tau_P),\end{aligned} \quad (5.5.37)$$

$$\begin{aligned}\frac{d}{dt}E^*(t) &= -\rho_E b_p^E M_V(t)H_E(t)\xi(m(t+\tau_E))E^*(t+\tau_E)\\&+ (b_p^E M_V(t)H_E(t) + b_{EC}C_V(t) + \alpha_E)E^*(t) + b_p^{H_E} M_V(t)H_E(t)H_E^*(t)\\&- nb_{CE}C_V(t)V_f^*(t) + b_{CE}C_V(t)(C_V^*(t) - m^*(t)),\end{aligned} \quad (5.5.38)$$

$$\begin{aligned}\frac{d}{dt}B^*(t) &= -\rho_B b_p^B M_V(t)H_B(t)\xi(m(t+\tau_B))B^*(t+\tau_B)\\&+ (b_p^B M_V(t)H_B(t) + \alpha_B)B^*(t) - b_p^P M_V(t)H_B(t)\xi(m(t+\tau_P))P^*(t+\tau_P)\\&+ b_p^{H_B} M_V(t)H_B(t)H_B^*(t),\end{aligned} \quad (5.5.39)$$

$$\frac{d}{dt}P^*(t) = \alpha_P P^*(t) - \rho_F F^*(t), \quad (5.5.40)$$

$$\frac{d}{dt}F^*(t) = \gamma_{VF} V_f(t)V_f^*(t) + (\gamma_{FV} V_f(t) + \alpha_F)F^*(t), \quad (5.5.41)$$

$$\begin{aligned}\frac{d}{dt}C_V^*(t) &= \left(\sigma V_f(t) + b_{CE}E(t) + b_m\right) C_V^*(t) - (b_{CE}E(t) + b_m)\, m^*(t)\\&+ b_{EC}E(t)E^*(t) - (\nu + nb_{CE}E(t) + \gamma_{VC}V_f(t))V_f^*(t),\end{aligned} \quad (5.5.42)$$

$$\begin{aligned}\frac{d}{dt}m^*(t) &= \alpha_m m^*(t) - \gamma_{VC} V_f(t) V_f^*(t) + \sigma V_f(t) C_V^*(t)\\ &- \frac{\partial \xi}{\partial m} \times (\rho_H^E b_H^E M_V(t-\tau_H^E) H_E(t-\tau_H^E) H_E^*(t)\\ &+ \rho_H^B b_H^B M_V(t-\tau_H^B) H_B(t-\tau_H^B) H_B^*(t)\\ &+ \rho_E b_p^E M_V(t-\tau_E) H_E(t-\tau_E) E(t-\tau_E) E^*(t)\\ &+ \rho_B b_p^B M_V(t-\tau_B) H_B(t-\tau_B) B(t-\tau_B) B^*(t)\\ &+ b_p^P M_V(t-\tau_P) H_B(t-\tau_P) B(t-\tau_P) P^*(t)).\end{aligned} \tag{5.5.43}$$

Initial data on $[T, T+\tau_i]$ are

$$\begin{aligned}&V_f^*(T)=0,\ M_V^*(T)=0,\ H_E^*(t)=0,\ T\le t\le T+\tau_H^E,\ H_B^*(t)=0,\ T\le t\le T+\tau_H^B,\\ &E^*(t)=0,\ T\le t\le T+\tau_E,\ B^*(t)=0,\ T\le t\le T+\max(\tau_P,\tau_B),\\ &P^*(t)=0,\ T\le t\le T+\tau_P,\ F^*(T)=0,\ C_V^*(T)=0,\ m^*(T)=0.\end{aligned} \tag{5.5.44}$$

5.5.5 HBV Infection: Chronic Versus Resolving Infection

We consider unique data that represent the dynamics of hepatitis B virus (HBV) in the blood of healthy male inmates of the U.S. Federal penitentiaries (1951–1954) ([27]). The volunteer subjects were inoculated with plasma samples from patients with HBV infection. The HBV load, the surface antigen (HBsAg) concentrations and the level of liver enzymes in patients' blood were examined every week. The primary objective of the study was to understand the factors that determine whether an individual with acute hepatitis B will resolve the illness and clear the virus below the detection level (about 10^5 and 10^8 HBV and HBsAg particles/ml, respectively) or will develop a chronic infection. The representative data sets from a patient who cleared the virus (we call him Patient 1) and a patient who became chronically infected (Patient 2) are summarized in Fig. 5.25(left) (the original data were reproduced manually from the graphs in [27]). Among the set of factors that were considered to be responsible for the chronic outcome in Patient 2 were the immune factors (e.g. a delay in the development of the antiviral response) as well as the virus factors (e.g. replication rate, immunogenicity). It was proposed that *high levels of viral replication during acute infection* predict progression to chronicity. In HBV chronic infection, the viral loads are generally high in the face of very low frequency of virus-specific CTL. However, it is not clear whether this reflects the higher initial virus replication levels or results from acute immune responses which appear to be weaker from the onset.

We further can analyse whether the above proposition is consistent with the data in hand. Specifically, we examine the sensitivity of the cumulative load functional

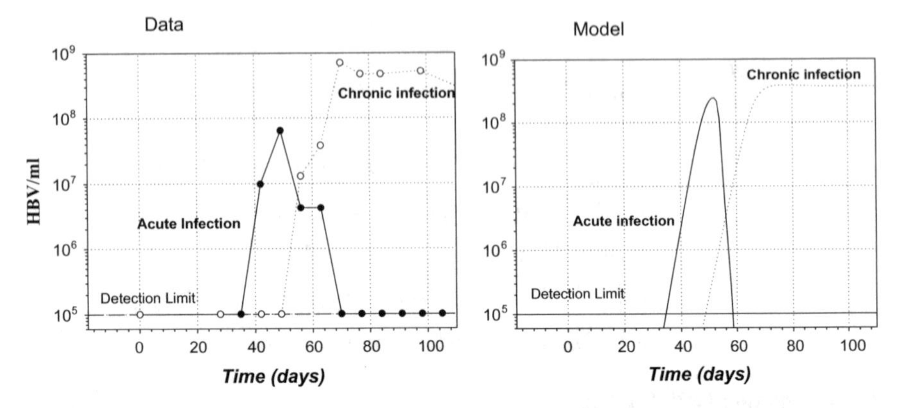

Fig. 5.25 Representative data on acute and chronic HBV infection and the model solutions reproducing the observed phenotypes, i.e. the kinetics of the viral load. (Reprinted from Journal of Computational and Applied Mathematics, Vol. 184, Marchuk et al., Adjoint equations and analysis of complex systems: Application to virus infection modelling, Pages 177–204, Copyright © 2005, with permission from Elsevier)

with respect to the parameters of the antiviral immune response model. Consider as an integral measure of the HBV infection the following linear functional ($T = 98$ days):

$$J_{[0,98]}(\mathbf{y}) \equiv J(V_f) = \int_0^T V_f(t)dt.$$

The data over days 0 to 98 suggest that for patients 1 and 2 the functional takes the values $J_1(V_f) \sim 5.8 \times 10^8$ HBV/ml day and $J_2(V_f) \sim 1.7 \times 10^{10}$ HBV/ml day, respectively. A visual inspection of the kinetics of HBV and HBsAg in acute versus chronic infection (See Fig. 5.25(left)) suggests that there is little difference (if any) in the initial growth phase. The differences appear about day 48 post infection at the peak of the acute infection viral load. Notice, that the parallel rightward shift of the growth curve might simply be due to the initial dose in the chronic infection. Therefore, we assumed that the sensitivity of the cumulative viral load functional to the model parameters computed for 48 and 98 days should provide information for the selection of those parameters that might be different between patients 1 and 2. This should provide a basis for further inference about the pathogenetic mechanisms of chronicity.

Using the adjoint equation technique, we compute the gradients of the viral load functional

$$J(V_f) = \int_0^{98} V_f(t)dt = \int_0^{48} V_f(t)dt + \int_{48}^{98} V_f(t)dt$$

for the whole observation interval and the first 48 days. The components of the gradient vector are ranked in Table 5.3 by their absolute values. According to the

Table 5.3 Sensitivity of the cumulative viral load functional evaluated over [0, 98] days and [0, 48] days (value in brackets) intervals. (Reprinted from Journal of Computational and Applied Mathematics, Vol. 184, Marchuk et al., Adjoint equations and analysis of complex systems: Application to virus infection modelling, Pages 177–204, Copyright © 2005, with permission from Elsevier)

Parameter	Gradient component $\frac{\partial J}{\partial \mathbf{p}_i} \cdot \mathbf{p}_i$ [$\times 10^9$]	Parameter	Gradient component $\frac{\partial J}{\partial \mathbf{p}_i} \cdot \mathbf{p}_i$ [$\times 10^9$]
ν	2.3 (3.9)	ρ_H^B	−1.7 (−0.0002)
σ	1.2 (3.8)	γ_{VM}	−1.2 (−1.6)
α_M	0.4 (−0.0006)	γ_{MV}	−0.6 (−0.0006)
α_H^B	0.2 (0.00003)	ρ_B	−0.4 (−0.000008)
γ_{FV}	0.06 (0.001)	γ_{VF}	−0.3 (−0.1)
α_P	0.05 (0.00006)	ρ_F	−0.3 (−0.06)
α_F	0.05 (0.06)	H_B^0	−0.3 (−0.0002)
α_m	0.01 (0.001)	b_p^P	−0.2 (−0.0002)
α_B	0.01 (0.0000008)	b_m	−0.2 (−0.31)
ρ_H^E	0.008 (−0.0001)	b_H^B	−0.2 (−0.00005)
b_{EC}	0.006 (0.0005)	B^0	−0.2 (−0.0001)
α_H^E	0.0006 (0.00003)	b_p^B	−0.1 (−0.000006)
$b_p^{H_B}$	0.000006 (<10^{-9})	P^0	−0.04 (−0.05)
		ρ_E	−0.01 (−0.04)
		b_p^E	−0.005 (−0.04)
		H_E^0	−0.005 (−0.0004)
		b_{CE}	−0.003 (−0.004)
		E^0	−0.003 (−0.004)
		b_H^E	−0.001 (−0.00004)
		α_E	−0.0006 (−0.00008)
		γ_{VC}	−0.00001 (−0.00005)
		$b_p^{H_E}$	−0.00000001 (<10^{-10})

following criteria of parameters selection, (i) a low sensitivity of $J_{[0,48]}$, (ii) a high sensitivity of $J_{[0,98]}$, (iii) the biological feasibility of such variation in the parameter value, which could lead to an increase in the viral load functional by two orders, the group of candidate parameters reduces to γ_{MV}, H_B^0, B^0, b_H^B, b_p^B, b_p^P out of the total set of 35 parameters. Further numerical simulations suggest that the transition from acute to chronic infection can be caused by about 10^2 fold decrease in the value of γ_{MV}, representing the efficacy of virus antigen processing and presentation by antigen-presenting cells as shown in Fig. 5.26. This change might be considered as a most parsimonious explanation of the chronic outcome of HBV infection in Patient 2.

A finer tuning of the solution via fitting of the acute and chronic viral load data (the corresponding solutions of the model is shown in Fig. 5.25(right)) suggests the following differences in the virus–host interaction parameters between Patient 1 and

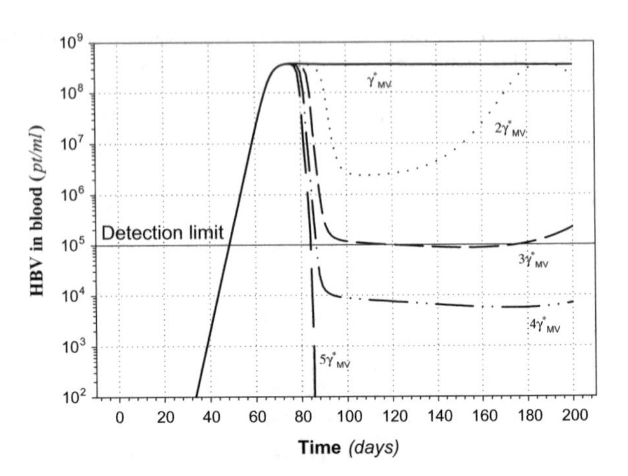

Fig. 5.26 Model prediction for HBV infection dynamics for different values of the efficacy function of antigen-presenting cells (the model parameter γ_{MV})

Patient 2: (i) antigen presentation rate is reduced by γ_{MV} 10^2; (ii) virus clearance by macrophages γ_{VM} reduced by 25%; (iii) virus replication rate ν is reduced by 20%; (iv) infection rate of target cells σ is 30% higher; (v) total number of target cells is 30% smaller.

Using the results of the above analysis, we propose that ($prop1$) the efficacy of antigen presentation might be the main cause of the chronic HBV infection in Patient 2; ($prop2$) the parameters of virus infection seem to be rather similar for the acutely and chronically infected patients. Overall, the explanation of HBV pathogenesis generated via the sensitivity analysis states that observed chronicity in the analysed HBV infection data results from a poor antigen presentation rather than high initial level of viral replication.

An important task in mathematical immunology is to explain and predict the changes in the patterns of infectious disease dynamics between various patents. Sensitivity analysis provides a means to establish causality relationships between basic parameters of the process kinetics and the clinically measured characteristics of infection. Various approaches to characterize the sensitivity can be used. In those cases when the modelled system is highly complex (e.g. the number of kinetics parameters is large) and integrative information about solutions in the form of functionals is appropriate, the adjoint equations framework allows one to compare the sensitivity of the system performance to deviations of the parameter values from their 'norms'. This should provide the way to individualize the disease dynamics and to use the mathematical models to design a treatment for an individual patient rather than for an abstract 'averaged' individual.

References

1. Zinkernagel R.M., Hengartner H., Stitz L.. On the role of viruses in the evolution of immune responses. *British Medical Bulletin* , **41** (1985) 92–97.
2. Grossman Z., Paul, W.E. Autoreactivity, dynamic tuning and selectivity, *Curr. Opin. Immunol.*, **13** (2001) 687–698.
3. Sester U, Sester M, Kohler H, Pees HW, Gartner BC, Wain-Hobson S, Bocharov G, Meyerhans A. Maintenance of HIV-specific central and effector memory CD4 and CD8 T cells requires antigen persistence. AIDS Res Hum Retroviruses. 2007; 23(4):549–553.
4. Moskophidis D, Lechner F, Pircher H, Zinkernagel RM: Virus persistence in acutely infected immunocompetent mice by exhaustion of antiviral cytotoxic effector T cells. Nature 1993, 362:758–761.
5. Germain, R. N. 2001. The art of the probable: system control in the adaptive immune system. Science 293:240–245.
6. Marchuk GI. Mathematical Models in Immunology. New York, Optimization Software, Inc., 1983. 378 pp.
7. B. Rehermann, C. Ferrari, C. Pasquinelli, and F. V. Chisari. The hepatitis B virus persists for decades after patients' recovery from acute viral hepatitis despite active maintenance of a cytotoxic T-lymphocyteresponse. *Nature Med.*, 2:1104–1108, 1996.
8. G. J. Webster, S. Reignat, M. K. Maini, S. A. Whalley, G. S. Ogg, A. S. King, D. Brown, P. L. Amlot, R. Williams, G. M. Dusheiko, and A. Bertoletti. Incubation phase of acute hepatitis B in man: dynamics of cellular immune mechanisms. *Hepatology*, 32:1117–1124, 2000.
9. Whalley, S. A., J. M. Murray, D. Brown, G. J. Webster, V. C. Emery, G. M. Dusheiko, and A. S. Perelson. 2001. Kinetics of acute hepatitis B virus infection in humans. J. Exp. Med. 193:847–854.
10. R. M. Zinkernagel. What is missing in immunology to understand immunity? *Nature Immunology*, 1:181–185, 2000.
11. A. Altinbas, Y. Yüksel, M. Pamukcu, F. Ekiz, Ö. Basar, O. Yüksel, Spontaneous HBsAg seroconversion after severe flare of chronic hepatitis B infection, Annals Hepatology. 9 (2010) 194–197.
12. Tatyana Luzyanina, Dirk Roose and Gennady Bocharov Numerical bifurcation analysis of immunological models with time delays J. Comput. Appl. Math. 184 (2005): 165–176.
13. Tatyana Luzyanina, Gennady Bocharov. Stochastic modelling of the impact of random forcing on persistent hepatitis B virus infection. Mathematics and Computers in Simulation, 96 (2014): 54–65.
14. G. Bocharov, B. Ludewig, A. Bertoletti, P. Klenerman, T. Junt, T. Luzyanina, C. Fraser, R.M. Anderson, Underwhelming the immune response: effect of slow virus growth rates on CD8+ T lymphocyte responses, J. Virol. 78 (2004) 2247–2254.
15. F.V. Chisari, M. Isogawa, S.F. Wieland, Pathogenesis of hepatitis B virus infection, Pathol. Biol. (Paris). 58 (2010) 258-266.
16. S.W. Cho, J.Y. Cheong, Y.S. Ju, D.H. Oh, Y.J. Suh, K.W. Lee, Human leukocyte antigen class II association with spontaneous recovery from hepatitis B virus infection in korenas: Analysis at the haplotype level, J. Korean Med. Sci. 23 (2008) 838–844.
17. D. J. Higham, P. E. Kloeden, MAPLE and MATLAB for stochastic differential equations in finance, in S. S. Nielsen (Ed.), Programming Languages and Systems in Computational Economics and Finance, volume 18 of *Advances in Computational Economics*, Kluwer Academic Publishers, 2002, pp. 233–270.
18. P. E. Kloeden, E. Platen, Numerical Solution of Stochastic Differential Equations, Springer, 1999.
19. M.K. Maini, C. Boni, C.K. Lee, J.R. Larrubia, S. Reignat, G.S. Ogg, A.S. King, J. Herberg, R. Gilson, A. Alisa, R. Williams, D. Vergani, N.V. Naoumov, C. Ferrari, A. Bertoletti, The role of virus-specific CD8(+) cells in liver damage and viral control during persistent hepatitis B virus infection, J. Exp. Med. 191 (2000) 1269–1280.

20. Matlab http://www.mathworks.com/.
21. M.A. Nowak, C.R. Bangham, Population dynamics of immune responses to persistent viruses, Science, 272 (1996) 74–79.
22. U. Picchini, SDE-Toolbox: Simulation and Estimation of Stochastic Differential Equations with MATLAB, Technical Report, 2007. Available from http://sdetoolbox.sourceforge.net.
23. A.T. Tan, S. Koh, W. Goh, H.Y. Zhe, A.J. Gehring, S.G. Lim, A. Bertoletti, A longitudinal analysis of innate and adaptive immune profile during hepatic flares in chronic hepatitis B, J. Hepatol. 52 (2010) 330–339.
24. G.J. Webster, S. Reignat, D. Brown, G.S. Ogg, L. Jones, S.L. Seneviratne, R. Williams, G. Dusheiko, A. Bertoletti, Longitudinal analysis of CD8+ T cells specific for structural and nonstructural hepatitis B virus proteins in patients with chronic hepatitis B: implications for immunotherapy, J Virol. 78 (2004) 5707–5719.
25. Bocharov G.A., Marchuk G.I., Romanyukha A.A. Numerical solution by LMMs of stiff delay differential systems modelling and immune response. *Numer. Math.* (1996) **73**, 131–148.
26. Doherty P.C. Some problem areas in the interaction between viruses and the immune system. *Immunol. Cell Biol.* (1986) **65**, 279–286.
27. Fong T., Di Bisceglie A.M., Biswas R., Waggoner J., Wilson L., Clagget J., Hoofnagle J.H. High levels of viral replication during acute hepatitis B infection predict progression to chronicity. *J. Med. Virol.* (1994) **43**, 155–158.
28. Grossman Z., Min B., Meier-Schellersheim M., and Paul W. Concomitant regulation of T-cell activation and homeostasis. *NATURE REVIEWS Immunology* (2004) **4**, 7–15.
29. Marchuk G.I. *Mathematical models of immune response in infectious diseases*. – Dordrecht: Kluwer Press, 1997.
30. Marchuk G.I. Adjoint equations and the sensitivity of functionals. *Earth. Obs. Rem. Sens.* (1999) **15**, 645–677.
31. Marchuk G.I., Petrov R.V., Romanyukha A.A., Bocharov G.A. Mathematical model of antiviral immune response I. Data analysis, generalized picture construction and parameters evaluation for hepatitis B. *J. Theoret. Biol.* (1991) **151**, 1–40.
32. Marchuk G.I., Romanyukha A.A., Bocharov G.A. Mathematical model of antiviral immune response II. Parameter indentification for acute viral hepatitis B. *J. Theoret. Biol.* (1991) **151**, 41–70.
33. G.I. Marchuk, V. Shutyaev and G. Bocharov. Adjoint equations and analysis of complex systems: application to virus infection modeling. J. Comput. Appl. Math. 184 (2005): 177–204.
34. Pontryagin, L. S., *Selected Works*, Nauka, Moscow, 1988 (in Russian).
35. Pupko, V. Ya., Zrodnikov, A. V., and Likhachev, Yu. I., *The Adjoint Function Method in Physics and Engineering*, Energoatomizdat, Moscow, 1984 (in Russian).
36. Schrödinger E. Quantisierung als Eigenwertproblem. *Ann. Phys.* (1926) **80**, 437–490.
37. Shutyaev V. An algorithm for computing functionals for a class of nonlinear problems using the adjoint equation. *Sov. J. Numer. Anal. Math. Modelling* (1991) **6**, 169–178.
38. Shutyaev V. *Control operators and iterative algorithms for variational data assimilation problems.*, Nauka, Moscow, 2001 (in Russian).
39. Bell G.I. Predator-prey equations simulating an immune response. *Mathematical Biosciences.* 1973; 16 (3–4), 291–314.

Chapter 6
Spatial Modelling Using Reaction–Diffusion Systems

Mathematical immunology is dealing with increasingly complex models of immune phenomena formulated with ODEs or DDEs. Except for few studies, mathematical models of the immune response against virus infections conventionally consider the infected whole organism as a single homogenous compartment. Thus, they do not take into account that the dynamics of infection spread differs between tissues, organs and blood. The spatial aspects of the immune processes can be partly taken into account by a compartmental view of the space. A more detailed description however should include this spatial heterogeneity with respect to virus propagation and immune response development. Models based upon PDEs are still rare in mathematical immunology. In this chapter, we present basic foundations of spatio-temporal modelling using reaction–diffusion (RDE) systems. The application of one-dimensional RDEs with time lag for predicting the qualitative regularities of the virus infection spreading in target tissues will be presented in the first sections. Then, a computational approach to study the cytokine distribution in LN will be illustrated. The results of this chapter are based on our earlier work published in [1, 2].

6.1 Reaction–Diffusion Equations for Immunology

6.1.1 Spatial Models of Infection Development

The virus distribution in tissue can be heterogeneous due to its non-homogeneous initial distribution and the emergence of spatio-temporal patterns determined by the interaction of virus reproduction, transport and the immune response. Regulation of

G. Bocharov et al., *Mathematical Immunology of Virus Infections*,
https://doi.org/10.1007/978-3-319-72317-4_6

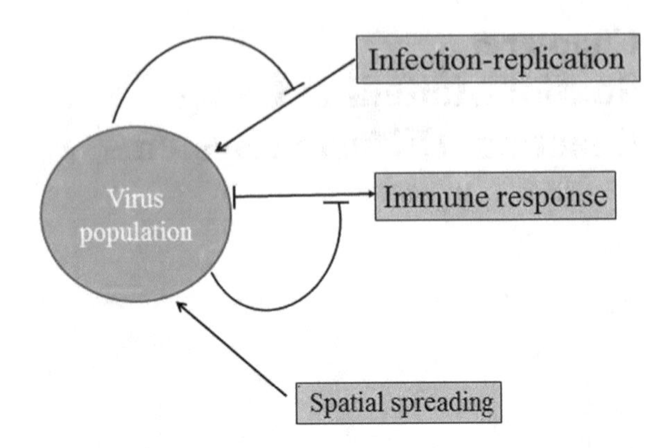

Fig. 6.1 Schematic representation of virus infection dynamics regulation (see the explanation in the text). (The figure is reprinted from Bocharov et al., PLOS ONE, 2016)

the virus population is schematically presented in Fig. 6.1.[1] There are three main processes which determine the distribution of virus in tissue: its reproduction, its elimination by immune cells and its transport. In the simplest case, virus reproduction can be described by the logistic function where the reproduction rate is proportional to its concentration v and to the concentration of uninfected host cells. In the dimensionless variables, the latter is proportional to $1 - v$. Therefore, the reproduction rate becomes proportional to the conventional logistic term $v(1 - v)$.

Virus elimination by immune cells is proportional to the product of their concentrations, cv. The concentration of immune cells at time t depends on the concentration of virus at some time $t - \tau$ since the immune cells need some time to proliferate and differentiate after being stimulated by the infection. Therefore $c(x, t) \sim f(v(x, t - \tau))$, where the function $f(v)$ describes the immune response depending on the infection level. This function has a specific bell shape: for small viral loads, the immune response is an increasing function of v while for large viral loads, it decreases since infection can downregulate the immune cells via anergy and activation-induced apoptosis.

Finally, virus population can spread in the tissue by different transport mechanisms. It can spread either by either direct cell–cell transmission or a random motion in the extracellular matrix [3, 4]. Under these assumptions, we obtain the following delay reaction–diffusion equation describing the virus distribution:

$$\frac{\partial v}{\partial t} = D\,\frac{\partial^2 v}{\partial x^2} + kv(1 - v) - f(v_\tau)v. \tag{6.1.1}$$

We will begin the investigation of virus spreading in tissue with this equation. In the end of this section, we will consider more complete models. In order to study dynamics of solutions of this model, we recall some basic definitions and results from the theory of reaction–diffusion waves.

[1]Material of Sects. 6.1 and 6.2 uses the results of our studies from Bocharov et al., Spatiotemporal dynamics of virus infection spreading in tissues, PlosOne, 2016, 11(12):e0168576.

6.1.1.1 Reaction–Diffusion Waves

The reaction–diffusion equation

$$\frac{\partial u}{\partial t} = D\frac{\partial^2 u}{\partial x^2} + F(u) \tag{6.1.2}$$

describes the evolution of the variable $u(x, t)$ in space and in time. In immunological models, it can be the concentrations of viruses, of cells or biochemical substances. The diffusion term in the right-hand side of this equation describes their random motion, and the function $F(u)$ shows their production (birth) and consumption (death) rates. A typical example is given by the following function:

$$F(u) = au^n(1 - u) - b(u). \tag{6.1.3}$$

It will allow us to describe specific features of solutions which are also observed for other nonlinearities considered below. For the dimensionless (normalized) concentration of cells or viruses, the first term in (6.1.3) describes their birth rate and second term their death rate. If $n = 1$, then the birth rate is proportional to the concentration of viruses (cells) and it is limited by the available resources $(1 - u)$, for example, host cells where virus multiplies. The death rate function $b(u)$ will be specified in the following section. At the moment, it will be taken in the simplest linear form $b(u) = \sigma u$ implying that the death rate is proportional to the concentration. In a more general and biologically interesting case, $b(u)$ can be a nonlinear function which takes into account specific properties of the immune response. The value $n > 1$ corresponds to a possible self-amplification of the birth rate.

We will consider Eq. (6.1.2) on the real axis, $-\infty < x < \infty$, with an initial condition $u(x, 0)$ assuming that it is a bounded non-negative and piecewise continuous function (Cauchy problem). The existence and uniqueness of solution of this problem is well known (see, e.g. [5]) and we will not discuss it here. If $n = 1$ and $b(u) = \sigma u$, then there are two different cases, $\sigma > a$ and $\sigma < a$ (for the sake of simplicity we consider only strict inequalities). In the first case, the equilibrium point $u = 0$ is a unique non-negative stationary solution, and it is globally asymptotically stable. The latter means that solution $u(x, t)$ of Eq. (6.1.2) uniformly converges to 0 as time t increases:

$$\sup_{x\in\mathbb{R}} |u(x, t)| \to 0, \quad t \to \infty.$$

This mathematical conclusion has a clear biological interpretation: if the virus death rate is greater than its birth rate, then infection disappears with time.

In the second case, where $\sigma < a$, there are two non-negative equilibrium points, $u_0 = 0$ and $u_1 = (a - \sigma)/a$. Considered as solutions of the ordinary differential equation

$$\frac{du}{dt} = F(u), \tag{6.1.4}$$

the former is unstable and the latter is stable. Therefore, any small positive concentration of viruses leads to their multiplication, and their concentration grows towards the positive equilibrium u_1.

Behavior of solutions of Eq. (6.1.2) can be described in this case by *travelling wave* solutions. Travelling wave is a solution of this equation of the form $u(x,t) = w(x-ct)$. It satisfies the second-order ordinary differential equation

$$Dw'' + cw' + F(w) = 0 \tag{6.1.5}$$

and the limits at infinity

$$w(-\infty) = u_1, \quad w(\infty) = u_0. \tag{6.1.6}$$

Here c is the wave speed. It is a constant, which is a priori unknown and which should be found as a solution of the problem. Hence, we search for the values of c for which problem (6.1.5), (6.1.6) has a solution.

Travelling wave solutions for reaction–diffusion equations were introduced by Fisher [6] and Kolmogorov–Petrovskii–Piskunov [5] for the models of population dynamics and by Zeldovich and Frank-Kamenteskii [9] in combustion theory (see also [10]). These works initiated a wide area of research with numerous applications [8].

6.1.2 *Existence and Stability of Waves*

6.1.2.1 Existence of Waves

In the example considered above, travelling waves describe a transition from an unstable equilibrium u_0 to a stable equilibrium u_1. Existence of such solutions for the scalar reaction–diffusion equation can be easily studied by the phase space analysis of the system of two first-order equations

$$w' = p, \quad p' = \frac{1}{D}(-cp - F(w)) \tag{6.1.7}$$

obtained from the second-order Eq. (6.1.5).

Existence of waves in the monostable case. In a more general setting, let the function $F(w)$ be positive on the interval $u_0 < w < u_1$ and $F(u_0) = F(u_1) = 0$. This is the so-called monostable case where one of the two non-negative equilibria of Eq. (6.1.4) is stable and another one is unstable. In this case, problem (6.1.5), (6.1.6) has a solution for all values of the speed c greater than or equal to the minimal speed

c_0. These solutions are monotonically decreasing functions of the space variable x. Their monotonicity is important for the stability of waves (see below).

The properties of the function $F(u)$ in (6.1.3) are different if $n > 1$. For simplicity, let $n = 2$. If $4\sigma > a$, then $F(u)$ has only one non-negative zero, $u_0 = 0$. It is globally asymptotically stable as a stationary solution of Eq. (6.1.2). If $4\sigma < a$, then $F(u)$ has three non-negative zeros, $u_0 = 0$, u_1 and u_2; the last two zeros are found as solutions of the equation $u(1-u) = \sigma/a$, $u_2 < u_1$. Stationary solutions u_0 and u_1 of Eq. (6.1.4) are stable, and u_2 is unstable.

Existence of waves in the bistable case. For a more general function $F(u)$, not necessarily given by equality (6.1.3), suppose that $F(u_0) = F(u_1) = F(u_2) = 0$ for some $u_0 < u_2 < u_1$, $F(w)$ is negative on the interval $u_0 < w < u_2$ and positive on the interval $u_2 < w < u_1$. This is the so-called bistable case since the zeros u_0 and u_1 are stable as stationary solutions of Eq. (6.1.4). In this case, there exists a unique value of c such that problem (6.1.5), (6.1.6) has a solution. It is a monotonically decreasing function of the space variable x.

6.1.2.2 Wave Speed

An important characterization of travelling waves is the speed of propagation. If $n = 1$ in (6.1.3), then the minimal speed is given by the formula $c_0 = 2\sqrt{DF'(0)}$. In general, explicit formula for the wave speed does not exist but there are various analytical approximations. In the bistable case, the wave speed can be positive or negative depending on the nonlinearity. Indeed, multiplying Eq. (6.1.5) by w' and integrating, one obtains

$$c = \int_{-\infty}^{\infty} F(u)du \Big/ \int_{-\infty}^{\infty} (w'(x))^2 dx \, .$$

Though it is not an explicit formula for the wave speed since the function $w'(x)$ is unknown, it allows the determination of the sign of c: it is positive (zero, negative) if and only if the integral of F is positive (zero, negative). This condition gives a simple and useful criteria of propagation, in particular, of virus spread which occurs under the positive wave speed.

6.1.2.3 Convergence to Waves

By definition, solution $u(x,t)$ of Eq. (6.1.2) with some initial condition $u(x,0) = u_0(x)$ uniformly converges to the travelling wave solution $w(x)$ if there exists a constant h such that

$$\sup_{x\in\mathbb{R}} |u(x,t) - w(x - ct - h)| \to 0, \quad t \to \infty. \tag{6.1.8}$$

This convergence characterizes behavior of solutions, and it shows that solution of the Cauchy problem propagates as a travelling wave solution. A more general type of convergence, called convergence in form occurs if instead of (6.1.8) one has

$$\sup_{x \in \mathbb{R}} |u(x,t) - w(x - m(t))| \to 0, \quad t \to \infty \tag{6.1.9}$$

for some function $m(t)$. Clearly, the uniform convergence is a particular case of the convergence in form where $m(t) = ct + h$. Therefore from the uniform convergence it follows the convergence in form. However, the inverse may not be true. If the function $m(t)$ is such that $m'(t) \to c$ as $t \to \infty$, then convergence (6.1.9) is convergence in form and in speed. However, convergence in form and in speed does not imply the uniform convergence neither. The function

$$m(t) = ct + k \log t + h \tag{6.1.10}$$

satisfies the property of convergence in speed but not the uniform convergence.

Having introduced these definitions, one can now present the main results on the convergence to waves in the monostable and in the bistable cases.

Convergence to the wave with the minimal speed in the monostable case. Let $u_0(x)$ be a non-negative monotone function such that $u_0(x) \equiv 0$ for $x \geq x_0$ and some x_0 and $u(-\infty) > 0$. Then, the solution $u(x,t)$ of Eq. (6.1.2) with the initial condition $u(x,0) = u_0(x)$ converges to the wave $w(x)$ with the minimal speed c_0 in form and in speed. In the case of function $F(u)$ given by (6.1.3) with $n = 1$, the function $m(t)$ in the definition of convergence has the form (6.1.10).

Convergence to the wave in the bistable case. Let $u_0(x)$ be a non-negative monotone function such that $u_0(\infty) = u_0$ and $u(-\infty) = u_1$. Then the solution $u(x,t)$ of Eq. (6.1.2) with the initial condition $u(x,0) = u_0(x)$ uniformly converges to the wave $w(x)$.

More general results about convergence to wave are known including convergence to the waves with the speed greater than the minimal one [7]. Convergence to waves is related to the wave stability determined by the location of the spectrum of the corresponding linearized operator. It should be noted that monotone waves are stable in appropriate classes of perturbations and non-monotone waves are not stable.

6.1.2.4 Systems of Waves

The existence of waves was discussed above for some particular functions $F(w)$. In what follows, some other functions will also be considered for which the waves may not exist and for which behavior of solutions of Eq. (6.1.2) can be different.

It was assumed above that the function $F(w)$ has two non-negative zeros u_0 and u_1 in the monostable case, and it is positive on the interval $u_0 < w < u_1$. Suppose now that it has two additional zeros u_* and u^* such that $u_0 < u_* < u^* < u_1$ and

$$F(w) > 0 \text{ for } u_0 < w < u_*, \quad u^* < w < u_1; \qquad F(w) < 0 \text{ for } u_* < w < u^*. \tag{6.1.11}$$

This case is also monostable since u_0 is unstable with respect to Eq. (6.1.4), and point u_1 is stable. However, the behavior of solutions in this case is more complex.

The limiting values of waves at infinity can now be different from (6.1.6). The waves with the limits $w(-\infty) = u_*$, $w(\infty) = u_0$ will be called the $[u_0, u_*]$-waves. Similarly, we define the $[u_*, u_1]$-waves and the $[u_0, u_1]$-waves.

The $[u_0, u_*]$-waves correspond to the monostable case considered in Sect. 6.1.2.1 since the function $F(w)$ is positive on the interval $u_0 < w < u_*$. Such waves exist for all values of the speed greater than or equal to the minimal speed c_0. The $[u_*, u_1]$-wave corresponds to the bistable case. It is unique, and its speed will be denoted by c_1. The existence of $[u_0, u_1]$-waves depends on the relation between the speeds c_0 and c_1.

Existence of waves in the monostable–bistable case. If $c_1 > c_0$, then $[u_0, u_1]$-waves exist for all speeds c in the interval $c_* \le c < c_1$ for some c_*. If $c_1 \le c_0$, then such waves do not exist.

An interesting question arises then, how to describe the behavior of solutions of Eq. (6.1.2) if the waves do not exist. In this case, not only waves but also systems of waves should be considered. The system of waves consists of two (or more) waves propagating one after another. Propagation of a single wave is shown in Fig. 6.2 (left), propagation of a system of waves in Fig. 6.2 (right). Here the $[u_0, u_*]$-wave with the minimal speed c_0 is faster than the $[u_*, u_1]$-wave. Therefore, the first waves runs away followed by a slower second wave.

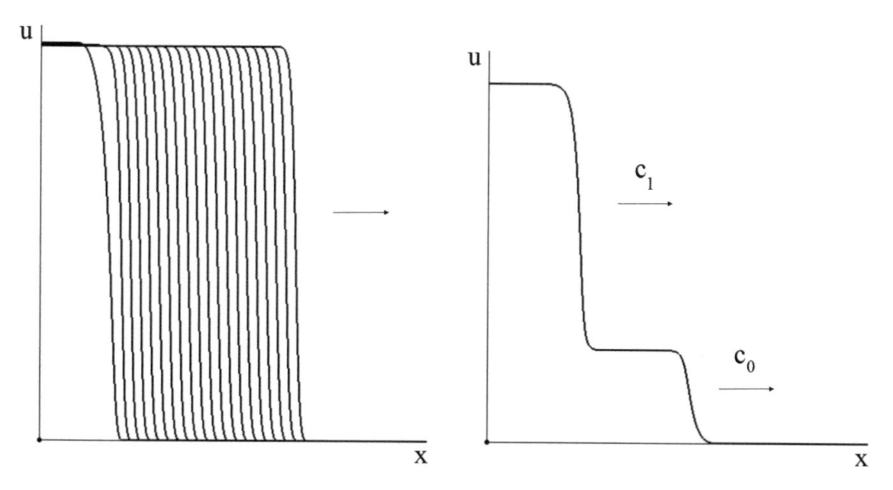

Fig. 6.2 Propagation of a single wave (left) and of system of waves (right). Different curves in the left figure show the solution $u(x, t)$ of Eq. (6.1.2) as a function of x at different moments of time. (The figure is modified from Bocharov et al., PLOS ONE, 2016)

6.2 Virus Spreading in Tissue

6.2.1 *The Model Without Time Delay*

The virus distribution in tissue such as spleen or lymph node can be described by the reaction–diffusion equation

$$\frac{\partial v}{\partial t} = D\,\frac{\partial^2 v}{\partial x^2} + kv(1-v) - f(v)v. \tag{6.2.1}$$

Here $v = v(x,t)$ is the dimensionless virus concentration, the first term in the right-hand side of this equation describes virus diffusion, the second term its production and the last term its elimination by CTLs. We suppose that the function $f(v)$ is non-negative and it is continuous together with its second derivatives. By a change of variables this equation can be reduced to the same equation with $D = 1, k = 1$.

The function $f(v)$ describes the concentration of immune cells which eliminate infection. It depends on the concentration of virus. According to its biological meaning, we suppose that it is growing for small v and decreasing for large v. Indeed, small viral load stimulates immune response while large viral load downregulates it. The qualitative form of this function is shown in Fig. 6.3. There are two different cases depending on the strength of the immune response for low viral load: $f(0) < 1$ (Fig. 6.3, left) or $f(0) > 1$ (Fig. 6.3, right).

The nonlinearity $F(v) = v(1 - v - f(v))$ corresponds to the model function considered in Sect. 6.1.2 with $n = 1$ and $b(v) = f(v)v$. We will consider this equation on the whole axis with a non-negative initial condition. Depending on the form of the function $f(v)$, we will get different behavior of solutions. Let us note that $F(0) = 0$. This function can have other zeros for $v > 0$.

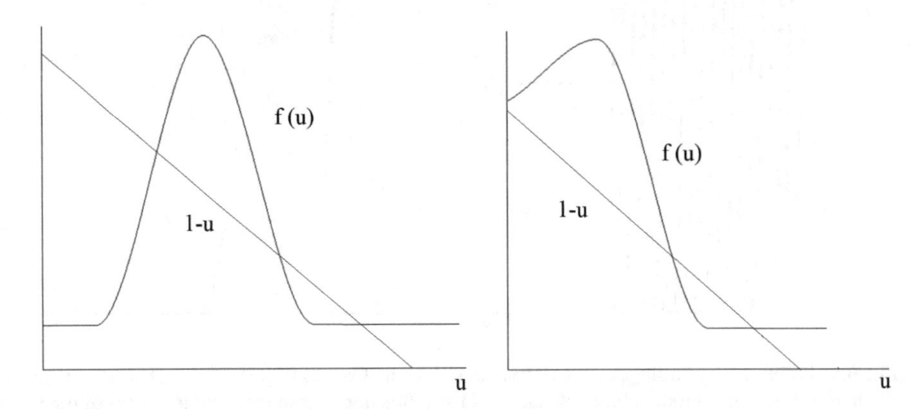

Fig. 6.3 Qualitative form of the function $f(v)$ with $f(0) < 1$ (left) and $f(0) > 1$ (right)

6.2.1.1 Stable Virus Free State

Suppose that $f(0) > 1$. Then $F'(0) < 0$, and the virus-free state $v = 0$ is locally asymptotically stable as a solution of Eq. (6.1.1). If $F(v) < 0$ for all $v > 0$, then it is globally asymptotically stable. This condition on the function $F(v)$ signifies that virus elimination due to the immune response is stronger than the virus reproduction for all values of v.

The form of the function $f(v)$ implies that $F(v)$ can have two other zeros for positive v. We suppose that they are different from each other and denote them by v_0 and v_-, $v_0 < v_-$. Then $F(v) < 0$ for $0 < v < v_0$ and $F(v) > 0$ for $v_0 < v < v_-$ (bistable case). In this case, there is a travelling wave solution of Eq. (6.1.1), $v(x,t) = w(x - ct)$. The function $w(x)$ satisfies the equation

$$w'' + cw' + F(w) = 0 \tag{6.2.2}$$

and it has the limits

$$w(-\infty) = v_-, \quad w(+\infty) = 0 \tag{6.2.3}$$

at infinity. Behavior of solutions of the Cauchy problem is determined by the wave speed and by the initial conditions.

Wave speed

The sign of the wave speed depends on the sign of the integral

$$I(F) = \int_0^{v_-} F(v)dv.$$

The wave speed c is positive (zero, negative) if $I(F)$ is positive (zero, negative). Positive wave speed corresponds to infection spreading. If the speed is negative, infection retreats.

Initial condition

Assume, first, that the initial condition $v(x, 0)$ is a monotonically decreasing function with the limits

$$v(-\infty, 0) = v^*, \quad v(+\infty) = 0.$$

If $v^* < v_0$, then the solution converges to 0 uniformly on the whole axis. If $v_- > v^* > v_0$, then it converges to the travelling wave uniformly on the whole axis. If $F'(v_-) < 0$, then the rate of this convergence is exponential.

Next, consider the case where the initial condition is non-negative and it has zero limits at infinity. If the function $v(x, 0)$ is sufficiently large, then $v(x,t) \to v_-$ as $t \to \infty$ locally uniformly. The transition between 0 and v_- is provided by travelling waves moving in the opposite directions. If $v(x, 0)$ is sufficiently small, then the solution converges to 0 uniformly on the whole axis. The conditions on $v(x, 0)$ under which one of these cases occurs can be formulated more precisely.

Thus, even if the wave speed is positive, infection propagation depends on the initial condition. The initial infection load should be sufficiently large to provide infection spreading.

6.2.1.2 Unstable Virus Free State

If $f(0) < 1$, then $F'(0) > 0$, and $v = 0$ is an unstable stationary solution of Eq. (6.1.1). For any non-negative initial condition $v(x, 0)$, which is not identically 0, the solution $v(x, t)$ does not converge to 0. It grows to some positive value.

We will discuss behavior of solution in more detail. Since $F(1) < 0$, then there is at least one zero of this function for $v > 0$. Taking into account the form of the function $f(v)$, the function $F(v)$ can have from one to three positive zeros. Let $F(v) = 0$ for $v = 0, v_0, v_1, v_-$, where $0 < v_0 < v_1 < v_-$. It is a monostable case with intermediate zeros (cf. Sect. 1.2.4):

$$F(v) > 0 \text{ for } 0 < v < v_0,\ v_1 < v < v_- \,;\ F(v) < 0 \text{ for } v_0 < v < v_1 \,.$$

There exist travelling waves with the limits $v(-\infty) = v_0$, $v(+\infty) = 0$, the $[0, v_0]$-waves (monostable case). Moreover, there is a $[v_0, v_-]$-wave with the limits $v(-\infty) = v_-$, $v(+\infty) = v_0$ (bistable case).

Denote by c_0 the minimal speed of the $[0, v_0]$-waves and by c_1 the speed of the (unique) $[v_0, v_-]$-wave. If $c_1 > c_0$, then there are $[0, v_-]$-waves with the speeds $c_* \le c < c_1$, where $c_0 \le c_* < c_1$. If $c_1 \le c_0$, then such wave does not exist. In this case, there is a system of waves consisting of two waves. First there is a $[0, v_0]$-wave followed by the $[v_0, v_-]$-wave propagating with a lesser speed.

Convergence to these waves and systems of waves depends on the initial conditions. The following cases of infection spread can be identified:

1. Weak infection v_0 is established after the infection spread in the tissue ($c_1 < 0$ or $c_1 > 0$ but the initial condition is sufficiently small). Infection spreads as $[0, v_0]$-wave with the minimal speed c_0,
2. Strong infection v_- is established after wave propagation ($c_1 > c_0$, initial condition is sufficiently large). Infection spreads as $[0, v_-]$-wave with some speed $c \ge c_0$,
3. Weak infection v_0 is followed by strong infection v_- ($0 < c_1 < c_0$, initial condition is sufficiently large). Infection spreads as two consecutive waves with different speeds (cf. Fig. 6.2, right).

6.2.2 *The Model with Time Delay*

In this section, we take into account time delay in the immune response and we use the results of the work [1]. The concentration of immune cells $f(v(x, t - \tau))$ at time t depends on the concentration of virus at time $t - \tau$. Therefore, we consider the equation

$$\frac{\partial v}{\partial t} = D\frac{\partial^2 v}{\partial x^2} + kv(1 - v) - f(v_\tau)v, \tag{6.2.4}$$

where the first term in the right-hand side describes virus transport, the second term its reproduction and the last one its elimination by the immune response, $v_\tau(x, t) = v(x, t - \tau)$. As above, we set $k = 1$ but we keep the diffusion coefficient as a parameter.

6.2.2.1 Time Oscillations

Time delay can lead to the instability of the homogeneous in space stationary solution, and it can also influence wave propagation. Let the following equality hold $f(v_0) = 1 - v_0$ for some v_0. Then $v(x, t) = v_0,\ x \in (-\infty, +\infty), t \in (-\infty, +\infty)$ is a stationary solution of Eq. (6.2.4). In order to study its local asymptotic stability with respect to small perturbations, we look for the solution of this equation in the form

$$v(x, t) = v_0 + \epsilon e^{\lambda t + iax},$$

where a, ϵ are real numbers, ϵ is a small parameter, and λ is an eigenvalue. Substituting the above function in (6.2.4) and equating the terms with the first power of ϵ, we get the following characteristic equation:

$$\lambda = -Da^2 - v_0\left(1 + f'(v_0)e^{-\lambda\tau}\right) \tag{6.2.5}$$

(we do not assume here that $D = 1$). The stability boundary of the steady-state solution can be computed by considering the characteristic roots in the form of purely imaginary eigenvalues $\lambda = i\phi$. Then we have

$$i\phi = -Da^2 - v_0\left(1 + f'(v_0)e^{-i\phi\tau}\right).$$

Therefore, the following equalities must hold for the real and imaginary parts:

$$Da^2 + v_0 + v_0 f'(v_0)\cos(\phi\tau) = 0, \tag{6.2.6}$$

$$v_0 f'(v_0)\sin(\phi\tau) = \phi. \tag{6.2.7}$$

Set $z = \phi\tau$. Then from (6.2.6), (6.2.7) we obtain

$$\cos z = -\frac{Da^2 + v_0}{v_0 f'(v_0)}, \quad \tau = \frac{z}{(v_0 f'(v_0) \sin z)}. \tag{6.2.8}$$

The first equation has a solution if the right-hand side is greater than -1. In particular, if $D = 0$, then the condition reduces to $f'(v_0) \geq 1$. Using z determined from the first equation, we find τ from the second equation.

Proposition 6.1 *If $f'(v_0) > 1 + Da^2/v_0$, then for all $\tau > z/(v_0 f'(v_0) \sin z)$ the solution v_0 of Eq. (6.2.4) is unstable. Here z is determined from the first equation in (6.2.8).*

Our analysis suggests that if the initial condition $v(x, 0)$ does not depend on the space variable, then the solution $v(x, t)$ of Eq. (6.2.4) is also homogeneous in space. In this case, depending on the values of parameters, the solution of the model either convergence to the stationary solution v_0 or to stable periodic time oscillations both being spatially homogeneous.

However, this behavior can be different in the case of travelling wave propagation. If we fulfil the linear stability analysis of the homogeneous in space stationary solution v_0 in the moving coordinate frame attached to the wave, we obtain the same stability conditions as before. It follows from the first equation in (6.2.8) that the onset of stability depends on the wave number a of the spatial perturbation and on the diffusion coefficient. The steady-state solution v_0 appears to be more stable with respect to spatially non-uniform perturbations ($a \neq 0$) than with respect to perturbations which are homogeneous in space ($a = 0$). The frequency of the spatial perturbations depends on the ratio of wave speed and the frequency of the time oscillations, $a = \phi/c$.

We illustrate the regimes of wave propagation in the case of the linear function $f(v) = rv$ (Fig. 6.4). In the first one, both types of perturbations, i.e. the spatial perturbations and time perturbations homogeneous in space, decay with time. They

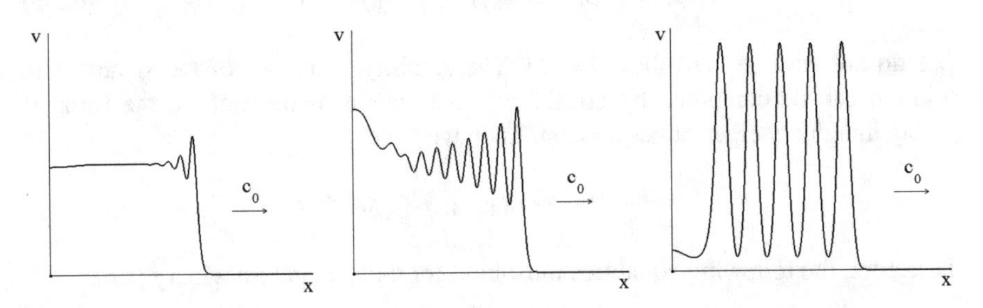

Fig. 6.4 Numerical simulations of Eq. (6.2.4) with the function $f(v) = rv$ ($r = 2$, $D = 10^{-4}$). Wave propagation for three different values of time delay, $\tau = 1.4, 2, 4$, respectively. For small time delay (left) space and time oscillations decay, for intermediate time delay (middle) space oscillations decay while time oscillations persists, for large time delay (right) both of them persist. (The figure is reprinted from Bocharov et al., PLOS ONE, 2016)

manifest themselves as decaying spatial oscillations behind the wave front followed by a spatially constant solution (Fig. 6.4, left). Another regime of wave propagation is characterized by the decaying spatially heterogeneous perturbation and the persisting homogeneous in space perturbations (Fig. 6.4, middle).

Finally, if the spatial perturbations persist, then the travelling wave propagation takes place with a (moving) periodic structure established behind the wave front (Fig. 6.4, right). Note that the wave speed in all these cases equals c_0.

6.2.3 *Full-Scale Viral Regulation of the Immune Response*

We consider now a full-scale viral regulation of the immune response, i.e. the whole function $f(v),\ v \geq 0$ as shown in Fig. 6.3 (left). The corresponding behavior of solutions will be assessed using numerical simulations and insight gained via the analyses of the simplified model problem with a linear function $f(v)$.

Let us recall that equation $f(v) = 1 - v$ has three roots (solutions), v_0, v_1 and v_-. In the model without delay (Sect. 2.1), Eq. (6.2.4) has wave solutions with the limits $w(-\infty) = v_0$, $w(\infty) = 0$ for all values of the speed greater than or equal to the minimal speed $c_0 = 2\sqrt{D(1 - f(0))}$. Suppose that the stationary point v_0 is stable (cf. Sect. 2.2.1), and the initial condition has limits 0 and v_0 at $\pm\infty$. Numerical simulations show that solution of Eq. (6.2.4) converges to the wave with the minimal speed. If the time delay is sufficiently large, then the wave is not monotone, similar to the model problems considered in Sect. 2.2.1.

The travelling wave with the limits $w(-\infty) = v_-$, $w(\infty) = v_0$ corresponds to the bistable case. It exists for a unique value c_1 of wavefront speed in the case without delay $\tau = 0$ and for the model problem with non-zero delay $\tau > 0$. Let us consider the case where $c_1 < c_0$. Then in the case without delay ($\tau = 0$) the wave with the limits $w(-\infty) = v_-$, $w(\infty) = 0$ does not exist. The behavior of solutions of Eq. (6.2.4) is described by a system of two waves propagating one after another with the speeds c_0 and c_1. A similar behavior is observed when the delay is present in the regulation of the immune response but it is sufficiently small (Fig. 6.5). The solution can be monotone behind the first wave or non-monotone depending on the value of τ.

It appears that the speed of the bistable travelling wave increases with time delay. This increase occurs because the bistable wave is preceded by a monostable wave characterized by a lower level of infection. Therefore, the values of the function $f(v)$ at the bistable wave also decrease due to time delay. Hence the speed c_1 increases as a function of τ, while the speed c_0 of the monostable wave does not depend on τ. For τ sufficiently large, the two waves merge and form a single wave with the limits $w(-\infty) = v_-$, $w(\infty) = 0$ (Fig. 6.5). This is specific for the system of waves where the bistable wave follows the monostable one. If τ is large enough, this resulting wave becomes non-monotone.

Figure 6.6 shows the last series of simulations in which the slope (sensitivity of immune response) of $f'(v_1)$ is large enough. In this case, we observe the existence of a monostable wave with spatial oscillations behind it. This wave can be separated

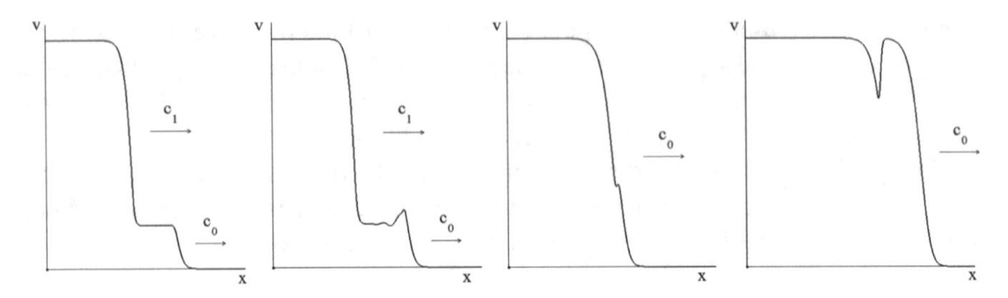

Fig. 6.5 Numerical simulations of different regimes of infection spreading depending on time delay, $\tau = 0.4, 0.95, 1.5, 10$; $D = 0.0001$. For small time delay (two left figures: $\tau = 0.4, 0.95$), there are two consecutive waves of infection propagating with different speeds. The first wave can be non-monotone. For large time delay (two right figures: $\tau = 1.5, 10$), the second wave propagates faster and they finally merge forming a single wave which can be either monotone or non-monotone. (The figure is reprinted from Bocharov et al., PLOS ONE, 2016)

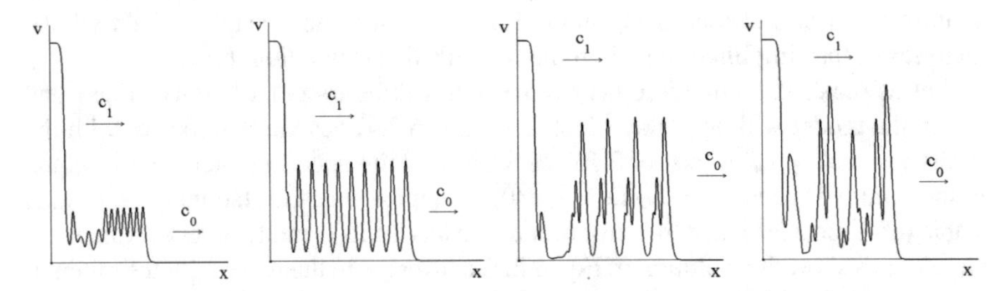

Fig. 6.6 Existence of a monostable wave with spatial oscillations behind it. This wave is separated from the bistable wave by a zone of irregular oscillations. Increase of the delay value results in a qualitative change of the spatial patterns of the infection spread. The two travelling waves do not merge and the monostable wave is not followed by steady space oscillations. Aperiodic oscillations are observed behind the wave front which propagates at a speed c_0. The values of time delay are, respectively, $\tau = 1, 2, 3, 4$; $D = 0.0001$. (The figure is reprinted from Bocharov et al., PLOS ONE, 2016)

from the bistable wave by a zone of irregular oscillations. Further increase of the delay results in a qualitative change of the spatial patterns of the infection spread. The two travelling waves do not merge any longer and the monostable wave is not followed by steady space oscillations. Aperiodic oscillations are observed behind the wave front which propagates, as before, at a constant speed c_0.

6.3 Spatial Model of Virus and Immune Cells Dynamics

In the previous section, we studied the simplest model of infection dynamics in which the concentration of immune cells in virus elimination term at time t is implicitly determined by the infection level at time $t - \tau$. The simplest model can be extended to reaction–diffusion system of equations

$$\frac{\partial v}{\partial t} = D_1 \frac{\partial^2 v}{\partial x^2} + kv(1-v) - \sigma vc, \tag{6.3.1}$$

$$\frac{\partial c}{\partial t} = D_2 \frac{\partial^2 c}{\partial x^2} + \phi(v_\tau)c(1-c) - \psi(v_\mu)c \tag{6.3.2}$$

which describes the spatio-temporal distributions of virus and immune cell concentrations. We suppose that $\phi(v)$ and $\psi(v)$ are some growing function with saturation.

We present here some examples of numerical simulations. Figure 6.7 shows two regimes of infection spreading. In the first one (without time delay, left image), virus distribution represents a sharp peak moving in space. Virus concentration vanishes behind the peak due to its elimination by immune cells whose concentration remains positive. The second example (Fig. 6.7, right) considers the case with time delay leading to the oscillations of persistent infection behind the propagating infection front.

Two other regimes are shown in Fig. 6.8. The first infection front propagates with a low infection level behind it and the presence of immune cells (left image). It is followed by a second front with high infection level and immune cell exhaustion. For a higher virus multiplication rate, transition to the high-level infection can occur directly, without the intermediate stage (Fig. 6.8, right).

The last example illustrates the influence of the virus diffusion coefficient characterizing the intensity of its random motion. If it is sufficiently high, then the development of the infection can begin as before (Fig. 6.7, left). However, after some time, the infection peak widens and forms two wave fronts propagating in the opposite directions. One of them leads to the elimination of immune cells (Fig. 6.9).

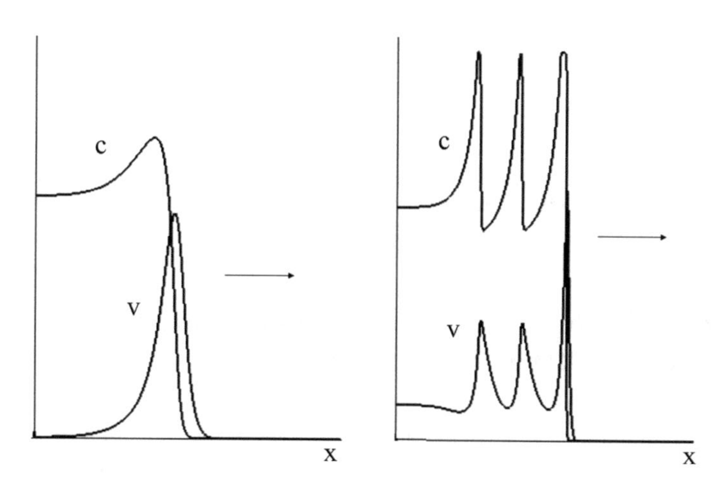

Fig. 6.7 Wave propagation without time delay (left). Virus peak is followed by its complete elimination by immune cells. Wave propagation with time delay (right) and with an oscillating persisting infection

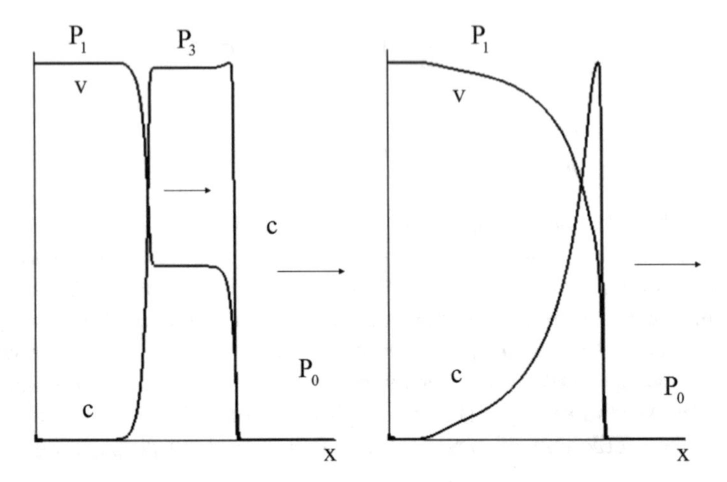

Fig. 6.8 Two stage infection propagation (left). First, a low-level infection is established followed by the second infection front with a high infection level and immune cell exhaustion. Transition to the high-level infection can occur directly (right)

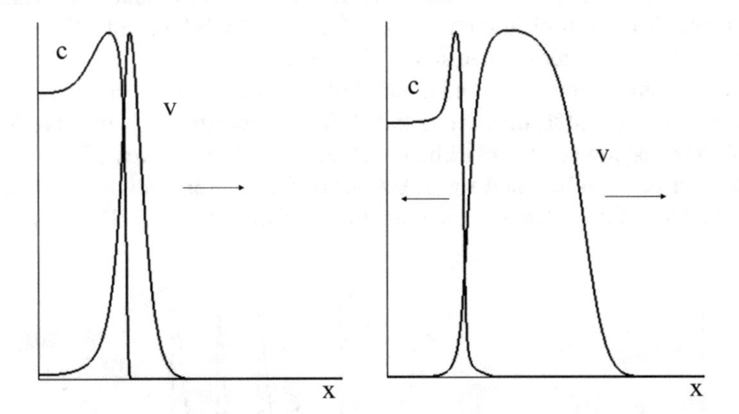

Fig. 6.9 Two snapshots of solution in the case of large virus diffusion coefficient. In the beginning, we observe the usual immune cell front accompanied by an infection spike (left). After some time, infection forms a two side propagating front eliminating immune cells (right)

6.4 Predicting the Type I IFN Field in Lymph Nodes During a Cytopathic Virus Infection

Secondary lymphoid organs (SLO) are tissues where the induction of innate and adaptive immune responses takes place.[2] The SLO microenvironment provides the

[2]Material of Sect. 6.4 uses the results from Mathematical Modelling of Natural Phenomena, Vol. 6, Bocharov et al., Reaction-Diffusion Modelling of Interferon Distribution in Secondary Lymphoid Organs, Pages 13–26, Copyright © 2011, with permission from EDP Sciences.

structural basis for the induction of T- and B-cell responses. The efficacy of immune responses relies on a productive interaction between antigen-presenting cells and lymphocytes. APCs protection by interferon requires that the in situ IFN-I concentration should exceed a certain threshold [14]. The interferon produced by pDCs is spatially distributed via convection and diffusion processes to various SLO compartments. The microanatomy of a paradigmatic SLO e.g. the lymph node can be essentially reduced to three zones: the subcapsular zone, which is an APC-rich area, the B-cell activation zone (B-cell follicle) and the T-cell activation zone [17]. Although direct measurements of the hydraulic conductivity within the T-cell zone are missing, it is considered that both diffusion and convection are extremely low [19]. Therefore, the spatial distribution of IFN in SLO deserves special investigation.

6.4.1 Reaction–Diffusion Model of IFN Dynamics

Delay differential equations considered in Chap. 4 in order to model the type I interferon reaction was developed following a compartmental approach, i.e. assuming the secondary lymphoid organ (spleen) in which the reaction takes place to be spatially homogeneous with instantaneous mixing. However, the secondary lymphoid organs have an elaborated architecture. The details of their structural organization allow us to suggest that an intensity of the transport processes, in particular, the diffusion, mediating the spread of soluble factors, in particular, IFN-I, differs essentially depending on the location in the specific anatomical SLO compartments [19]. To examine the effect of diffusion on the spatial distribution of IFN-I, we consider the concentration of the interferon I at time t to be also dependent on position $\mathbf{x} \in \Omega \subset R^3$, where Ω represents the spatial region occupied by the lymphoid organ. In the case of a paradigmatic SLO, it consists of a number (N) of non-overlapping subdomains Ω_i: $\Omega = \bigcup_{i=1}^{N} \Omega_i$. The spatio-temporal dependence of the variable $I(\mathbf{x}, t)$ is treated as a continuum field of concentrations evolving according to the reaction–diffusion equation with a source term:

$$\frac{\partial I}{\partial t}(\mathbf{x}, t) = \nabla \cdot (D \nabla I(\mathbf{x}, t)) - d_I I(\mathbf{x}, t) + \sum_{l=1}^{L} F_l(\mathbf{x}). \tag{6.4.1}$$

Here, D stands for the diffusion mass transfer tensor which is assumed to be a scalar constant coefficient depending on the subdomain ($D = D_i \cdot E$, $i = 1, \dots, N$) since diffusion is considered to be isotropic. Interferon degradation is described by the term $-d_I I(\mathbf{x}, t)$. IFN-I secretion by different types of activated cells located at some positions $\left(\mathbf{x}_k^m\right)_{k=1,K_l}^{l=1,L}$ is represented by the source term. It is the sum over Dirac delta functions $F_l(\mathbf{x}) = \sum_{k=1}^{K_l} \rho^{(l)} \delta(\mathbf{x} - \mathbf{x}_k^{(l)})$ with $\rho^{(l)}$ representing the per capita cell type specific secretion rate. Due to the singularity of $F_l(\cdot)$, the equations are understood in the weak sense [20].

The production–degradation parameters were estimated based on the actual data using the compartmental model in [14]. In particular, the production rates of IFN-I by activated pDC and macrophage are $\rho^{pDC} = 4.4 \times 10^{-4}$ pg/h/cell and $\rho^{M\varphi} = 10^{-6}$ pg/h/cell, respectively. The estimated decay rate of IFN-I in a cell-free medium is $d_I = 0.012\,\mathrm{h}^{-1}$. The last value does not consider the consumption of interferon and can be increased by up to tenfold to account for the internalization of free IFN-I by various cells in SLO.

The diffusion characteristics of the subdomains representing various compartments of a paradigmatic secondary lymphoid organ and corresponding boundary conditions are discussed in the next section.

6.4.2 3D Approximation of a Paradigmatic Lymph Node

Secondary lymphoid organs have a highly elaborate structure and organization to facilitate the interaction between the immune cells and the lymph-borne pathogens derived from distant tissues. Following a building block approach, a paradigmatic lymph node synthesis is presented in [17]. Functionally, the lymph node (domain Ω) consists of three major subdomains:

- an outer antigen-sampling zone (subcapsular sinus, trabecular sinuses, conduit tubes), referred to as subdomain Ω_1,
- B-cell follicles which make subdomain Ω_2,
- T-cell zone (cortex and paracortex) denoted as subdomain Ω_3.

A paradigmatic LN schematic view is presented in Fig. 6.10. The subcapsular sinus in the LN contains aggregates of macrophages and dendritic cells which trap soluble antigens and serve both innate and adaptive immune responses. Once the pDCs detect viral RNA or DNA, they pass through ordinary activation programs to subsequently start secreting IFN-I. Conduits represent an important system of distribution channels for small soluble antigens and immune modulators (with molecular weight below 70 kDa). They extend from subcapsular sinus floor through the T-cell zone and form a contiguous lumen with fluid channels around the high endothelial venules (HEVs), thus making a network highly connected with the cortex capillaries and venules [17, 21]. The bulk flow of water and tracers passing via conduits is documented [19]. As the IFN-I molecules mass is rather low, $\sim$17 kDa, they could also enter the conduit system.

The paradigmatic lymph node domain Ω is schematically described in Fig. 6.10 using a constructive solid geometry representation by combining three subdomains $\Omega_i, i = 1, 2, 3$, defined below. Each subdomain is made of composition of geometric primitives, such as spheres and cylinders differing in their sizes and orientation. The Open CASCADE technology (see http://www.opencascade.org) was used to construct a 3D geometric model of a paradigmatic lymph node. The first domain Ω_1 is topologically defined as follows: $(S_{\text{out}} \backslash S_{\text{inn}}) \cup C_{TS} \cup C_{CT}$, where S_{out} and

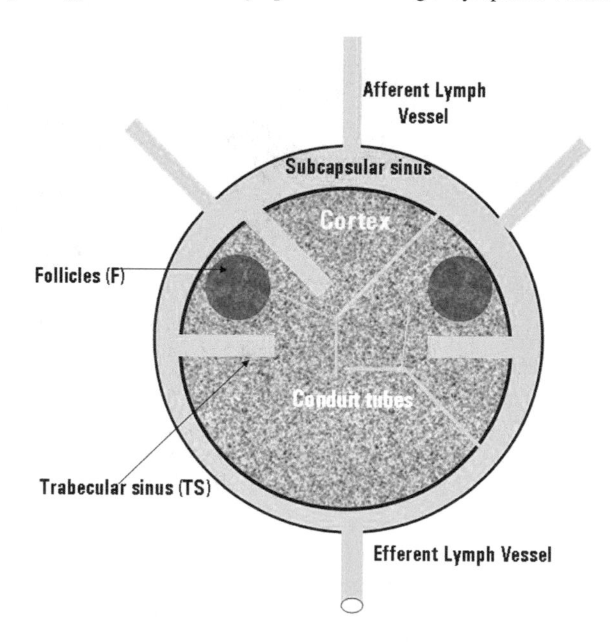

Fig. 6.10 Schematic representation of a paradigmatic lymph node (LN). LN major functional building blocks are subcapsular sinus, trabecular sinus, conduits, B-cell follicles, T-cell zone. (Reprinted from Mathematical Modelling of Natural Phenomena, Vol. 6, Bocharov et al., Reaction-Diffusion Modelling of Interferon Distribution in Secondary Lymphoid Organs, Pages 13–26, Copyright © 2011, with permission from EDP Sciences)

S_{inn} are the outer and inner spheres of diameters 2 mm and 1.9 mm, respectively; $C_{TS} = \bigcup_{i=1}^{4} C_i$ stands for the union of four right circular cylinders with 0.05 mm diameter and 0.5 mm length; and $C_{CT} = \bigcup_{i=1}^{2} Y_i$ is a set of two conduits having Y-shape and 0.0005 mm diameter. The second domain Ω_2, representing the B-cell follicles, is a union of four disconnected spheres $\bigcup_{i=1}^{4} S_{F,i}$ with a diameter of 0.2 mm located inside S_{inn}. Finally, the third domain Ω_3 topologically can be described as $S_{\text{inn}} \setminus (\Omega_2 \cup C_{TS} \cup C_{CT})$. The final Open CASCADE-based 3D geometric model of the lymph node consists of 50 vertices, 62 curved edges and 30 curved faces. The solid geometry model and its mesh approximation are presented in Fig. 6.11. This geometry requires multiple length scales resolution for the representation of the conduits and overall major domains. To this end, we used the mesh approximation based on unstructured-mesh approach.

The CAD-designed geometric model was imported into the Ani3D mesh generation toolkit. Ani3D is a solid modeller-based preprocessing tool for robust generation of three-dimensional unstructured tetrahedral meshes [15]. Ani3D algorithms control and automate much of the meshing process. The tetrahedrization technology implements advancing front methods supplemented by Delaunay meshing technique to deal with void faces. Notice, that the mesh is a fine-grained closer to conduits in order to properly represent the structure of conduits, whose diameter is about three

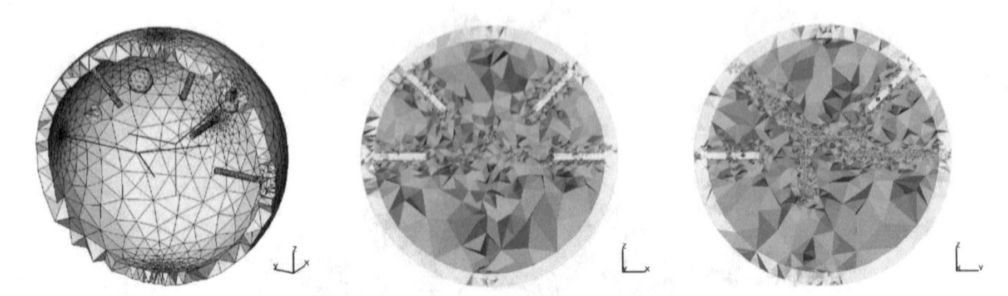

Fig. 6.11 Reconstruction of the 3D geometry of the LN. Slice planes showing the tetrahedral spatial discretization of the LN: left: slice plane for the subdomain Ω_1; middle: slice plane xz; right: slice plane yz. Refining of mesh approximating conduits is an important feature of the meshing algorithm. The domain Ω_1 is shown in yellow, the follicles domain Ω_2, consisting of small spheres, in green, and the subdomain Ω_3 in grey. (Reprinted from Mathematical Modelling of Natural Phenomena, Vol. 6, Bocharov et al., Reaction-Diffusion Modelling of Interferon Distribution in Secondary Lymphoid Organs, Pages 13–26, Copyright © 2011, with permission from EDP Sciences)

orders of magnitude smaller than the lymph node one. The lymph node surface approximating triangle mesh consists of 24720 vertices and 95732 triangles. The generated tetrahedral mesh model of the lymph node containing 103891 vertices and 619691 tetrahedrons was further improved using mesh smoothing implemented in aniMBA toolkit of the Ani3D package. The resulting mesh is characterized by the minimal cell quality of $q_{min} = 0.08$ and the total number of tetrahedra is reduced to 594898.

Experimental studies reviewed in [19] indicate that a small molecular tracer was quickly distributed from the subcapsular sinus into the B-cell follicle, where it located diffusely between the lymphocytes. In contrast, within the T-cell zone, the tracer distribution was restricted to conduits. It has been proposed that the differences in the distribution result from the biophysical characteristics of the T-cell zone, in particular, a hydraulic conductivity. Although direct measurements of the hydraulic conductivity within the T-cell area are missing, it is likely that both diffusion and convection are extremely low. The B-cell zone is considered to have a larger hydraulic conductivity that the T-cell zone. The above semi-quantitative observations have direct implications for the diffusion properties of soluble molecules like interferon. We assumed that the diffusion in domains Ω_1 to Ω_3 is characterized by the rates obeying the following diffusion coefficients ranking: $D_1 \gg D_2 \gg D_3$. As the molecular weight of IFN-I is very close to that of myoglobin, we used the following estimate of the diffusion coefficient $D^* = 0.16\ \text{mm}^2/\text{h}$ [18] as a baseline value.

6.4.3 Numerical Results

The reaction–diffusion partial differential equation governing the interferon dynamics was used to analyse the steady-state distribution of IFN-I across the subdomains of the 3D lymph node geometry. The corresponding reduced model (second-order elliptic equation) reads

$$-D\left(\frac{\partial^2}{\partial x^2}+\frac{\partial^2}{\partial y^2}+\frac{\partial^2}{\partial z^2}\right)I(\mathbf{x},t)+d_I I(\mathbf{x},t)=F(\mathbf{x}), \tag{6.4.2}$$

where

$$D=D_i\cdot E\ \text{ in } \mathbf{x}\in\Omega_i,\ i=1,\ldots,3,\quad F(\mathbf{x})=\sum_{k=1}^{K}\rho\delta(\mathbf{x}-\mathbf{x}_k). \tag{6.4.3}$$

The equation is understood in the weak sense. It considers interferon diffusion in the lymph node (the first term on the left side), its production and loss due to degradation and uptake by various cells (the first and the second terms on the right side, respectively). As the major IFN-I producers are pDCs, we neglected the contribution of macrophages to the source term.

We considered the scenario, where the individual pDCs are located randomly, mainly in domain Ω_1 around the polar region of the subcapsular sinus, although there was a no-zero probability of them to be placed in the upper half of the cortex and follicles. The algorithm for specifying the individual pDCs position with respect to the lymph node centre in cylindrical coordinates, $\mathbf{x}_k=(z_k,r_k,\theta_k)$, $k=1,\ldots,K$, makes use of random numbers ($\varepsilon_i\sim\mathscr{U}(0,1)$) generated from a standard continuous uniform distribution as follows: $z_k=R_{LN}(1-\varepsilon_1\varepsilon_2\varepsilon_3)$, $r_k=R_{z_k}(1-\varepsilon_4\varepsilon_5\varepsilon_6)$, $\theta_k=2\pi\varepsilon_7$, where R_{LN} is the radius of the outer sphere approximating the LN, and $R_{z_k}=\sqrt{(R_{LN}^2-z_k^2)}$.

The boundary conditions were specified as follows. At the outer boundary of the domain Ω_1 and the part of its inner boundary, overall defined as $\partial^*\Omega_1=\partial S_{\text{out}}\cup\partial S_{\text{inn}}\backslash(S_{\text{inn}}\cap C_{TS})\backslash(S_{\text{inn}}\cap C_{CT})$ the homogeneous Neumann boundary condition was used

$$\mathbf{n}\cdot D\nabla I(\mathbf{x},t)=0 \text{ on } \partial^*\Omega_1. \tag{6.4.4}$$

The vector fields are not differentiable with respect to spatial coordinates at the other interfaces between the lymph node domains with different diffusion properties. Therefore, we impose the boundary conditions, which describe the continuity of both the IFN-I concentration and the diffusion flux. If $\partial_{123}\Omega$ denotes the interface between the domains Ω_2 and Ω_3, Ω_1 and Ω_3, except for the part $\partial S_{\text{inn}}\backslash(S_{\text{inn}}\cap C_{TS})\backslash(S_{\text{inn}}\cap C_{CT})$, then we require that

$$I(\mathbf{x},t)\text{ continuous across }\partial_{123}\Omega,\qquad \mathbf{n}\cdot D\nabla I(\mathbf{x},t)\text{ continuous across }\partial_{123}\Omega.$$

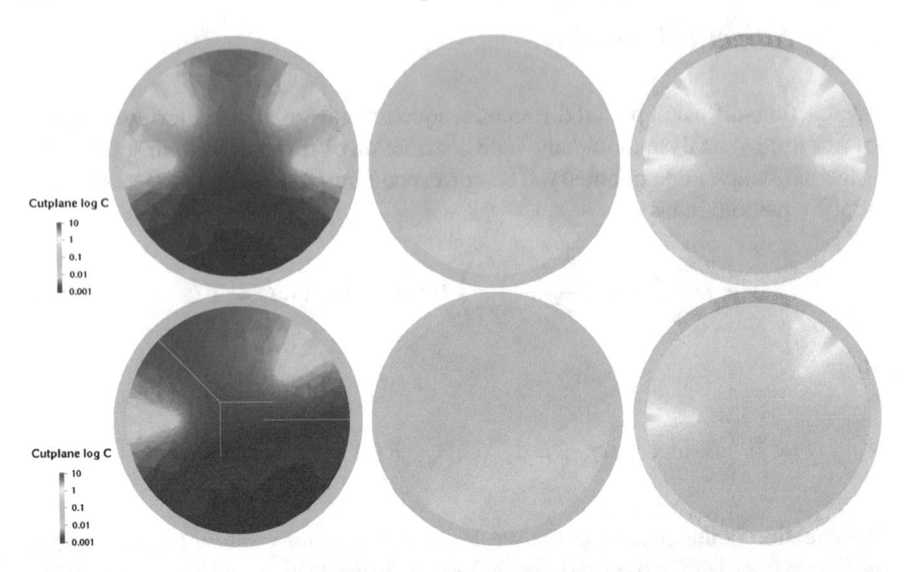

Fig. 6.12 Simulated stationary distribution of IFN-I across the paradigmatic lymph node for various number of source cells, pDCs, located randomly, mainly in the subcapsular sinus around the upper pole. Upper panel: OXZ cross section; left: 1 pDC, middle: 10 pDCs, right: 100 pDCs. Lower panel: OYZ cross section; left: 1 pDC, middle: 10 pDCs, right: 100 pDCs. The logarithmic colour bars represent the concentration of the interferon (pg/mm^3). (Reprinted from Mathematical Modelling of Natural Phenomena, Vol. 6, Bocharov et al., Reaction-Diffusion Modelling of Interferon Distribution in Secondary Lymphoid Organs, Pages 13–26, Copyright © 2011, with permission from EDP Sciences)

The numerical simulations of the full three-dimensional model were carried out using a monotone nonlinear cell-centred finite volume method developed for diffusion equations on conformal polyhedral meshes in [16]. The practically observed convergence order was about two. The diffusion coefficients in subdomains 1 to 3 we set as follows: $D_1 = D^*$, $D_2 = 0.1D^*$ and $D_3 = 0.01D^*$.

We started by studying the interferon spatial distribution in the LN for different number of activated plasmacytoid dendritic cells. Figure 6.12 shows the stationary IFN-I distribution across the LN for the source size (parameter M in Eq. (4.23)) consisting of 1, 10 and 100 pDCs. These numbers correspond to the fraction of interferon secreting cells covering the range of about 0.15 to 15% of the total pDCs in the lymph node. The IFN-I distribution appears to be highly inhomogeneous with the differences in concentration ranging over two orders of magnitude between different regions. The subdomain Ω_1 (subcapsular sinus, trabecular sinuses and conduits) is characterized by much higher concentration of the IFN-I than subdomain Ω_3 representing T-cell zone. A pDCs number increase from 1 to 100 results in the rise of the concentration field of IFN-I so that larger parts of the LN become protected against virus infection. In Fig. 6.13 the subdomains are shown, where the IFN-I concentration is below the threshold values of 0.01, 0.1 and 1 pg/mm^3, for the corresponding numbers of pDCs. The results suggest that as few as 10 pDCs in the

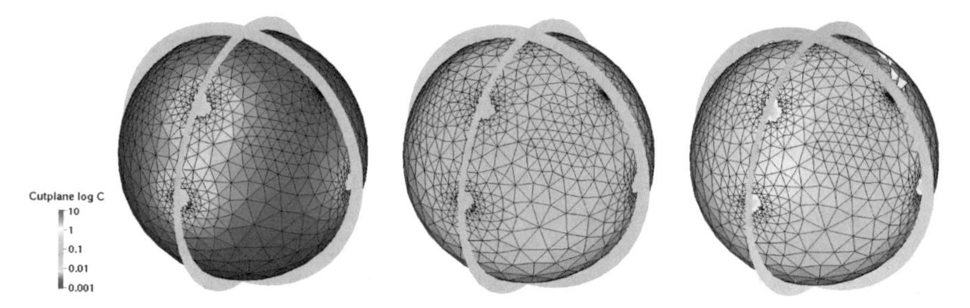

Fig. 6.13 Simulated stationary distribution of IFN-I across the paradigmatic lymph node for various number of source cells, pDCs, located randomly, mainly in the subcapsular sinus around the upper pole, with the tetrahedra showing the regions in which the IFN concentration is below some threshold θ. OXZ and OYZ cross sections are shown. Left: 1 pDC with $\theta = 0.01$ pg/mm^3, middle: 10 pDCs with $\theta = 0.1$ pg/mm^3, right: 100 pDCs with $\theta = 1.0$ pg/mm^3. The logarithmic colour bar represents the concentration of the interferon (pg/mm^3). (Reprinted from Mathematical Modelling of Natural Phenomena, Vol. 6, Bocharov et al., Reaction-Diffusion Modelling of Interferon Distribution in Secondary Lymphoid Organs, Pages 13–26, Copyright © 2011, with permission from EDP Sciences)

Fig. 6.14 Simulated stationary distribution of IFN-I across the paradigmatic lymph node for various number of source cells, pDCs, located randomly, mainly in the subcapsular sinus around the upper pole with the interferon loss rate d_I increased by tenfold . OYZ cross section is shown. Left: 1 pDC, middle: 10 pDCs, right: 100 pDCs. The logarithmic colour bar represents the concentration of the interferon (pg/mm^3). (Reprinted from Mathematical Modelling of Natural Phenomena, Vol. 6, Bocharov et al., Reaction-Diffusion Modelling of Interferon Distribution in Secondary Lymphoid Organs, Pages 13–26, Copyright © 2011, with permission from EDP Sciences)

lymph node secrete enough interferon to ensure the protection of APCs (such as macrophages) in the whole subdomain Ω_1.

The sensitivity of the stationary interferon concentration to the rate of loss d_I has been studied. The simulation results presented in Fig. 6.14 show that tenfold increase leads to about twofold decrease of the maximum concentration in subcapsular domain and conduits but dramatically reduces the amount of available interferon in the B- and T-cell zones.

This study results suggest that the spatial stationary distribution of IFN-I is essentially heterogeneous across the lymph node. Highly protected subdomains such as sinuses, conduits, coexist with the regions, where the stationary concentration of IFN-

I is lower by about 100-fold. This implies that for some infections, the pathogens can escape the IFN-I effect if the infected target cell is localized/migrated into poorly protected SLO regions.

The study allows one to put forward a hypothesis on the process of establishment of viral persistence which is related to the heterogeneity of spatial distribution of IFN-I. The morphology of secondary lymphoid organs can lead to the formation of poorly protected areas. In these areas, the localization of the cells infected by viruses, such as $CD4^+$ T lymphocytes in the case of HIV infection or macrophages provides the conditions for the continuation of active infectious processes. The predicted diffusion-mediated compartmentalization of cytokines, chemokines and drugs heavily depends on the assumed variations in hydraulic conductivity of various zones of the secondary lymphoid organs which deserves further experimental investigation. However, there are clinical observations on HIV ART-based treatment failure related to insufficient drug concentration in LT [11] suggesting that HIV-1 persists in tissues during drug therapy allowing ongoing, low-level replication in tissues [12, 13].

References

1. G. Bocharov, A. Meyerhans, N. Bessonov, S. Trofimchuk, V. Volpert. Spatiotemporal dynamics of virus infection spreading in tissues. PlosOne, 2016 Dec 20;11(12):e0168576.
2. Bocharov G., Danilov A., Vassilevski Yu., Marchuk G.I., Chereshnev V.A., Ludewig B. (2011) Reaction-Diffusion Modelling of Interferon Distribution in Secondary Lymphoid Organs. Math. Model. Nat. Phenom., 6(7): 13–26.
3. Mikael Boulle, Thorsten G. Muller, Sabrina Dahling, Yashica Ganga, Laurelle Jackson, Deeqa Mahamed, Lance Oom, Gila Lustig, Richard A. Neher, Alex Sigal. HIV Cell-to-Cell Spread Results in Earlier Onset of Viral Gene Expression by Multiple Infections per Cell. PLOS Pathogens, https://doi.org/10.1371/journal.ppat.1005964 November 3, 2016.
4. Walther Mothes, Nathan M. Sherer, Jing Jin, Peng Zhong. Virus Cell-to-Cell Transmission. Journal of Virology, Sept. 2010, Vol. 84, No. 17, 83608368.
5. A.N. Kolmogorov, I.G. Petrovsky, N.S. Piskunov. A study of the diffusion equation with increase in the amount of substance, and its application to a biological problem. Bull. Moscow Univ., Math. Mech., 1:6 (1937), 1–26. In: Selected Works of A.N. Kolmogorov, Vol. 1, V.M. Tikhomirov, Editor, Kluwer, London, 1991.
6. R. A. Fisher, *The wave of advance of advantageous genes*, Ann. Eugenics 7 (1937), 355–369.
7. A. Volpert, Vit. Volpert, Vl. Volpert. Traveling wave solutions of parabolic systems. Translation of Mathematical Monographs, Vol. 140, Amer. Math. Society, Providence, 1994.
8. V. Volpert. Elliptic partial differential equations. Volume 2. Reaction-diffusion equations. Birkhäuser, 2014.
9. Ya. B. Zeldovich, D. A. Frank-Kamenetskii. A theory of thermal propagation of flame, Acta Physicochim. USSR 9 (1938), 341–350; Zhurnal Fizicheskoi Himii, 9 (1939), no. 12, 1530 (Russian)
10. Ya. B. Zeldovich, G. I. Barenblatt, V. B. Librovich, G. M. Makhviladze. The mathematical theory of combustion and explosion. Plenum Publishing Co., New York, 1985.
11. Fletcher CV, Staskus K, Wietgrefe SW, Rothenberger M, Reilly C, Chipman JG, Beilman GJ, Khoruts A, Thorkelson A, Schmidt TE, Anderson J, Perkey K, Stevenson M, Perelson AS, Douek DC, Haase AT, Schacker TW. Persistent HIV-1 replication is associated with lower antiretroviral drug concentrations in lymphatic tissues. Proc Natl Acad Sci U S A. 2014;111(6):2307–12

12. Rose R, Lamers SL, Nolan DJ, Maidji E, Faria NR, Pybus OG, Dollar JJ, Maruniak SA, McAvoy AC, Salemi M, Stoddart CA, Singer EJ, Mcgrath MS. 2016. HIV maintains an evolving and dispersed population in multiple tissues during suppressive combined antiretroviral therapy in individuals with cancer. J Virol 90:89848993. https://doi.org/10.1128/JVI.00684-16.
13. Lorenzo-Redondo R, Fryer HR, Bedford T, Kim EY, Archer J, Pond SLK, Chung YS, Penugonda S, Chipman J, Fletcher CV, Schacker TW, Malim MH, Rambaut A, Haase AT, McLean AR, Wolinsky SM. Persistent HIV-1 replication maintains the tissue reservoir during therapy. Nature. 2016;530(7588):51–56. https://doi.org/10.1038/nature16933.
14. G. Bocharov, R. Zust, L. Cervantes-Barragan, T. Luzyanina, E. Chiglintcev, V.A. Chereshnev, V. Thiel, B. Ludewig. A systems immunology approach to plasmacytoid dendritic cell function in cytopathic virus infections. PLoS Pathogens, 6(7) (2010), e1001017. https://doi.org/10.1371/journal.ppat.1001017, 1–14.
15. A.A. Danilov. Unstructured tetrahedral mesh generation technology. Comput. Math. Math. Phys., 50 (2010), 146–163.
16. A.A. Danilov, Yu.V. Vassilevski. A monotone nonlinear finite volume method for diffusion equations on conformal polyhedral meshes. Russ. J. Numer. Anal. Math. Modelling, 24 (2009), 207–227.
17. T. Junt, E. Scandella, B. Ludewig. Form follows function: lymphoid tissues microarchitecture in antimicrobial immune defense. Nature Rev. Immunol., 8 (2008), 764–775.
18. J. Keener, J. Sneyd. Mathematical physiology. Springer-Verlag, New York, 1998.
19. T. Lammermann, M. Sixt. The microanatomy of T cell responses. Immunol. Reviews, 221 (2008), 26–43.
20. G.I. Marchuk. Methods of Numerical Mathematics. Springer-Verlag, New York, 1982.
21. F. Pfeiffer, V. Kumar, S. Butz, D. Vestweber, B.A. Imhof, J.V. Stein, B. Engelhardt. Distinct molecular composition of blood and lymphatic vascular endothelial cell junctions establishes specific functional barriers within the peripheral lymph node. Eur. J. Immunol., 38 (2008), 2142–2155.

Chapter 7
Multi-scale and Integrative Modelling Approaches

7.1 Multi-scale Models

The term multi-scale models is used with different meanings in different sciences. In mathematics, it implies the presence of one or several small parameters, homogenization and averaging techniques. In physics, it is understood in the sense of microscopic–macroscopic scales (e.g. molecular dynamics versus continuum mechanics). These questions are exhaustively discussed in particular in [39]. Multi-scale modelling in biology has been intensively developed during the last decade. It implies that the model includes different biological scales: cells, intracellular regulation, extracellular matrix, the tissue under study and other organs (not necessarily all of them in the same model). It is in this sense that we understand multi-scale modelling in this chapter.

Multi-scale models in physiology imply the presence of different levels of the description of biological processes: intracellular (molecular), cellular, extracellular, tissue and organ, the whole organism. The corresponding mathematical models can include ordinary and partial differential equations, agent-based models and their combinations. A more detailed description can be found in [1, 2]. In this section, we introduce an emerging multi-scale approach to modelling in immunology.

7.2 Multi-scale Approaches in Mathematical Immunology

The immune system is regulated by multiple processes at various levels of biological organization including the genetic, cellular, tissue, organ and the whole organism levels. The resulting structural and functional complexity of the immune system called for a major shift towards information-rich, systems-based approaches in immunological research. High-throughput technologies generate vast amounts of data that facilitate dissection of the immunological processes at ever finer resolution. The

G. Bocharov et al., *Mathematical Immunology of Virus Infections*,
https://doi.org/10.1007/978-3-319-72317-4_7

need to embed immune processes into their spatial context both at the molecular and cellular level is a hallmark of the systems immunology approach. In fact, there are many examples of how the fate decisions in the immune system depend on the spatial–temporal dynamics of cytokines, e.g. the interleukin-2 (IL-2) [5] and type I interferon (IFN) [6, 7]. The physical scales of the processes underlying functioning of immune system during infections are represented in Fig. 7.1.

Although the spatial and temporal scales of the immune and pathogen dynamics and regulation are well appreciated, the overall modelling practice is rather modest. A multi-scale framework turned out to be insightful for understanding the mechanisms and identifying potential therapeutic targets for human infection with *Mycobacterium tuberculosis* [40–42]. Other specific examples of immune system analysis based on multi-scale models are the studies of early CD8$^+$ T-cell immune responses in lymph nodes (LN) [43, 44], the NF-κB signalling pathway [45] and immune processes in lymph nodes [46, 47]. The gained experience led to the formulation of some more general principles for developing and computationally implementing integrative models of immune responses [41, 48, 49]. A recent study integrating the spatial structure of the T-cell zone of lymph nodes and the dynamics of T-cell responses in

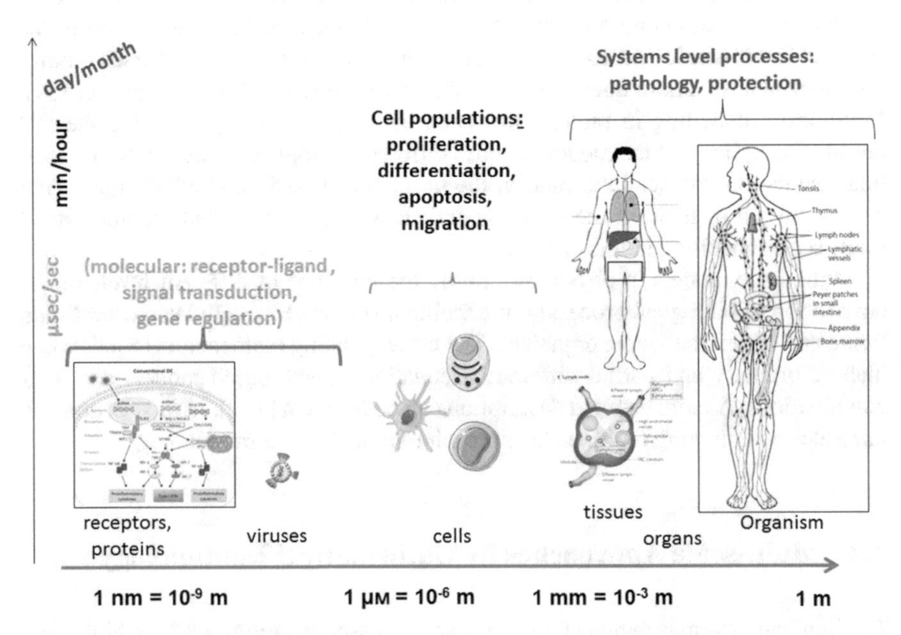

Fig. 7.1 Multi-scale nature of the immune system structure and function as defence system against virus infections. The graphical elements shown in the figure are adapted and modified using figures from (1) the website http://anatomybody101.com/the-immune-responses-of-the-lymphatic-system/the-immune-responses-of-the-lymphatic-system-3d3979405d904f69a2f42156fa6a7973/, (2) Junt et al., Nat Rev Immunol. 2008, 8(10):764–75, and (3) from Akira S et al., Cell 124, 783–801 (2006)

HIV infection has quantified the effect of the destruction of the fibroblastic reticular cell (FRC) network on T-cell reactivity [50].

This chapter aims to present a computational approach to setting up a multi-scale mathematical model of immune response to infection and apply it to examine the impact of heterogeneity of spatial signalling on the kinetics of immune responses.

7.2.1 Equations of Cell Kinetics and Cell Dynamics

The behaviour of biological cells can be characterized by three aspects: cell fate, cell motion and cell interaction with other cells and molecules. Cell fate includes their self-renewal (proliferation without differentiation), differentiation (with or without proliferation) and apoptosis (programmed cell death). Suppose that cells A can self-renew with the rate coefficient k_s, differentiate, k_d and die by apoptosis, k_a. If they are uniformly distributed in space, then their concentration is described by the ordinary differential equation

$$\frac{dA}{dt} = I + (k_s - k_d - k_a)A, \qquad (7.2.1)$$

where I is the cell influx. Assuming that the coefficients of these equation are constant, we conclude that the concentration $A(t)$ exponentially grows if $k_s > k_d + k_a$ and $A(0) > 0$ or $I > 0$. If the relation between the coefficients is opposite, $k_s < k_d + k_a$, then $A(t)$ converges to the stationary value $I/(k_d + k_a - k_s)$.

This description of cell population dynamics is oversimplified. In general, the coefficients of the equation are not constant, and they can depend on the concentrations of other cells and on various molecular substances. Moreover, spatial cell distribution and their motion can also be essential to be taken into account.

Equation (7.2.1) is an example of a cell kinetics equation. In the case of n cell types, we get a system of equations: each of them can self-renew, differentiate or die by apoptosis with certain rates. We have the following equations for their concentrations:

$$\frac{dA_i}{dt} = 2k_i^s A_i - k_i^m A_i - A_i \sum_{j=1}^{n} k_{ij}^d + \sum_{j \neq i} k_{ji}^d A_j, \quad i = 1, ..., n. \qquad (7.2.2)$$

We do not specify the form of differentiation (with or without proliferation, symmetric or asymmetric). This can be taken into account in the coefficients of the system. Here, k_i^s is the rate of self-renewal of cells A_i, k_i^m the rate of their apoptosis and k_{ij}^d the rate of differentiation of cells A_i into cells A_j.

If the coefficients of system (7.2.2) are constant, then it is a system with non-negative off-diagonal elements. The point $A_i = 0$, $i = 1, ..., n$ is a stationary solution of this system. Its stability is determined by the eigenvalues of the matrix $P = (p_{ij})$, where

$$p_{ii} = 2k_i^s - k_i^m - \sum_{j=1}^{n} k_{ij}^d, \quad p_{ij} = \sum_{j \neq i} k_{ji}^d \geq 0.$$

According to the well-known criterion of stability of matrices with non-negative off-diagonal elements, this stationary point is stable if $p_{i1} + ... + p_{in} < 0$ for all i and unstable if $p_{ii} > 0$ for some i.

Taking into account random cell motion, we obtain the system of equations of cell dynamics:

$$\frac{\partial A}{\partial t} = D\Delta A + F(A). \tag{7.2.3}$$

Here, $A = (A_1, ...A_n)$, $F = (F_1, ..., F_n)$, D is the diagonal matrix of diffusion coefficients supposed to be constant. The functions F_i, $i = 1, ..., n$ are determined by the right-hand sides of Eq. (7.2.2). Thus, equations of cell dynamics represent reaction–diffusion equations. In a more detailed description, convective cell motion can be taken into account when it is appropriate. This equation should be completed by the boundary and initial conditions. In particular, if cells cannot cross the boundary, then we have the Neumann boundary condition

$$\frac{\partial A}{\partial n} = 0. \tag{7.2.4}$$

Here, n is the outer normal vector to the boundary which is supposed to be sufficiently smooth. The initial condition $A(x, 0)$ is some non-negative function.

Density-dependent cell proliferation.

In the case of self-renewal, $A \rightarrow 2A$, the proliferation rate k can decrease with the local cell density, $k(A) = k_0(A_0 - A)$. Then, we get the following equation for the cell concentration:

$$\frac{\partial A}{\partial t} = D\Delta A + k_0 A(A_0 - A) \tag{7.2.5}$$

(A here is a scalar variable and D is a number). We considered similar equations in the previous chapter. Let us recall that they have travelling wave solutions describing growth of cell populations (including tumour growth). If the initial condition is positive and the boundary condition is given by (7.2.4), then the solution converges to the value A_0 everywhere in the domain.

Differentiation with feedback.

Suppose that cells A can make a choice between self-renewal and differentiation into cells B:

$$A \rightarrow 2A, \quad A \rightarrow 2B,$$

and cells B can die by apoptosis. The corresponding rate coefficients are, respectively, k_s, k_d and k_a Then, the cell concentrations are described by the equations

$$\frac{\partial A}{\partial t} = D\Delta A + (k_s - k_d)A, \tag{7.2.6}$$

$$\frac{\partial B}{\partial t} = D\Delta B + 2k_d A - k_a B. \tag{7.2.7}$$

If the rate coefficients are constant, and $k_s \neq k_d$, then this system has a unique homogeneous in space stationary solution $A = B = 0$.

Assume now that differentiated cells B can influence the rate coefficients, upregulate differentiation and downregulate self-renewal of cells A. Then, k_s is a decreasing function of B and k_d an increasing function. If the equation $k_s(B) = k_d(B)$ has a solution $B = B_0$, then (A_0, B_0), where $A_0 = k_a B_0/(2k_d(B_0))$, is a positive stationary point. Solutions of system (7.2.6) and (7.2.7) with no-flux boundary conditions and with positive initial conditions converge to this stationary solution for large time.

In the models of the immune response, cells A correspond to naive lymphocytes and cells B to mature lymphocytes. The latter produce various cytokines stimulating differentiation of the naive immature cells.

7.2.2 Global Extracellular Regulation

The total number of cells A_i in the domain Ω is given by the integral

$$J_i(t) = \int_\Omega A_i(x, t)dx.$$

If cells produce some biochemical substances, such as cytokines, hormones and growth factors, then their quantity depends on the total number of cells J_i in the tissue. If the rate coefficients in the cell kinetics Eq. (7.2.2) are constant (do not depend on cell concentrations), then integrating cell dynamics Eq. (7.2.3) with boundary conditions (7.2.4), we obtain the system of equations for total cell numbers:

$$\frac{dJ_i}{dt} = F(J), \quad i = 1, ..., n, \tag{7.2.8}$$

where $J = (J_1, ..., J_n)$. We obtain a similar system of equations if the rate coefficients depend on J_i, that is, on the concentrations of substances whose production is determined by the total number of cells. Here, cells A_i can belong to different tissues and organs. Thus, we obtain a model of global regulation of cell dynamics.

Global regulation of the immune response.

Let us complete the system of Eqs. (7.2.6) and (7.2.7) for naive and differentiated T cells by the equation for the concentration of virus V in blood:

$$\frac{dV}{dt} = kV(V_0 - V) - \phi(T)V. \tag{7.2.9}$$

Here, the first term in the right-hand side describes virus multiplication with a logistic growth, and the second term its elimination by T cells (CTLs). The rate coefficients in the equations of lymphocyte production depend on the virus concentration:

$$\frac{\partial A}{\partial t} = D\Delta A + (k_s(V) - k_d(V))A, \tag{7.2.10}$$

$$\frac{\partial B}{\partial t} = D\Delta B + 2k_d(V)A - k_a(V)B. \tag{7.2.11}$$

Furthermore, the concentration of T cells in blood is proportional to the total number of differentiated lymphocytes in the tissue:

$$T(t) = \alpha \int_{\Omega} B(x,t)dx,$$

where α is some positive coefficient. Integrating Eqs. (7.2.10) and (7.2.11) with the no-flux boundary conditions, we obtain the ODEs:

$$\frac{dJ}{dt} = (k_s(V) - k_d(V))J, \tag{7.2.12}$$

$$\frac{dT}{dt} = 2\alpha k_d(V)J - k_a(V)T. \tag{7.2.13}$$

Systems (7.2.9), (7.2.12) and (7.2.13) describe the interaction of a virus infection with the immune response. A similar model with a single type of immune cells was considered in the previous chapter (with or without delay and with logistic growth of the number of cells).

7.2.3 *Local Extracellular Regulation*

We will assume in this section that cell density is sufficiently small such that cells do not prevent random motion of each other and convective motion of the medium does not occur. Cell adhesion is also neglected, so that they move independently of each other. Let $C = (C_1, ..., C_n)$ be the vector of cell concentrations and $u =$

$(u_1, ..., u_m)$ the vector of concentrations of extracellular substances. Assuming that the rates of self-renewal, differentiation and apoptosis depend on cell concentrations and extracellular variables, we get the reaction–diffusion system

$$\frac{\partial C}{\partial t} = D_c \Delta C + F(C, u), \tag{7.2.14}$$

$$\frac{\partial u}{\partial t} = D_u \Delta u + G(C, u), \tag{7.2.15}$$

where $F = (F_1, ..., F_n)$ are the rates of cell production, $G = (G_1, ..., G_n)$ are the rates of production of extracellular species. By the rate of cell production, we understand the overall rate of change of cell concentration taking into account their self-renewal, proliferation, apoptosis. Similarly, the functions G_i take into account production, consumption and destruction of the corresponding biochemical species. We will assume for simplicity that the matrices of diffusion coefficients are diagonal.

7.2.4 Intracellular Regulation

Cell fate, that is, the choice between self-renewal, differentiation and apoptosis is determined by the intracellular regulation through the concentrations of intracellular proteins and other molecules. Continuous models of cell dynamics, where cells are described by their concentrations, imply that a small space volume contains a large number of cells. The cells of the same type in this small volume can differ by the concentrations of the intracellular substances. Hence, there is no one-to-one correspondence between space points and intracellular concentrations, and the latter cannot be considered as functions of space and time. Therefore, we need to introduce another description of cell dynamics with the intracellular regulation.

Let $p = (p_1, ..., p_m)$ be a vector of intracellular concentrations. We will consider them as independent variables together with space and time. Cell concentrations depend now on all these variables, $A_i = A_i(p, x, t)$. The evolutions of the concentrations are described by the equations

$$\frac{\partial A_i}{\partial t} + \sum_{j=1}^{m} \frac{\phi_j \partial A_i}{\partial p_j} = D_p \Delta_p A + D \Delta_x A_i + \Phi_i(A, p), \tag{7.2.16}$$

where

$$D_p \Delta_p A_i = \sum_{j=1}^{m} D_j \frac{\partial^2 A_i}{\partial p_j^2}, \quad \Delta_x A_i = \sum_{j=1}^{3} \frac{\partial^2 A_i}{\partial x_j^2},$$

$\phi_j(p, u)(= \frac{dp_j}{dt})$ is the rate of production of the corresponding intracellular concentrations. It can depend on both intracellular variables p and extracellular variables u. Let us note that besides the usual diffusion in the physical space (second term in the right-hand side of this equation), we can also consider diffusion in the state space which describes random fluctuations of the intracellular variables. If we omit the diffusion terms, we obtain conventional transport equations considered in structured population dynamics. The last term in the right-hand side of Eq. (7.2.16) describes cell proliferation, differentiation and death which depend on their concentrations and on the intracellular variables.

The interval of variation of the variables p_j depends on cell properties. In particular, if cells A_i divide for $p_j = p_j^0$, give cells of the same type, and the daughter cells have a half of the concentration $p_j^0/2$ in comparison with the mother cell before division, then the variable p_j changes in the interval $p_j^0/2 \leq p_j \leq p_j$. In this case, the boundary conditions for the function A_i become as follows: $A_i(p_j^0/2) = 2A_i(p_j^0)$. Similarly, we can introduce the boundary conditions in the case of cell differentiation and death.

7.3 Hybrid Multi-scale Models

Hybrid models represent a combination of different models used to describe some complex process. Each of these models is more appropriate for some particular part of this process, and they should be considered in their interaction [1, 2]. We will present here in more detail a particular type of hybrid models well adapted for the description of physiological processes.

Biological cells are considered as individual-based objects. They can move, divide, change their type, produce and consume various biochemical substances, or die. In the latter case, they are removed from the computational domain. In the simplest representation, cells can be considered as elastic spheres.

Moving from the molecular and cellular level to a multi-scale model requires the development of novel modelling methodologies for an iterative integration of data from different biological levels into mechanism-based modular mathematical models [8, 41, 48]. So far, very few mathematical models have been proposed to describe the multi-scale spatial regulation of immune responses in a genuine hybrid manner [40, 43, 46, 47].

The model presented here takes into account: (1) spatial aspects of the immune response in the lymph node by means of cell and concentration distributions, (2) regulation of T lymphocytes in the lymph node including their asymmetric division and their interaction with extracellular cytokine concentrations, (3) the intracellular regulation of T cells depending on IL-2 and type I IFN and (4) the interaction of the tissue level processes and the systemic infection dynamics (blood) via inter-compartmental cell fluxes as shown in Fig. 7.2.

Conventional models of the immune response are based on ordinary differential equations, and they do not take into account spatial distributions of cells and concentrations in the lymph node. The multi-scale models previously developed use a similar agent-based cell description. However, they do not take into account processes such as asymmetric cell division, their interaction with IFN or the interaction of the tissue level with the organism level which is one of the key features of our multi-scale model. The presentation below follows our recent works [3, 4].

7.3.1 Hybrid Models of Immune Response

To formulate the mathematical model, we consider a part of the lymph node, i.e. the T-cell zone, which contains various cell types, mainly the antigen-presenting cells (APCs) and subsets of T lymphocytes (see Fig. 7.2). Naive T cells and some APCs (such as plasmacytoid dendritic cells, pDCs) enter the node with blood flow via the high endothelial venules (HEVs), whereas effector and/or memory T cells, and mainly DCs and macrophages home to lymph nodes via afferent lymphatic vessels [16, 17]. Following activation with pathogens, APCs acquire a motile state that allows their translocation to the T-cell zone of draining lymph node with the afferent lymph flow [18, 19]. Therefore, we assume that the influx of APCs is proportional to the level of infection in the organism. Differentiation of naive T cells into $CD4^+$ and $CD8^+$ T cells occurs in the thymus from progenitor T cells [20]. We suppose that they enter lymph nodes already differentiated and that there is a given influx of each cell type.

The APCs bearing foreign antigens activate the clonal expansion of naive T lymphocytes. The activation of T-cell division and death is regulated by a set of signals coming from the interactions of the antigen-specific T-cell receptors (TCRs) with the MHC class I or class II presented peptides and IL-2 receptors binding IL-2. Naive

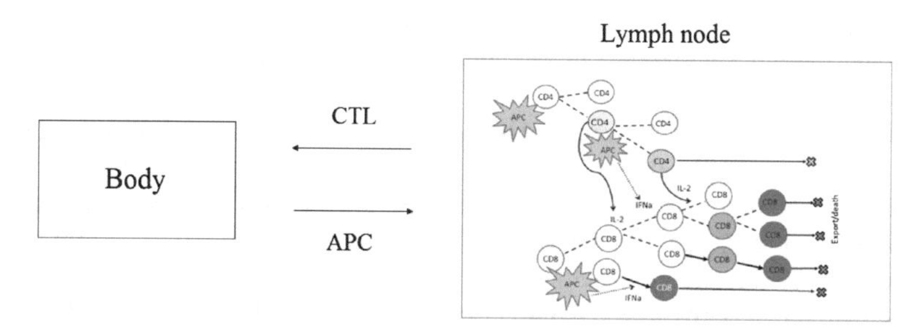

Fig. 7.2 Schematic representation of the two-scale model. Naive T cells and antigen-presenting cells (APC) enter the lymph node. Due to asymmetric cell division, some T cells differentiate. Mature $CD8^+$ T cells leave the lymph node and kill infected cells. Mature $CD4^+$ T cells produce IL-2 that influences cell survival and differentiation. APCs are shown in green, and naive T cells are white. Differentiated $CD4^+$ T cells are yellow and $CD8^+$ T cells are blue. Levels of yellow and blue indicate cell maturation. Infection level and immune cells in the body are described by ODEs. Cells in the lymph node are considered as individual objects, intracellular regulation is described by ODEs and extracellular substances by PDEs

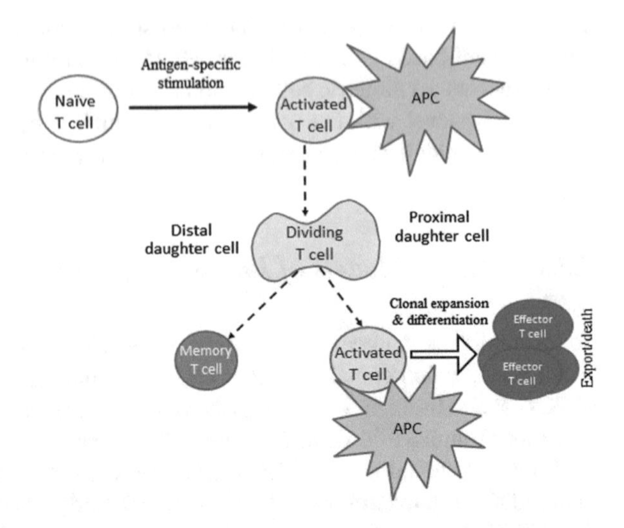

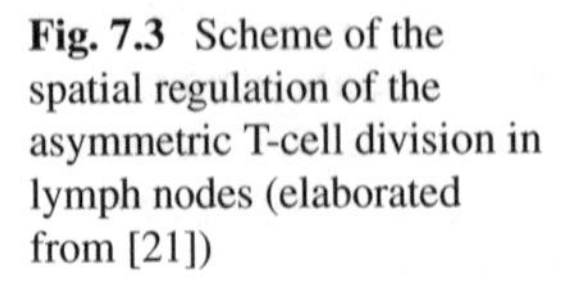
Fig. 7.3 Scheme of the spatial regulation of the asymmetric T-cell division in lymph nodes (elaborated from [21])

T cells undergo asymmetric division [21]. Some of the daughter cells continue to proliferate and differentiate. Mature $CD4^+$ T cells produce IL-2 [20, 22, 23] which influences survival and differentiation of both $CD4^+$ and $CD8^+$ T cells. The proliferation of $CD8^+$ T cells is stimulated by IL-2 [22]. They can expand their number many 1000-fold. In addition to IL-2 enhancing the proliferation of T cells, APCs start to secrete type I IFN (IFN-I) which has an antiviral and immunomodulatory function. In fact, the effect of IFN-I depends on the temporal sequence of the signals obtained by naive T cells [6]. It can change from a normal activation of T cells followed by their proliferation and differentiation to an already differentiated state followed by apoptosis. Overall, the regulated death of T cells by apoptosis depends on the availability and the timing of the T cell receptor (TCR), IL-2 and IFN signalling.

Mature $CD8^+$ T cells (effector cells) leave the lymph node and kill infected cells. Therefore, there is a negative feedback between production of mature $CD8^+$ T cells and the influx of APCs.

In the model, an asymmetric T-cell division is considered as shown in Fig. 7.3. Naive T cell entering the draining lymph node is recruited into the immune response after the contact interaction via the TCR with APC presenting the foreign antigen. The activation and prolonged contact with APC can result in polarity of the lymphocyte. The position of the contact with the APC determines the direction of cell division and the difference between the daughter cells in terms of their differentiation state. According to [21], the proximal daughter cell will undergo clonal proliferation and differentiation resulting in the generation of terminally differentiated effector cells (mature $CD8^+$ T cells) that leave the lymph node for peripheral tissues to search and kill infected cells. The distal daughter cell becomes a memory cell. The memory cells are capable of self-renewal by slowly dividing symmetrically in the absence of recurrent infection.

In our model of cell dynamics, cells are considered as individual objects that can move, divide, differentiate and die. Their behaviour is determined by the surrounding cells, by intracellular regulatory networks described by ordinary differential equations and by various substances in the extracellular matrix whose concentrations are described by partial differential equations. This approach was used to model hematopoiesis and blood diseases [9–15].

Cell motion and geometry.

The cells divide and can increase their number leading to their displacement in the lymph node. We describe cell displacement by the following model. Let us denote the centre of two cells by x_1 and x_2 and their radii by r_1 and r_2, respectively. Then, if the distance h_{12} between the two cells is less than the sum of their radii $(r_1 + r_2)$, there will be a repulsive force f_{12} between them. This force should depend on the difference between $(r_1 + r_2)$ and h_{12}. Let us consider the case of one cell interacting with different cells in the lymph node. The total force applied to this cell will be $F_i = \sum_{j \neq i} f_{ij}$. We describe the motion of the particles as the motion of their centres which can be found by the applying Newton's second law:

$$m\ddot{x}_i + m\mu\dot{x}_i - \sum_{j \neq i} f_{ij} = 0, \tag{7.3.1}$$

where m is the mass of the particle, μ is the friction factor due to contact with the surrounding medium. The potential force between two cells is given explicitly by:

$$f_{ij} = \begin{cases} K \frac{h_0 - h_{ij}}{h_{ij} - (h_0 - h_1)} , & h_0 - h_i < h_{ij} < h_0 \\ 0 , & h_{ij} \geq h_0 \end{cases} ,$$

where h_{ij} is the distance between the centres of the two cells i and j, h_0 is the sum of their radii, K is a positive parameter and h_1 is the sum of the incompressible part of each cell. The force between the particles tends to infinity if h_{ij} decreases to $h_0 - h_1$.

Cells and concentrations.

Cells in the lymph node:

1. $n_{APC}(\mathbf{x}, t)$—the density of APCs in T-cell zone;
2. $n_{CD4}(\mathbf{x}, t)$—the density of $CD4^+$ T cells in T-cell zone (with different levels of maturity);
3. $n_{CD8}(\mathbf{x}, t)$—the density of $CD8^+$ T cells in T-cell zone (with different levels of maturity);

Extracellular variables:

4. $I_e(\mathbf{x}, t)$—the concentration of IL-2 in T-cell zone;
5. $C_e(\mathbf{x}, t)$—the concentration of type I IFN in T-cell zone;

Intracellular variables:

6. $I_i(t)$—the intracellular concentration of IL-2-induced signalling molecules in the i-th cell;
7. $C_i(t)$—the intracellular concentration of type I IFN-induced signalling molecules in the i-th cell;

The state variables at the level of the whole organism:

8. $N_{ef}(t)$—the total number of effector CD8$^+$T cells in the body;
9. $N_{inf}(t)$—the total number of infected cells in the body;

Cell division and differentiation.

APC and naive T cells enter the computational domain with a given frequency if there is available space. Naive T cells move in the computational domain randomly. If they contact APC, they divide asymmetrically Fig. 7.3. The distant daughter cell is similar to the mother cell, and the proximal daughter cell becomes differentiated.

When the cell reaches the half of its life cycle, it will increase its size. When it divides, two daughter cells appear, and the direction of the axis connecting their centres is chosen randomly from 0 to 2π. The duration of the cell cycle is 18 h with a random perturbation of -3 to $+$ 3 h.

We consider two levels of maturity of CD4$^+$ T cells and three levels of CD8$^+$ T cells. If a differentiated cell has enough IL-2 (see the next paragraph), then it divides and gives two more mature cells. Finally, differentiated cells leave the lymph node. In the simulations, this means that they are removed from the computational domain.

Intracellular regulation.

The survival and differentiation of activated CD4$^+$- and CD8$^+$ T lymphocytes depends on the amount of signalling via the IL-2 receptor and the type I IFN receptor. It is controlled primarily by the concentration of the above cytokines in the close proximity of the respective receptors. The signalling events lead to the upregulation of the genes responsible for cell proliferation, differentiation and death. One can use similar type of equation to model qualitatively the accumulation of the respective intracellular signalling molecules linked to IL-2- and type I IFN receptors. The IL-2-dependent regulatory signal dynamics in individual cells can be described by the following equation:

$$\frac{dI_i}{dt} = \frac{\alpha_1}{n_T} I_e(\mathbf{x_i}, t) - d_1 I_i. \quad (7.3.2)$$

Here, I_i is the intracellular concentration of signalling molecules accumulated as a consequence of IL-2 signals transmitted through transmembrane receptor IL2R downstream the signalling pathway to control the gene expression in the i-th cell. The concentrations inside two different cells are in general different from each other. The first term in the right-hand side of this equation shows the cumulative effect of IL-2 signalling. The extracellular concentration I_e is taken at the coordinate $\mathbf{x_i}$ of the centre of the cell. The second term describes the degradation of IL-2-induced signalling molecules inside the cell. Furthermore, n_T is the number of molecules internalized by T-cell receptors.

In a similar way, the IFN-dependent regulatory signal dynamics in individual cells can be described by the following equation:

$$\frac{dC_i}{dt} = \frac{\alpha_2}{n_T} C_e(\mathbf{x_i}, t) - d_2 C_i. \tag{7.3.3}$$

Here, C_i is the intracellular concentration of signalling molecules accumulated as a consequence of IFN signals transmitted through transmembrane receptor IFNR downstream the signalling pathway to control the gene expression in the i-th cell. The concentrations inside two different cells are in general different from each other. The first term in the right-hand side of this equation shows the cumulative effect of IFN signalling. The extracellular concentration C_e is taken at the coordinate $\mathbf{x_i}$ of the centre of the cell. The second term describes the degradation of IFN-induced signalling molecules inside the cell.

To model the fate regulation of growth versus differentiation of the activated cells in relation to the timing of the IL-2 and type I IFN signalling, we implement the following decision mechanisms:

1. If the concentration of activation signals induced by type I IFN, C_i, is greater than some critical level C_i^* at the beginning of the cell cycle and that of I_i is smaller than the critical level I_i^*, then the cell will differentiate resulting in a mature cell.
2. If the concentration of activation signals induced by IL-2, I_i, is greater than some critical level I_i^* at the end of the cell cycle, then the cell will divide producing two more mature cells.
3. If $C_i < C_i^*$ at the beginning of cell cycle and $I_i < I_i^*$ at the end of cell cycle, then the cell will die by apoptosis and will be removed from the computational domain.

Stochastic aspects of the model.

As it is discussed above, mechanical interaction of cells results in their displacement described by Eq. (7.3.1) for their centres. In order to describe random motion of cells, we add random variables to the cell velocity in the horizontal and vertical directions. Duration of cell cycle is given as a random variable in the interval $[T - \tau, T + \tau]$.

Extracellular dynamics of cytokines.

Proliferation and differentiation of T cells in the lymph node depends on the concentration of IL-2 and type I IFN. These cytokines are produced by mature CD4$^+$ T cells and antigen-presenting cells, respectively. Spatial distribution of IL-2 is described by a reaction–diffusion equation as follows:

$$\frac{\partial I_e}{\partial t} = D_{IL} \Delta I_e + W_{IL} - b_1 I_e. \tag{7.3.4}$$

Here, I_e is the extracellular concentration of IL-2, D is the diffusion coefficient, W_{IL} is the rate of its production by CD4$^+$ T cells, and the last term in the right-hand side of this equation describes its consumption and degradation. The production rate W_{IL} is determined by mature CD4$^+$ T cells. We consider each such cell as a source term with a constant production rate ρ_{IL} at the area of the cell. Let us note that we do not take into account explicitly consumption of IL-2 by immature cells in order not to introduce an additional parameter. Implicitly, this consumption is taken into account in the degradation term.

For type I IFN, the equation and the terms in it have a similar interpretation:

$$\frac{\partial C_e}{\partial t} = D_{IFN} \Delta C_e + W_{IFN} - b_2 C_e. \tag{7.3.5}$$

Initial and boundary conditions for both concentrations IL-2 and IFN are taken zero. As before, the production rate W_{IFN} equals ρ_{IFN} at the area filled by APC cells and zero otherwise.

Infection.

Mature T cells leave the bone marrow. The level of CD8$^+$ T cells (effector cells) N_{ef} in the body is determined by the equation

$$\frac{dN_{ef}}{dt} = k_1 T - k_2 N_{ef}, \tag{7.3.6}$$

where T is their number in the lymph nodes. So the first term in the right-hand side of this equation describes production of effector cells in the lymph nodes and the second term their death in the body.

Let N_{inf} be the number of virus-infected cells. Its dynamics is described by the equation

$$\frac{dN_{inf}}{dt} = f(N_{inf}) - k_3 N_{ef} N_{inf}. \tag{7.3.7}$$

The first term in the right-hand side of this equation describes growth of the number of infected cells and the second term their elimination by effector cells. The function f will be considered in the form:

$$f(N_{inf}) = \frac{aN_{inf}}{1 + hN_{inf}},$$

where a and h are some positive constants.

Finally, the influx of APCs into the lymph nodes is proportional to the number of infected cells N_{inf}. This influx is limited by the place available in the lymph node. If there is a free place sufficient to put a cell, new cells are added. Let us also note that the lymph nodes can increase due to infection in order to produce more effector cells.

7.3.2 *Numerical Experiments*

We illustrate the model performance by considering two scenarios, reflecting different spatial patterns of IL-2 and type I IFN concentration fields. In the first one, both cytokines have the same diffusion coefficient $D_{IL2} = D_{IFN}$, whereas in the second case the diffusion rate of IFN is tenfold faster. The details of the numerical implementation of the hybrid model and the parameter values used for the simulations are presented in [3, 4]. Cell population densities and cytokine concentrations are scaled with respect to some reference values. These are determined by the cell density in the lymph node $\sim 10^5 - 10^6\,\text{mm}^{-3}$, the relative proportions of APCs, CD4$^+$ T cells and CD8$^+$ T cells [24–30] and the production rate of the cytokines [38]. The considered cell numbers correspond to a computational domain in the T-cell zone of about $100 \times 100 \times 100\mu\text{m}$.

The model presented above contains two compartments, the lymph node where effector cells are produced and the body where infection develops. The lymph node is described with the hybrid model while infection development in the organism by ordinary differential equations for infected cells and for effector cells. These two compartments are coupled by means of flux of effector cells from the lymph node to the body and by the flux of APC cells to the lymph node.

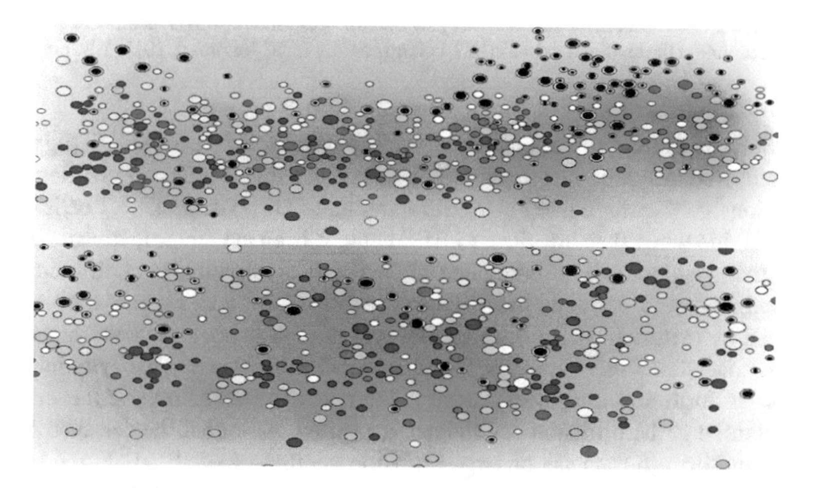

Fig. 7.4 Snapshot of numerical simulations of the cells and cytokines distribution in lymph node. Different cells are shown: APC (green), naive CD4$^+$ T cells (black), naive CD8$^+$ T cells (white), three maturity levels of differentiated CD8$^+$ T cells (blue), two maturity levels of CD4$^+$ T cells (yellow). Mature CD4$^+$ T cells produce IL-2 whose concentration in the extracellular matrix is shown by the level of green. APC produce IFN-I (red). The upper figure shows the simulation (day 8 post-infection) with equal diffusion coefficients of IL-2 and IFN-I, in the lower figure (day 80 post-infection) the diffusion coefficient of IFN is 10 times larger than the diffusion coefficient of IL-2. (Reprinted from Bouchnita et al., Hybrid approach to model the spatial regulation of T cell responses, BMC Immunol. 2017, 18(Suppl 1):29)

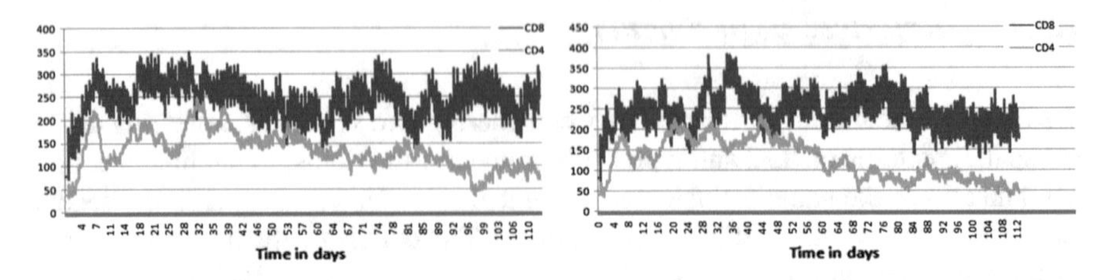

Fig. 7.5 The numbers of CD4$^+$ and CD8$^+$ T cells in time in the case of equal diffusion coefficients (left panel) and for the diffusion coefficient of IFN-I 10 times larger than the diffusion coefficient of IL-2 (right panel). (Reprinted from Bouchnita et al., Hybrid approach to model the spatial regulation of T cell responses, BMC Immunol. 2017, 18(Suppl 1):29)

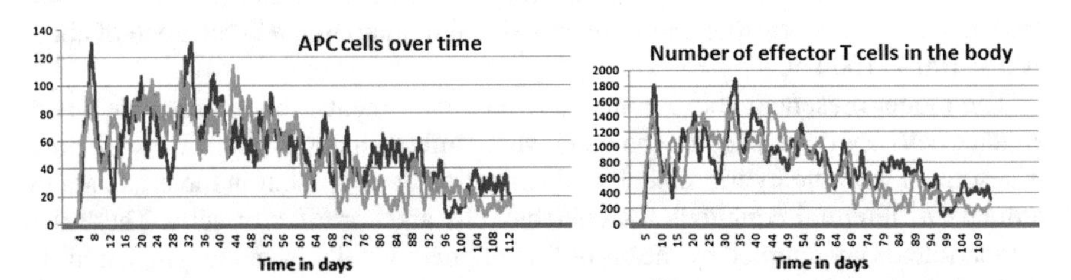

Fig. 7.6 The numbers of APC cells (left panel) and effector T cells (right panel) in time in the case of equal diffusion coefficients (black curve) and for the diffusion coefficient of IFN-I 10 times larger than the diffusion coefficient of IL-2 (grey curve). (Reprinted from Bouchnita et al., Hybrid approach to model the spatial regulation of T cell responses, BMC Immunol. 2017, 18(Suppl 1):29)

The results of the simulations are shown in Figs. 7.4, 7.5, 7.6 and 7.7. Figure 7.4 represents a snapshot of the lymph node T-cell zone with all cells participating in the simulations: APC cells, naive T cells, differentiated CD4$^+$ T and CD8$^+$ T cells. Naive T cells divide when they are close to APC cells. It is an asymmetric division where a proximal daughter cell differentiates while a distant cell remains undifferentiated. Differentiated cells continue their division and maturation in the presence of IL-2 produced by mature CD4$^+$ T cells. If the level of IL-2 is not sufficient, they die by apoptosis. Mature T cells leave the lymph node. One can see that the cytokine fields are non-uniform and their distribution patterns change essentially if the turnover parameters, e.g. the diffusion coefficient, are varied. Note that the cell distribution is more uniform in the case of large diffusion coefficient of IFN-I (Fig. 7.4 lower image) compared with the case of small diffusion coefficient (upper image).

The evolution of the total number of CD4$^+$ and CD8$^+$ T cells in the lymph node T-cell zone is shown in Fig. 7.5. The dynamics of APC cells in the lymph node T-cell zone and the effector cells in the body is shown in Fig. 7.6. The magnitude of the immune response is not sufficient to eradicate completely the infection. Indeed, the number of infected cells decreases but remains positive (Fig. 7.7). As virus infection is not cleared, the cell populations fluctuate around some constant values. Overall, the model reproduces the qualitative patterns of long-term persistent infection (experimental infections and in humans) dynamics (e.g. [31–35]). The pri-

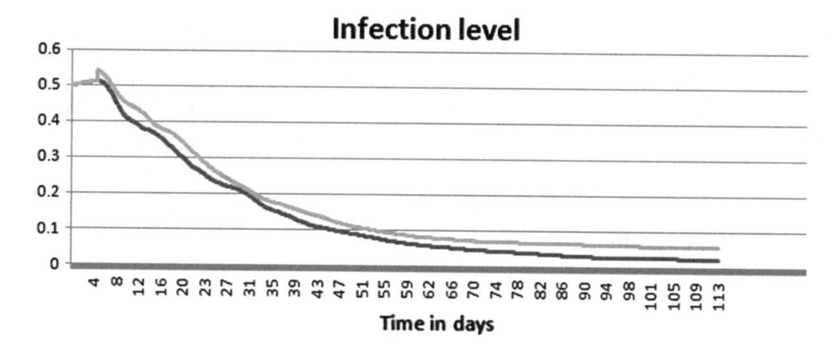

Fig. 7.7 The level of virus infection in the body in the case of equal diffusion coefficients (black curve) and for the diffusion coefficient of IFN-I 10 times larger than the diffusion coefficient of IL-2 (grey curve). (Reprinted from Bouchnita et al., Hybrid approach to model the spatial regulation of T cell responses, BMC Immunol. 2017, 18(Suppl 1):29)

Table 7.1 Cumulative numbers of key variables of the model over 113 days post-infection

	$D_{IFN} = D_{IL-2}$	$D_{IFN} = 10D_{IL-2}$
Number of CD4$^+$ T cells	27544	27040
Number of CD8$^+$ T cells	15194	14139
Number of APCs	4749	5293
The infection load	16.98	19.31
Number of N_{ef}	87849	80967

mary clonal expansion takes about 7 days and is followed by an enhanced long-term T-cell response to the persistent infection. The increase in the spread of type I IFN changes the relative distributions pattern of IL-2 and IFN-I, so that the resulting alteration in cytokine signalling reduces the clonal expansion and increases the overall level of virus infection.

The cumulative numbers of CD4$^+$ and CD8$^+$ T cells and virus infection load.

As single simulation runs of the stochastic model are characterized by a fluctuating and overlapping dynamics, we quantified integrative characteristics of the model behaviour. To describe the effect of the diffusion coefficient D_{IFN} on the T-cell production, we compared the cumulative numbers of CD4$^+$ and CD8$^+$ T cells as well as the infection load over the total time of the simulation for the two scenarios. We also show the cumulative numbers of effector T cells in the body N_{ef}. The results are shown in Table 7.1.

The net effect of the increase in the diffusion rate of type I IFN is a reduction in the clonal expansion of the T cells, in particular the effector T cells in the peripheral organs (by ~10%) and a rise in the infection level (by ~20%). The changes in the clonal T-cell expansion are the consequence of the differences in the cytokine concentration fields, which in turn alter the timing and the sequence of the IL-2 and type I IFN signalling.

The types and the relative densities of immune cells considered in the model essentially correspond to the clonal, APC-induced expansion of T cells activated by virus infections (see, e.g. [36]). The motility of T cells in the lymph node is determined by their random motion and mechanical cell–cell interactions [37]. The spatial distribution of cytokines considered in the model (IL-2 and type I IFN), though it requires more detailed investigation, corresponds to the actual understanding of the role of these cytokines.

7.3.3 *Infection Spreading in the Lymph Node*

Antigen-presenting cells can become infected and, when they enter the lymph node, initiate infection spreading in the lymphoid tissue. This question was discussed in the previous chapter where we studied reaction–diffusion models of the immune response. Here, we present an example of numerical simulations of the infection spreading with the hybrid model (Fig. 7.8).

The process begins with a single infected APC in the lymph node. Virus spreads to the lymph node through direct cell–cell interaction and by means of random motion in the extracellular matrix. More and more APC and $CD4^{+}$ T cells become infected. The number of uninfected cells decays resulting in the deceleration of their reproduction and weakening of the immune response.

7.4 Basis for Further Work

The aim of this chapter is to present a methodology for a hybrid modelling of immunological processes in their spatial context. A two-level hybrid mathematical model of immune cell migration and interaction integrating cellular and organ levels of regulation for a 2D spatial consideration of idealized secondary lymphoid organs is developed. It considers the population dynamics of antigen-presenting cells, $CD4^{+}$ and $CD8^{+}$ T lymphocytes in naive, proliferation and differentiated states. Cell division is assumed to be asymmetric and regulated by the extracellular concentration of interleukin-2 (IL-2) and type I interferon (IFN), together controlling the balance between proliferation and differentiation. The cytokine dynamics is described by reaction–diffusion PDEs, whereas the intracellular regulation is modelled with a system of ODEs. The mathematical model has been developed, calibrated and numerically implemented to study various scenarios in the regulation of T-cell immune responses to infection, in particular, the change in the diffusion coefficient of type I IFN as compared to IL-2. We have shown that a hybrid modelling approach provides an efficient tool to describe and analyse the interplay between spatio-temporal processes in the emergence of abnormal immune response dynamics.

Virus persistence in humans is often associated with an exhaustion of T lymphocytes. Many factors can contribute to the development of exhaustion. One of them

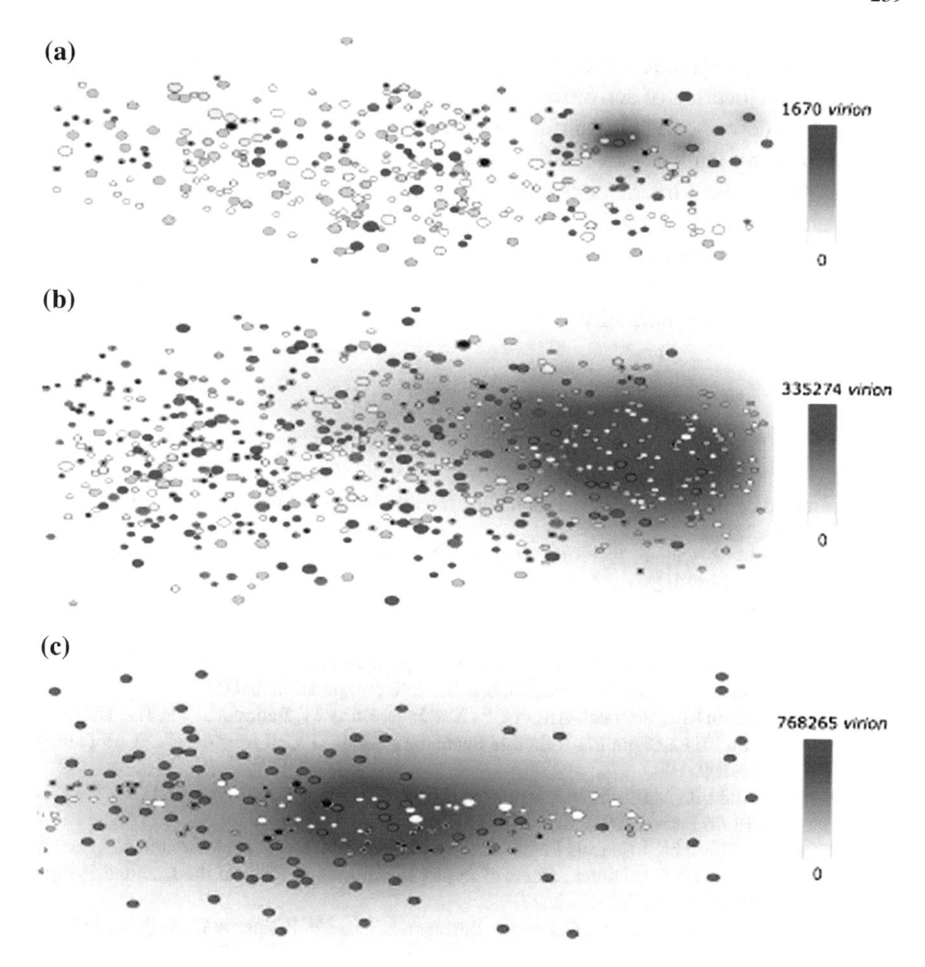

Fig. 7.8 Snapshots of simulations corresponding to 1 day after infection appearance in the lymph node (**a**), 20 days (**b**) and 60 days (**c**). Infection level is shown in blue, infected APCs in red and infected CD4$^+$ T cells in orange, naive uninfected CD4$^+$ T cells (black), naive CD8$^+$ T cells (white), three maturity levels of differentiated CD8$^+$ T cells (blue) and two maturity levels of uninfected CD4$^+$ T cells (yellow)

is associated with a shift from a normal clonal expansion pathway to an altered one characterized by an early terminal differentiation of T cells. We propose that an altered T-cell differentiation and proliferation sequence can naturally result from a spatial separation of the signalling events delivered via TCR, IL-2 and type I IFN receptors. Indeed, the spatial overlap of the concentration fields of extracellular IL-2 and IFN in lymph nodes changes dynamically due to different migration patterns of APCs and CD4$^+$ T cells secreting them.

The proposed hybrid mathematical model of the immune response represents a novel expandable and tunable analytical tool to examine challenging issues in the spatio-temporal regulation of cell growth and differentiation, in particular, the effect of timing and location of activation signals. Further application of hybrid modelling approach to describe the virus infection dynamics (HIV) can be found is [4].

The growing tendency in modern immunology to deal with the dynamic complexity of virus–host interactions needs advanced mathematical modelling tools and hence an interaction between experimentalists, theoreticians and mathematicians which is by no means easy. We hope that our experience shared in this chapter helps the research community to promote the coming of genuine integrative tools for the analysis of infection immunity.

References

1. A. Stephanou, V. Volpert. Hybrid modelling in cell biology. Math. Model. Nat. Phenom. 10 (2015), no. 1, 1-3.
2. A. Stephanou, V. Volpert. Hybrid modelling in biology: a classification review. Math. Model. Nat. Phenom. 11 (2016), no. 1, 37-48.
3. Bouchnita A, Bocharov G, Meyerhans A, Volpert V. Hybrid approach to model the spatial regulation of T cell responses. BMC Immunol. 2017; 18(Suppl 1):29.
4. A. Bouchnita, G. Bocharov, A. Meyerhans, V. Volpert. Towards a Multiscale Model of Acute HIV Infection. 2017, 5(1), 6; https://doi.org/10.3390/computation5010006.
5. Khan SH, Martin MD, Starbeck-Miller GR, Xue H-H, Harty JT, Badovinac VP. The Timing of Stimulation and IL-2 Signaling Regulate Secondary $CD8^{+}$ T Cell Responses. PLoS Pathog. 2015;11(10):e1005199.
6. Welsh RM, Bahl K, Marshall HD, Urban SL. Type 1 Interferons and Antiviral $CD8^{+}$ T-Cell Responses. PLoS Pathog. 2012;8(1):e1002352.
7. Sandra Hervas-Stubbs, Jose Luis Perez-Gracia, Ana Rouzaut, Miguel F. Sanmamed, Agnes Le Bon, and Ignacio Melero Direct Effects of Type I Interferons on Cells of the Immune System. Clin Cancer Res. 2011;17:2619–2627.
8. Ludewig B, Stein JV, Sharpe J, Cervantes-Barragan L, Thiel V. Bocharov G. A global imaging view on systems approaches in immunology. Eur J Immunol. 2012;42:3116–3125.
9. Bessonov N, Eymard N, Kurbatova P, Volpert V. Mathematical modelling of erythropoiesis in vivo with multiple erythroblastic islands. Applied Mathematics Letters. 2012;25: 1217–1221.
10. Fischer S, Kurbatova P, Bessonov N, Gandrillon O, Volpert V, Crauste F. Modelling erythroblastic islands: using a hybrid model to assess the function of central macrophage. J Theoretical Biol. 2012;298:92–106.
11. Kurbatova P, Bernard S, Bessonov N, Crauste F, Demin I, Dumontet C, Fischer S, Volpert V. Hybrid model of erythropoiesis and leukemia treatment with cytosine arabinoside. SIAM J Appl Math. 2011;71(6):2246–2268.
12. Volpert V, Bessonov N, Eymard N, Tosenberger A. Modèle multi-échelle de la dynamique cellulaire. In: Glade N, et dAngelique Stephanou, editeurs. Le vivant discret et continu. Editions Materiologiques; 2013.
13. Kurbatova P, Eymard N, V. Volpert V. Hybrid Model of Erythropoiesis. Acta Biotheoretica. 2013;61:305–315.
14. Eymard N, Bessonov N, Gandrillon O, Koury MJ, Volpert V. The role of spatial organization of cells in erythropoiesis. J Math Biol. 2015;70:71–97.
15. Stéphanou A, Volpert V. Hybrid modelling in biology: a classification review. Math Model Nat Phenom. 2016;11(1):37–48.
16. Mueller SN, Germain RN. Stromal cell contributions to the homeostasis and functionality of the immune system. Nat Rev Immunol. 2009;9(9):618–629.

17. Girard JP, Moussion C, Forster R. HEVs, lymphatics and homeostatic immune cell trafficking in lymph nodes. Nat Rev Immunol. 2012;12(11):762–773.
18. Junt T, Scandella E, Ludewig B. Form follows function: lymphoid tissue microarchitecture in antimicrobial immune defence.Nat Rev Immunol. 2008; 8(10):764–775.
19. Forster R, Braun A, Worbs T. Lymph node homing of T cells and dendritic cells via afferent lymphatics.Trends Immunol. 2012;33(6):271–280.
20. Goldsby RAA, Kuby J, Kindt ThJ. Immunology. W H Freeman & Co (Sd); 4th edition, 2000; Ch.10:222.
21. Chang JT, Reiner SL. Asymmetric Division and Stem Cell Renewal without a Permanent Niche: Lessons from Lymphocytes. Cold Spring Harb Symp Quant Biol. 2008;73:73–79.
22. Broere F, Apasov SG, Sitkovsky MV, van Eden W. T cell subsets and T cell-mediated immunity. In: Nijkamp FP, Parnham MJ, editors. Principles of Immunopharmacology: 3rd revised and extended edition. Basel: Springer AG; 2011.
23. Nelson BH. IL-2, Regulatory T Cells, and Tolerance. J Immunol. 2004;172:3983–3988.
24. Ganusov VV, De Boer RJ. Do most lymphocytes in humans really reside in the gut? Trends Immunol. 2007;28(12):514–518.
25. Scandella E, Bolinger B, Lattmann E, Miller S, Favre S, Littman DR, Finke D, Luther SA, Junt T, Ludewig B. Restoration of lymphoid organ integrity through the interaction of lymphoid tissue-inducer cells with stroma of the T cell zone. Nat Immunol. 2008;9(6):667–675.
26. Kumar V, Scandella E, Danuser R, Onder L, Nitschke M, Fukui Y, Halin C, Ludewig B, Stein JV. Global lymphoid tissue remodeling during a viral infection is orchestrated by a B cell-lymphotoxin-dependent pathway. Blood. 2010;115(23):4725–4733.
27. Bocharov G, Zust R, Cervantes-Barragan L, Luzyanina T, Chiglintsev E, Chereshnev VA, et al. A Systems Immunology Approach to Plasmacytoid Dendritic Cell Function in Cytopathic Virus Infections. PLoS Pathog. 2010;6(7):e1001017.
28. Bocharov G, Chereshnev V, Gainova I, Bazhan S, Bachmetyev B, Argilaguet J, Martinez J, Meyerhans A. Human Immunodeficiency Virus Infection: from Biological Observations to Mechanistic Mathematical Modelling. Mathematical Modelling of Natural Phenomena, 2012;7 (5): 78–104.
29. Cremasco V, Woodruff MC, Onder L, Cupovic J, Nieves-Bonilla JM, Schildberg FA, Chang J, Cremasco F, Harvey CJ, Wucherpfennig K, Ludewig B, Carroll MC, Turley SJ. B cell homeostasis and follicle confines are governed by fibroblastic reticular cells. Nat Immunol. 2014;15(10):973–981.
30. Giese C, Marx U. Human immunity in vitro - solving immunogenicity and more.Adv Drug Deliv Rev. 2014 69-70:103–122.
31. Moskophidis D, Lechner F, Pircher H, Zinkernagel RM. Virus persistence in acutely infected immunocompetent mice by exhaustion of antiviral cytotoxic effector T cells. Nature. 1993;362(6422):758–761.
32. Moskophidis D, Battegay M, van den Broek M, Laine E, Hoffmann-Rohrer U, Zinkernagel R. Role of virus and host variables in virus persistence or immunopathological disease caused by a non-cytolytic virus. J. Gen. Virol. 1995;76(2):381–391.
33. McMichael AJ, Borrow P, Tomaras GD, Goonetilleke N, Haynes BF. The immune response during acute HIV-1 infection: clues for vaccine development. Nature reviews Immunology. 2010;10(1):11–23.
34. Cervantes-Barragan L, Lewis KL, Firner S, Thiel V, Hugues S, Reith W, Ludewig B, Reizis B. Plasmacytoid dendritic cells control T-cell response to chronic viral infection. Proceedings of the National Academy of Sciences of the United States of America. 2012;109(8):3012–3017.
35. Hansen SG, Piatak M, Ventura AB, Colette M., Hughes CM, Gilbride RM, Ford JC, Oswald K, Shoemaker R, Li Y, Lewis MS, Gilliam AN, Xu G, Whizin N, Burwitz BJ, Planer SL, Turner JM, Legasse AW, Axthelm MK, Nelson JA, Fruh K, Sacha JB, Estes JD, Keele BF, Edlefsen PT, Lifson JD, Picker LJ. Immune clearance of highly pathogenic SIV infection. Nature. 2013;502(7469):100–104.
36. Kaech SM, Wherry EJ. Heterogeneity and cell-fate decisions in effector and memory CD8+ T cell differentiation during viral infection. Immunity. 2007;27(3):393–405.

37. Fricke GM, Letendre KA, Moses ME, Cannon JL. Persistence and Adaptation in Immunity: T Cells Balance the Extent and Thoroughness of Search. PLoS Computational Biology. 2016;12(3):e1004818.
38. Baker CTH, Bocharov GA, Paul CAH. Mathematical Modelling of the Interleukin-2 T-Cell System: A Comparative Study of Approaches Based on Ordinary and Delay Differential Equation. J Theor Medicine. 1997;1(2):117–128.
39. Weinan, E. Principles of Multiscale Modelling; Cambridge University Press: Cambridge, UK, 2011.
40. Fallahi-Sichani, M.; El-Kebir, M.; Marino, S.; Kirschner, D.E.; Linderman, J.J. Multi-scale computational modeling reveals a critical role for TNF receptor 1 dynamics in tuberculosis granuloma formation. J. Immunol. 2011, 186, 3472–3483.
41. Cilfone, N.A.; Kirschner, D.E.; Linderman, J.J. Strategies for efficient numerical implementation of hybrid multi-scale agent-based models to describe biological systems. Cell. Mol. Bioeng. 2015, 8, 119–136.
42. Simeone Marino, S.; Kirschner, D.E. A multi-compartment hybrid computational model predicts key roles for dendritic cells in tuberculosis infection. Computation 2016, 4, 39.
43. Prokopiou, S.A.; Barbarroux, L.; Bernard, S.; Mafille, J.; Leverrier, Y.; Arpin, C.; Marvel, J.; Gandrillon, O.; Crauste, F. Multiscale Modeling of the Early CD8 T-Cell Immune Response in Lymph Nodes: An Integrative Study. Computation 2014, 2, 159–181.
44. Gao, X.; Arpin, C.; Marvel, J.; Prokopiou, S.A.; Gandrillon, O.; Crauste, F. IL-2 sensitivity and exogenous IL-2 concentration gradient tune the productive contact duration of CD8(+) T cell-APC: A multiscale modeling study. BMC Syst. Biol. 2016 , 10, doi:10.1186/s12918-016-0323-y.
45. Williams, R.A.; Jon Timmis, J.; Qwarnstrom, E.E. Computational Models of the NF-KB Signaling Pathway. Computation 2014 , 2, 131–158.
46. Baldazzi, V.; Paci, P.; Bernaschi, M.; Castiglione, F. Modeling lymphocyte homing and encounters in lymph nodes. BMC Bioinform. 2009, 10, https://doi.org/10.1186/1471-2105-10-387.
47. Gong, C.; Mattila, J.T.; Miller, M.; Flynn, J.L.; Linderman, J.J.; Kirschner, D. Predicting lymph node output efficiency using systems biology. J. Theor. Biol. 2013, 335, 169–84.
48. Palsson, S.; Hickling, T.P.; Bradshaw-Pierce, E.L.; Zager, M.; Jooss, K.; OBrien, P.J.; Spilker, M.E.; Palsson, B.O.; Vicini, P. The development of a fully-integrated immune response model (FIRM) simulator of the immune response through integration of multiple subset models. BMC Syst. Biol. 2013, 7, https://doi.org/10.1186/1752-0509-7-95.
49. Germain, R.; Meier-Schellersheim, M.; Nita-Lazar, A.; Fraser, I. Systems biology in immunology—A computational modeling perspective. Annu. Rev. Immunol. 2011, 29, 527–585.
50. Donovan, G.M.; Lythe, G. T cell and reticular network co-dependence in HIV infection. J. Theor. Biol. 2016, 395, 211–220.

Chapter 8
Current Challenges

The outcome of a viral infection is determined by the race between an expanding virus and the respective antiviral immune response of the infected host. This dynamic interaction lends itself to a quantitative description inherent in mathematical modelling. In this book, we have been covering several basic aspects of virus–host interactions, in particular,

- the kinetic regulation of antigen-specific T-cell responses,
- the quantitative characterization of T-cell proliferation,
- the causal links between biological process parameters and the pathogenesis of chronic viral infections.

The description of our work on these immunologically relevant issues covers diverse computational modelling technologies and specifies practical aspects of model building, parameter estimation and sensitivity analysis. However, all this represents just a start in a meaningful analysis of virus–host interactions and many more complex questions need to be addressed in the future. The major issue is how to deal efficiently with the complexities of the immune system using mathematics as an exploratory tool. These complexities are revealed by modern high-throughput measurements and visualization technologies generating overwhelming big-data sets from which the multi-level regulation of the existing, phenotypically different infection dynamics may be derived and understood in mechanistic terms. Questions arising are as follows:

(1) What is the key information needed to mechanistically understand the control of a given infection dynamics?
(2) What constitutes the functional network of processes that regulate homeostasis of the immune system?
(3) How is the redundancy of the immune system elements used to generate robust immune system functioning?

G. Bocharov et al., *Mathematical Immunology of Virus Infections*,
https://doi.org/10.1007/978-3-319-72317-4_8

(4) What is the limit in the prediction of immune system functions when starting from quantitative characteristics of its components, e.g. genetic-, cellular- or tissue characteristics?
(5) Can one finally link antiviral immune responses including its limits and failures to the first physical–chemical principles of living systems in terms of energy balance, genetic make-up and reaction kinetics?
(6) How can one assess the control and regulation of immune processes that are impacted by the spatial structures of various organs dynamically evolving during their functioning?

In order to handle the above issues and to identify the immune systems regulation rules, it is clear that mathematical modelling has to go beyond a routine development of low-resolution phenomenological models, and exploit the full knowledge available in mathematical theory and technology of automatic control. Furthermore, the reaction systems that describe the elementary immune functions have to be embedded into the physiologically distinct compartments and the morphological constraints inherent to chromosomes, cells, tissues and the whole organism. This will then allow the research community not only to get a better quantitative understanding of immune system functioning in infections, but also enable to build predictive pharmacokinetics and pharmacodynamics models for antiviral and immunomodulatory drugs of various physical and chemical nature and ideally, to specify the necessary criteria for drug performance.

From the view of a mathematician, a number of challenging issues remain that deserve further analyses, both analytical and numerical. The spatial distribution of virus and immune cells in tissues can influence the dynamics of immune responses. If we compare ODE models (without diffusion) and the corresponding RDE models (with diffusion), then they may have the same equilibria. However, basins of attraction of these equilibria and transitions between them are likely to be different. For example, if we have two stable stationary points within an ODE model, then we cannot pass from one to another one due to a small perturbation. In contrast, this can easily happen within the frame of an RDE model via a localized in space perturbation that can initiate a reaction–diffusion wave of transition to another stable equilibrium. Hence, localized spatial perturbations can be important from the point of view of infection development and the corresponding immune response, and the correct model choice might critically impact its predictive power.

Additional properties of spatial models of the immune response in comparison with ODE models are related to the diffusion terms describing random motion of virus and immune cells. Virus displacement in tissue is an important issue in biological research. Tissue invasion happens by virus moving either directly between host cells or through the intercellular matrix. The mechanisms of this motion are not yet sufficiently elucidated but it is clear that a possible control of these mechanisms would provide additional means to eliminate infections. Similarly, as we discussed above, the intensity of motion of immune cells can strongly influence the efficacy of immune responses. However, the models considered up to now are based on simplified mechanisms of virus and cell motion assuming that it is random and can be described by conventional diffusion terms. In fact, displacement of biological objects can be of complex nature. There are numerous evidence that anomalous diffusion models can be appropriate. In the case of virus and immune cells, this question is yet completely open.

Another dimension in modelling of immune responses is provided by multi-scale models. Their development is in the very beginning but the necessity of more precise quantitative descriptions of immune responses will determine their development. An immune response is a complex systemic process, which involves different organs and tissues of a host organism. Clearly, we cannot neglect this aspect and restrict ourselves to a single-level average model. Bone marrow, thymus, lymph nodes and the lymphatic system, spleen and blood circulation are all involved in an immune response and they have very different structures and functions. Thus considering a human organism as a one-compartment entity is an oversimplification, which becomes largely outdated. One of the modern approaches to multi-scale modelling is provided by hybrid models, where various discrete and continuous models are chosen to describe the corresponding physiological process in the most appropriate way. However, issues like approximation, stability, consistency, and efficacy of the numerical implementation of such models need further rigorous analyses.

To conclude, we expect that mathematical immunology is entering the phase of a genuinely mutual inspiration of immunological and mathematical studies. Novel challenges will emerge on the way to master the full potential of the immune system to controlling unfavorable dynamics of virus infections. An exciting time is lying ahead.

Printed in the United States
By Bookmasters